Functional and Operatorial Statistics

Sophie Dabo-Niang · Frédéric Ferraty

Functional and Operatorial Statistics

Springer

Dr. Sophie Dabo-Niang
Laboratoire GREMARS-EQUIPPE
Université Charles de Gaulle
Lille 3
Maison de la Recherche
Domaine du Pont de Bois
BP 60149
F-59653 Villeneuve d'ascq cedex
France
sophie.dabo@univ-lille3.fr

Dr. Frédéric Ferraty
Institut de Mathématiques de
Toulouse Equipe LSP
Université Paul Sabatier
F-31062 Toulouse Cedex 9
France
ferraty@math.univ-toulouse.fr

ISBN: 978-3-7908-2061-4 e-ISBN: 978-3-7908-2062-1

Library of Congress Control Number: 2008928585

Cover design: WMXDesign GmbH, Heidelberg

Printed on acid-free paper

9 8 7 6 5 4 3 2 1

springer.com

Preface

An increasing number of problems and methods involve infinite-dimensional aspects. This is due to the progress of technologies which allow us to store more and more information while modern instruments are able to collect data much more effectively due to their increasingly sophisticated design. This evolution directly concerns the statisticians who have to propose new methodologies while taking into account such high-dimensional data (e.g. continuous processes, functional data, etc.). The numerous applications (microarrays, paleo-ecological data, radar waveforms, spectrometric curves, speech recognition, continuous time series, 3-D images, etc.) in various fields (biology, econometrics, environmetrics, the food industry, medical sciences, paper industry, speech recognition, etc.) make researching this statistical topic very worthwhile. New challenges emerge both from theoretical and practical point of views. This First International Workshop on Functional and Operatorial Statistics (IWFOS) aims to emphasize this fascinating field of research and this volume gathers the contributions presented in this conference. It is worth noting that this volume mixes applied works (with original datasets and/or computational issues) as well as fundamental theoretical ones (with deep mathematical developments). Therefore, this book should cover a large audience, like academic researchers (theoreticians and/or practitioners), graduate/PhD students and should appeal to anyone working in statistics with industrial companies, research institutes or software developers.

This Workshop covers a wide scope of statistical aspects. Numerous works deal with classification (see for instance chapters 6, 7, 17, 21 or 41), functional PCA-based methods (see for instance chapters 2, 16, 30 or 37), mathematical toolbox (see for instance chapters 11, 13, 24 or 29), regression (see for instance chapters 3, 4, 5, 8, 10, 18, 19, 23, 26, 27, 33 or 34), spatial statistics (see for instance chapters 9, 22, 35, 36 or 42), time series (12, 28, 39, 40 or 44). Other topics are also present as subsampling (see chapter 38) as well as transversal/explorative methodologies (see for instance chapters 14, 20, 31 or 43). In addition, interesting works focus on original/motivating applications (see for instance chapters 15, 25 or 32). This splitting into topics (classification, functional PCA-based methods, ...) is introduced just for giving an idea on the contents but most of the time, one can assign a same work to several subjects. It is worth noting that numerous contributions deal with statistical methodologies for functional data which is certainly the main common denominator of IWFOS (see chapters 2, 3, 4, 5, 6, 7, 8, 9, 10, 12, 13, 14, 16, 17, 18, 19, 20, 21, 22, 23, 25, 26, 29, 30, 31, 32, 33, 34, 35, 36, 37, 40, 41, 43, 44).

The scientific success of this event is obviously linked with the wide variety of participants coming from about 20 countries covering all the continents. One would like to thank them gratefully and specially the invited speakers and all the contributors for the high quality of their submitted works.

Of course the hard core is the STAPH members (see chapter 1), which managed and coordinated this Workshop both from a scientific and organizational point of view. But this international conference would not exist without the help of many people. In particular, K. Benhenni (France), B. Cadre (France), H. Cardot (France), A. Cuevas (Spain), A. Dahmani (Algeria), A. Goia (Italia), W. Gonzalez-Manteiga (Spain), W. Härdle (Germany), A. Kneip (Germany), Ali Laksaci (Algeria), A. Mas (France), E. Ould-saïd (France), M. Rachdi (France), E. Salinelli (Italia) and I. Van Keilegom (Belgium) have greatly contributed to the high quality of IWFOS'2008 and are gratefully thanked.

The final thanks go to Marie-Laure Ausset which took charge of secretary tasks as well as the institutions which have supported this Workshop via grants or administrative supports (CNRS, Conseil Général de la Haute-Garonne, Conseil Régional Midi-Pyrénées, Laboratoire de Statistique et Probabilités, Institut de Mathématiques de Toulouse, Université Paul Sabatier, Laboratoire Gremars-Equippe of university Charles De Gaulle, Équipe de Probabilités-Statistique of University Montpellier 2, laboratoire LMPA (Littoral) and University del Piemonte Orientale (Italy)).

Toulouse, France
May 2008

Sophie Dabo-Niang
Frédéric Ferraty

Contents

1.2 The STAPH group

The STAPH group has been created in 1999 and pursues two main scientific aims. The first one is still to go on with the developments of both historical topics described just before: multivariate analysis and nonparametric estimation. As outlined before, interactions between these two wide areas of research in statistics are more and more frequent. This includes in particular various extensions to the functional data setting. So, the second aim is to bridge the numerous gaps between these two important fields of research. In particular, one can expect that the mixing of multivariate technics with functional estimation can allow the building of new pertinent statistical methods with a special emphasis on infinite-dimensional data. In this infinite-dimensional spirit, the three main statistical topics on which the STAPH group focuses are:

⇒ The operatorial background of statistics and its connected subjects,
⇒ The univariate/multivariate nonparametric functional estimation,
⇒ Statistical methods for infinite-dimensional data.

The infinite-dimensional setting is certainly one of the most heavy challenge since it combines original theoretical and practical problems but this will be the price to pay for developing new methodologies. A sample of main recent contributions involving STAPH's members in these fields can be found in Ferraty and Vieu (2006) who make the gap between nonparametric statistics and functional data analysis, in Crambes, Kneip and Sarda (2008) who produce deep asymptotics study in functional linear regression, in Boudou and S. Viguier-Pla (2006), who study the specificities of PCA of times series, that implies to work in frequency domain. Other various study domains are for example the properties of some tensor operators by Romain (2002) or the definition of operator-based random measures in Banach spaces with applications to stationary series by Benchik et al. (2007).

STAPH is also a way of thinking the research. There is no "universally better" way of making research in that sense that there is no hierarchy between practical or theoretical aspects. Said differently, there is no universal method of producing pertinent research; deep theoretical problems can have potential impacts on real applications whereas fundamental work can emerge from practical developments, and all intermediate situations can occur. This is why it is important to consider theoretical as well as practical aspects, without any predominance. Throughout informal but deep discussions, each participant in STAPH brings its own brick in such a way that, brick by brick, the collective scientific knowledge progresses. In this spirit, the human component plays a major role, which is the key of the success. This can be summarized by this quotation of François Rabelais (1533): *Science sans conscience n'est que ruine de l'âme. (i.e. Science without consciousness is nothing but a ruin of soul).* This is the foundation of the STAPH group and one aims to develop and share more and more the knowledge on these topics not only inside our institution (Mathematical Institute of Toulouse) but also with the French community and finally the international one.

1.3 The first IWFOS'2008

It is now something like ten years that STAPH is developping its activities, and details of activities can be founded in Staph (2008). After several publications on its privileged topics, regular seminars have been organized. Step by step, strong links have been built with various statistical french laboratories (Grenoble, Lille, Montpellier, Paris,...). When the STAPH network grew up sufficiently, regular meetings were organized in France (Toulouse, Grenoble, Lille) with more and more participants coming not only from french universities, but also from foreign countries (Algeria, Germany, Spain,...). The international network of STAPH was also expanded via its participation in the organization of sessions in international meetings. It is clear that the main common interest of various participants to this scientific events was statistical modelling in high dimensional setting with a special emphasis on functional data. At the same time, the number of international publications in this area became larger and larger. The popular success of this recent field of statistics is certainly motivated by the numerous domains of application (chemometrics, econometrics, environmetrics, high technologies, medical sciences, industries,...) as well as by the new theoretical challenges arising from the infinite-dimensional setting.

Given that, the necessity of organizing an international workshop around these topics, which can be gathered under the label "Statistical methods and problems in infinite-dimensional spaces", became an evidence and here borned the first International Workshop on Functional and Operatorial Statistics. As attested by the abstracts collected in this document, this meeting gathers the most influent statisticians actually active on these fields. More information on this event can be found in IWFOS (2008).

According to the STAPH spirit, this workshop is the opportunity to present as well the most recent theoretical works on infinite dimensional statistics as various practical case studies. Moreover, still keeping an eye on the future, a special attention has been given to young researchers who have been voluntarily mixed to the most confirmed ones.

This scientific event is the result of fruitful collaborations. Of course, the hard core are the STAPH members, which managed and coordinated this Workshop both from a scientific and organizational point of view. But this international conference would not exist without the help of many people. Among them one would like to thank especially to S. Dabo-Niang for the realization of these proceedings and for its participation to the Scientific Committee.

Naturally, the various scientific contacts developed around the STAPH's activities during the last decade play a key role as well for the high scientific level of IWFOS'2008 as for their material implications. K. Benhenni (France), B. Cadre (France), H. Cardot (France), A. Cuevas (Spain), S. Dabo-Niang (France), A. Dahmani (Algeria), A. Goia (Italia), W. Gonzalez-Manteiga (Spain), W. Härdle (Germany), A. Kneip (Germany), Ali Laksaci (Algeria), A. Mas (France), E. Ould-saïd (France), M. Rachdi (France), E. Salinelli (Italia) and I. Van Keilegom (Belgium) have greatly contributed to the high quality of IWFOS'2008. They are gratefully thanked for accepting the invitation to join the organizing/scientific committee and for their help in reviewing the contributions contained in this proceedings. Denis Bosq and Jim Ramsay have high scientific authority in infinite dimensional statistics (see for instance Bosq, 2000 or Ramsay and Silverman, 1997). We would like to express them our strong gratitude as well for the continuous interest shown for STAPH's activities as for their various contributions to this workshop. One also grateful thanks the invited speakers as well as all contributors whose the high quality of the submitted works ensured the scientific success of this event.

References

[1] Benchikh, T. Boudou, A. Romain, Y.: Mesures aleatoires operatorielle et banachique. Application aux series stationnaires. (French) [Operatorial and Banach space-valued random measures. Application to stationary series] C. R. Math. Acad. Sci. Paris, **345** (6), 345-348 (2007).

[2] Bosq, D.: Linear processes in function spaces, theory and applications. Lecture notes in statistics, **149**, Springer-Verlag, New York. (2000).

[3] Boudou, A., Viguier-Pla, S.: On proximity between PCA in the frequency domain and usual PCA. Statistics, **40**, 447-464. (2006).

[4] Crambes, C., Kneip, A., Sarda, P.: Smoothing splines estimators for functional linear regression. Ann. Statist. (to appear). (2008).

[5] Collomb, G.: Méthodes non paramétriques en régression, analyse de séries temporelles, prédiction et discrimination. Univesité Paul Sabatier, Thèse d'Etat. (1983).

[6] De Boor, C.: A practical guide to splines. Springer, New York. (1978).

[7] De Leeuw, J. and J. Van Rijckvorsel.: Component and Correspondence analysis: dimension reduction by functional approximation. Wiley, New York. (1988).

[8] Dauxois, J. and Pousse, A.: Une extension de l'analyse canonique. Quelques applications. Annales de l'Institut Henri Poincare, Vol. XI , **4**, 355-379. (1975).

[9] Ferraty, F., Vieu, P.: Nonparametric functional data analysis: theory and practice. Springer, New York. (2006).

[10] Hastie, T. and Tibshirani, R. J.: Generalized additive models. Chapman and Hall, New York. (1990).

[11] IWFOS.: http://www.lsp.ups-tlse.fr/staph/IWFOS2008 (2008).

[12] Ramsay, J. and Silverman, B.: Functional data analysis. Springer, New York. (1997).

[13] Romain, Y.: Perturbation of functional tensors with applications to covariance operators. Stat. Proba. Letters, **58**, 253-264 (2002).

[14] Staph .: http://www.lsp.ups-tlse.fr/staph. (2008).

Chapter 2
Solving Multicollinearity in Functional Multinomial Logit Models for Nominal and Ordinal Responses

Ana Aguilera and Manuel Escabias

Abstract Different functional logit models to estimate a multicategory response variable from a functional predictor will be formulated in terms of different types of logit transformations as base-line category logits for nominal responses or cumulative, adjacent-categories or continuation-ratio logits for ordinal responses. Estimation procedures of functional logistic regression based on functional PCA of sample curves will be generalized to the case of a multicategory response. The true functional form of sample curves will be reconstructed in terms of basis expansions whose coefficients will be estimated from irregularly distributed discrete time observations.

2.1 Introduction

The functional logistic regression model is the most used method to explain a binary variable in terms of a functional predictor as can be seen in many applications in different fields. Ratcliffe *et al.* (2002) used this model for predicting if human foetal heart rate responses to repeated vibroacoustic stimulation. The relation between the risk of drought and time evolution of temperatures has been modeled in Escabias *et al.* (2005). With respect to the problem of estimation of this model, Escabias *et al.* (2004) proposed different functional principal component approaches for solving multicollinearity by providing an accurate parameter function estimation. An alternative estimation procedure based on PLS logit regression has been recently considered (Aguilera *et al.*, 2007).

Ana Aguilera
Department of Statistics and O.R. University of Granada, Spain, e-mail: `aaguiler@ugr.es`

Manuel Escabias
Department of Statistics and O.R. University of Granada, Spain, e-mail: `escabias@ugr.es`

In the general context of generalized functional linear models, James (2002) assumes that each predictor can be modeled as a smooth curve from a given functional family. Then, the functional model can be equivalently seen as a generalized linear model whose design matrix is given by the unobserved basis coefficients for the predictor and the EM algorithm is used for estimating the model from longitudinal observations at different times for each individual. On the other hand, Müller and StadtMüller (2005) considered an orthonormal representation of sample curves and used as predictor variables of the functional model a finite number of coefficients of such orthonormal expansion. Asymptotic tests and simultaneous confidence bands for the parameter function have been obtained by using this dimension reduction approach. An estimation procedure based on B-splines expansion maximizing the penalized log-likelihood has been studied in Marx and Eilers (1999) for a functional binomial response model and in Cardot and Sarda (2005) for the general case of functional generalized linear models.

The natural generalization of the functional logit model is the functional multinomial regression model where the response variable has a finite set of categories and the predictor is a functional variable. An initial work on this issue has been developed by Cardot *et al.* (2003) where a functional baseline-category logit model has been considered for predicting land use with the temporal evolution of coarse resolution remote sensing data. In this paper we propose a different approach comparing different methods of estimation based on the approximation of the functional predictor and the parameter functions in a finite space generated by a basis of functions what turns the functional model into a multiple one. Model estimation will be improved by developing several functional principal component approaches and selecting the predictor principal components according to their ability to provide the best possible estimation of the parameter functions.

2.2 Functional multinomial response model

Let us consider a functional predictor $\{X(t) : t \in T\}$, whose sample curves belong to the space $L^2(T)$ of square integrable functions on T, and a categorical response random variable Y with S categories.

Given a sample of observations of the functional predictor $\{x_i(t) : t \in T, i = 1, \ldots, n\}$, the sample of observations of the response associated to them is a set of n vectors $(y_{i1}, \ldots, y_{iS})'$ of dimension S defined by

$$y_{is} = \begin{cases} 1 & \text{if category } s \text{ is observed for } X(t) = x_i(t) \\ 0 & \text{other case} \end{cases}$$

so that each observation is generated by a multinomial distribution $M(1; \pi_{i1}, \ldots, \pi_{iS})$ with $\pi_{is} = P[Y = s|X(t) = x_i(t)]$ and $\sum_{s=1}^{S} \pi_{is} = 1 \quad \forall i = 1, \ldots, n$.

Let us observe that y_{iS} is redundant. Then, if we denote by $y_i = (y_{i1}, \ldots, y_{i,S-1})'$ the vector response for subject i, with mean vector $\mu_i = E[Y_i] = (\pi_{i1}, \ldots, \pi_{i,S-1})'$, the multinomial response model is a particular case of generalized linear model $y_{is} = \pi_{is} + \varepsilon_{is}$ with

$$g_s(\mu_i) = \alpha_s + \int_T \beta_s(t) x_i(t) dt, \quad s = 1, \ldots, S-1, \tag{2.1}$$

where the link function components g_s can be defined in different ways, ε_{is} are independent and centered errors and α_s and $\beta_s(t)$ a set of parameters to be estimated. In this paper we are going to generalize the functional logit model for a binary response to the case of a multinomial response. Because of this we will consider as link functions different types of logit transformations $l_{is} = g_s(\mu_i)$ (see Agresti (2002) for a detailed explanation).

2.2.1 Nominal responses

Baseline-category logits for nominal response pair each response with a baseline category

$$l_{is} = \log\left[\pi_{is}/\pi_{iS}\right].$$

Then, the equation that expresses baseline-category logit models directly in terms of response probabilities is ($\alpha_S = 0, \beta_S(t) = 0$)

$$\pi_{is} = \frac{\exp\left\{\alpha_s + \int_T x_i(t)\,\beta_s(t)\,dt\right\}}{\sum_{s=1}^{S} \exp\left\{\alpha_s + \int_T x_i(t)\,\beta_s(t)\,dt\right\}}, \quad s = 1, \ldots, S,\ i = 1, \ldots, n. \tag{2.2}$$

2.2.2 Ordinal responses

When the response variable is ordinal the logit transformations l_{is} reflect ordinal characteristics such as monotone trend. Next, several types of ordinal logits will be studied.

Cumulative logits

Cumulative logits use category ordering by forming logits of cumulative probabilities. The most popular logit model for ordinal responses is the proportional odds model

$$l_{is} = \log \frac{P[Y \leq s|x_i(t)]}{1 - P[Y \leq s|x_i(t)]} = \frac{\sum_{j=1}^{s} \pi_{ij}}{\sum_{j=s+1}^{S} \pi_{ij}} = \alpha_s + \int_T \beta(t) x_i(t) dt,$$

$s = 1, \ldots, S-1$, that has the same effects $\beta(t) = \beta_s(t) \quad \forall s = 1, \ldots, S-1$ for each cumulative logit. Then, each response probability is obtained as $\pi_{is} = F_{is} - F_{i,s-1}$ with

$$F_{is} = P[Y \leq s|x_i(t)] = \frac{\exp\left(\alpha_s + \int_T \beta(t) x_i(t) dt\right)}{1 + \exp\left(\alpha_s + \int_T \beta(t) x_i(t) dt\right)}.$$

Adjacent-categories logits

Logits for ordinal responses do not need use cumulative probabilities. Alternative logits for ordinal responses are the adjacent-categories logits and the continuation-ratio logits.

Adjacent-categories logits are defined as $l_{is} = \log\left[\pi_{is}/\pi_{i,s+1}\right]$, $s = 1, \ldots, S-1$. Taking into account the relation between baseline-category logits and adjacent-categories, the adjacent-categories logit model with common effect $\beta(t)$ (equal odds model)

$$\log\left[\frac{\pi_{is}}{\pi_{i,s+1}}\right] = \alpha_s + \int_T \beta(t) x_i(t) dt,$$

can be expressed in terms of the response probabilities as

$$\pi_{is} = \frac{\exp\left[\sum_{j=s}^{S-1} \alpha_j + \int_T (S-s) \beta(t) x_i(t) dt\right]}{1 + \sum_{s=1}^{S-1} \exp\left[\sum_{j=s}^{S-1} \alpha_j + \int_T (S-s) \beta(t) x_i(t) dt\right]}.$$

Continuation-ratio logits

Continuation-ratio logits are $l_{is} = \log\left[P(Y=s)/P(Y>s)\right] = \log\left[\pi_{is}/\sum_{j=s+1}^{S} \pi_{ij}\right]$. Denoting by $p_{is} = \frac{\pi_{is}}{\pi_{is} + \cdots + \pi_{iS}}$ the probability of response s, given response s or higher, the continuation-ratio logit models can be seen as ordinary binary logit models

$$l_{is} = \log \frac{\pi_{is}}{\sum_{j=s+1}^{S} \pi_{ij}} = \log \frac{p_{is}}{1 - p_{is}} = \alpha_s + \int_T \beta_s(t) x_i(t) dt,$$

so that these conditional probabilities are modeled as

$$p_{is} = \frac{\pi_{is}}{\pi_{is} + \cdots + \pi_{iS}} = \frac{\exp\left(\alpha_s + \int_T \beta_s(t) x_i(t) dt\right)}{1 + \exp\left(\alpha_s + \int_T \beta_s(t) x_i(t) dt\right)}.$$

error of the parameter functions estimates. The relation of this minimum with special trends in other goodness of fit measures will be also investigated. An adequate model selection method based on the results will be proposed.

Acknowledgements This research has been funded by project P06-FQM-01470 from *"Consejería de Innovación, Ciencia y Empresa. Junta de Andalucía, Spain"* and project MTM2007-63793 from *Dirección General de Investigación, Ministerio de Educación y Ciencia, Spain.*

References

[1] Agresti, A.: Categorical Data Analysis. Second edition, Wiley: New York. (2002).
[2] Aguilera, A. M. Escabias, M. and Valderrama, M. J.: Using principal components for estimating logistic regression with high-dimensional multicollinear data. Computational Statistics and Data Analysis. **50**(8), 1905-1924 (2006).
[3] Cardot, H., R. Faivre and Goulard, M.: Functional approaches for predicting land use with the temporal evolution of coarse resolution remote sensing data. Journal of Applied Statistics. **30**(10), 1185-1199 (2003).
[4] Cardot, H. and Sarda, P.: Estimation in generalized linear models for functional data via penalized likelihood. Journal of Multivariate Analysis. **92**, 24-41 (2005).
[5] Escabias, M., Aguilera, A. M. and Valderrama, M. J.: Principal component estimation of functional logistic regression: discussion of two different approaches. Journal of Nonparametric Statistics. **16**(3-4), 365-384 (2004).
[6] Escabias, M., Aguilera, A. M. and Valderrama, M. J.: Modelling environmental data by functional principal component logistic regression. Environmetrics. **16** (1), 95-107 (2005).
[7] Escabias, M. Aguilera, A. M. and Valderrama, M. J.: Functional PLS logit regression model. Computational Statistics and Data Analysis. **51**(10), 4891-4902 (2007).
[8] James, G. M.: Generalized linear models with functional predictors. Journal of the Royal Statistical Society, Series B. **64**(3), 411-432 (2002).
[9] Marx, B.D. and P.H.C. Eilers.: Generalized linear regression on sampled signals and curves. A p-spline approach. Technometrics. **41**, 1-13 (1999).
[10] Muller, H-G. and StadtMüller, U.: Generalized functional linear models. The Annals of Statistics. **33**(2), 774-805 (2005).
[11] Ocaña, F.A., Aguilera, A.M. and Escabias, M.: Computational considerations in functional principal component analysis. Computational Statistics. **22**(3), 449-466 (2007).
[12] Ramsay, J. O. and Silverman, B. W.: Functional Data Analysis. Second edition, Springer-Verlag: New York. (2005).
[13] Ratcliffe, S. J., Leader, L. R. and Heller, G. Z.: Functional data analysis with application to periodically stimulated foetal heart rate data. II: functional logistic regression. Statistics in medicine. **21**(8), 1115-1127 (2002).

Chapter 3
Estimation of Functional Regression Models for Functional Responses by Wavelet Approximation

Ana Aguilera, Francisco Ocaña and Mariano Valderrama

Abstract A linear regression model to estimate a sample of response curves (realizations of a functional response) from a sample of predictor curves (functional predictor) is considered. Different procedures for estimating the parameter function of the model based on wavelets expansions and functional principal component decomposition of both the predictor and response curves are proposed. Wavelets coefficients will be estimated from discrete observations of sample curves at irregularly spaced time points that could be different among sample individuals.

3.1 Introduction

Functional data analysis is an emerging field in the statistical research designed for modeling a sample of curves that can be seen as realizations of a functional variable. A detailed study of the most common techniques in FDA can be found in the book of Ramsay and Silverman (2005). Theorethical and pratical aspects of nonparametric methods for FDA are collected in the recent book by Ferraty and View (2006). Functional regression models have been studied intensively in the recent literature on this topic. These are regression models where predictors or responses may be viewed as functional variables. The functional linear model to estimate an scalar response from a functional predictor has been studied in Cardot *et al.* (1999). The situation

---

Ana Aguilera
Department of Statistics and O.R., University of Granada, Spain, e-mail: `aaguiler@ugr.es`

Francisco Ocaña
Department of Statistics and O.R., University of Granada, Spain, e-mail: `focana@ugr.es`

Mariano Valderrama
Department of Statistics and O.R., University of Granada, Spain, e-mail: `valderra@ugr.es`

where the predictor is a vector or scalar and the response is functional was analized by Chiou *et al.* (2004).

The aim of this paper is to propose an estimation procedure for a functional regression model where both predictor and response variables are functions. This model has been recently studied by Yao *et al.* (2005) that proposed an estimation approach for sparse and irregular longitudinal data based on a nonparametric estimation of the eigenfunctions of the sample covariance operators associated to both predictor and response functional variables. As a particular case of this functional regression model, principal component prediction models were firstly introduced by Aguilera *et al.* (1999) to forecast a continuous time stochastic process on a future interval from its recent past. The same prediction problem have been solved by using wavelets methods on the notion of autoregressive Hilbert processes (Antoniadis and Sapatinas, 2003).

In this paper we propose a four step estimation procedure of such a functional regression model summarized as

1. Wavelets smoothing of predictor and response sample curves from irregularly spaced longitudinal data.
2. Functional principal component analysis of the orthonormal wavelets approximations of both predictor and response functional variables that is reduced to standart PCA of both matrices of wavelets coefficients.
3. Multivariate linear regression of each functional principal component (PC) of the response curves on an optimum set of PCs of the predictor curves.
4. Cross-validation variable selection procedure which takes into account both the variance explained by each PC of the predictor variable and its correlation with the PC of the response variable that we want to predict.

Finally, the predictive performance of the proposed functional regression model will be studied with real and simulated data.

3.2 Functional linear model for a functional response

Let us consider a functional predictor variable $\{X_w(t) : t \in T, w \in \Omega\}$ and a functional response variable $\{Y_w(s) : s \in S, w \in \Omega\}$ where $(\Omega, \mathcal{A}, \mathcal{P})$ is a probability space, T and S are intervals in $\mathbb{R}$, and both processes have square integrable sample paths.

The sample consists of pairs of random trajectories $\{(x_w(t), y_w(s)), w = 1, ..., n\}$ that can be seen as realizations of the functional predictor and response variables, respectively.

As an extension of the multivariate linear regression model, the functional linear regression model to estimate functional response $Y(s)$ from functional predictor $X(t)$ is

$$y_w(s) = \alpha(s) + \int_T \beta(t,s)x_w(t)dt + \varepsilon_w(s) \quad s \in S,$$

with ε_w being independent and centered random errors and β a square integrable bivariate regression function. We will suppose without loss of generality that predictor and response variables are centered ($\mu_X(t) = \mu_Y(s) = 0$), in other case centered sample curves $\tilde{x}_w(t) = x_w(t) - \mu_X(t))$ and $\tilde{y}_w(s) = y_w(s) - \mu_Y(s)$ will be used instead of $x_w(t)$ and $y_w(s)$, respectively.

Then, the problem is reduced to estimate the conditional mean function

$$E[Y(s)/x_w] = \int_T \beta(t,s)x_w(t)dt. \tag{3.1}$$

Our main aim is to estimate parameter function β. This is an ill-posed problem due to the infinite dimension of predictor and response realizations.

3.3 Model estimation

A first estimation of parameter function β can be obtained by assuming that both predictor and response sample curves belong to finite dimension spaces generated by two different basis $\{\vartheta_p : p = 1, \ldots, P\}$ and $\{\varphi_q : q = 1, \ldots, Q\}$. That is,

$$x_w(t) = \sum_{p=1}^{P} a_{wp}\vartheta_p(t) \quad y_w(s) = \sum_{q=1}^{Q} b_{wq}\varphi_q(s). \tag{3.2}$$

If we assume that the parameter function is expressed as $\beta(t,s) = \sum_{p=1}^{P}\sum_{q=1}^{Q} \beta_{pq}\vartheta_p(t)\varphi_q(s)$, then model (3.1) is equivalent to the following multivariate linear regression model:

$$b_{wq} = \sum_{p=1}^{P} \beta_{pq} \sum_{r=1}^{P} a_{wr}\psi_{pr} + \varepsilon_{wq} \quad q = 1\ldots, Q, \tag{3.3}$$

with $\psi_{pr} = \int_T \vartheta_p(t)\vartheta_r(t)dt$, and ε_{wq} independent and centered random errors.

In matrix form $B = A\Psi\beta + \Upsilon$, where $B = (b_{wq}), A = (a_{wp}), \Psi = (\psi_{pr})$ and $\Upsilon = (\varepsilon_{wq})$. Let us observe that the functional model has been reduced to a multivariate linear regression model of response sample curves coefficients on predictor sample curves coefficients multiplied by the matrix of inner products between predictor basis functions. Then, we can obtain the following estimation of the parameter function

$$\hat{\beta}(t,s) = \sum_{p=1}^{P}\sum_{q=1}^{Q} \hat{\beta}_{pq}\vartheta_p(t)\varphi_q(s),$$

from the least squares estimation of its coefficients matrix
$\hat{\beta} = ((A\Psi)'(A\Psi))^{-1}(A\Psi)'B$.

The problem is that the columns of the design matrix $(A\Psi)$ of this model are usually highly correlated (multicollinearity) so that the estimation of function β can be inaccurate despite the good predictive ability of the model. In this paper we propose an estimation of the parameter function based on functional principal component decomposition of predictor and response curves.

Let us consider the following orthogonal decomposition of predictor and response sample curves, respectively,

$$x_w(t) = \sum_{i=1}^{n-1} \xi_{wi} f_i(t) \quad y_w(s) = \sum_{j=1}^{n-1} \eta_{wj} g_j(s),$$

where ξ_i and η_j are the principal components (PCs) vectors of predictor and response curves, respectively, given by $\xi_{wi} = \int_T x_w(t) f_i(t) dt$ and $\eta_{wj} = \int_S y_w(s) g_j(s) ds$, with $f_i(t)$ and $g_j(s)$ being the principal component weights obtained as the eigenfunctions of the sample covariance operators of predictor and response curves, respectively.

Then, functional regression is equivalent to linear regression of each PC of $Y(s)$ in terms of all PCs of $X(t)$. That is model (3.1) can be written as

$$\eta_{wj} = \sum_{i=1}^{n-1} \xi_{wi} \nu_{ij} + \varepsilon_{wj}, \tag{3.4}$$

so that the parameter function is given by $\beta(t,s) = \sum_{i=1}^{n-1} \sum_{j=1}^{n-1} \nu_{ij} f_i(t) g_j(s)$.

It is known that least squares estimation of ν_{ij} is given by $\hat{\nu}_{ij} = \frac{\sigma_{ij}}{\sigma_i^2}$, with σ_{ij} the corresponding element of the sample cross-covariance matrix of predictor and response principal components and σ_i^2 the sample variance of predictor PC ξ_i.

A functional principal component estimation of parameter function β can be obtained by selecting an optimum number of pc's of the response variable and regress each of them in terms of and optimum number of pc's of the predictor variable.

Then, the following prediction equation can be used to forecast the response $y^*(s)$ associated to a new predictor curve x^*

$$E[y^*(s)/x^*] = \sum_{j=1}^{J} \eta_j^* g_j(s) = \sum_{j=1}^{J} \sum_{i \in I_j} \frac{\sigma_{ij}}{\sigma_i^2} \xi_i^* g_j(s),$$

where principal component ξ_i is predicted as $\xi_i^* = \int_T x^*(t) f_i(t) dt$.

In order to estimate principal components weights $f_i(t)$ and $g_j(s)$ we will assume that predictor and response curves can be expressed as in (3.2). Then, the results set out in Ocaña *et al.* (2007) allow to demonstrate that functional principal component analysis (FPCA) of the predictor sample paths (resp. the response sample curves) with respect to the usual inner product in $L^2(T)$ is equivalent to PCA of the data matrix $A\Psi^{1/2}$ (resp. $B\Pi^{1/2}$) with respect to the usual inner product in $\mathbb{R}^P$ (resp. $\mathbb{R}^Q$), with $\Pi = (\pi_{qr})$ being the matrix of inners products between basis functions defined by $\Pi_{qr} = \int_S \varphi_q(s)\varphi_r(s)ds$.

Let $\Gamma^X = (\xi_{wi})_{n\times P}$ (resp. $\Gamma^Y = (\eta_{wj})_{n\times Q}$) be the matrix whose columns are the PCs of the $A\Psi^{1/2}$ matrix (resp. $B\Pi^{1/2}$), and V^X (resp. V^Y) the one whose columns are the eigenvectors of the sample covariance matrix of $A\Psi^{1/2}$ (resp. $B\Pi^{1/2}$.) Then, $\Gamma^X = \left(A\Psi^{1/2}\right)V^X$ (resp. $\Gamma^Y = \left(B\Pi^{1/2}\right)V^Y$) and the PCs weight functions are

$$f_i(t) = \sum_{p=1}^{P} f_{pi}\vartheta_p(t),\ i = 1,\ldots,P, \quad g_j(t) = \sum_{q=1}^{Q} g_{qj}\varphi_q(t),\ j = 1,\ldots,Q,$$

with $F = (f_{pi})_{P\times P} = \Psi^{-1/2}V^X$ (resp. $G = (g_{qj})_{Q\times Q} = \Pi^{-1/2}V^Y$).

Then, basis coefficients of the parameter function (coefficients $\beta = (\beta_{pq})_{P\times Q}$ of model (3.3)) are estimated in terms of the estimates of the parameters $\nu = (\nu_{ij})_{P\times Q}$ of functional principal component regression model (3.4) as $\hat{\beta} = \Psi^{-1/2}V^X\hat{\nu}(V^Y)'\Pi^{-1/2}$.

Let us observe that when basis are orthonormals, FPCAs of $\{x_w(t)\}$ and $\{y_w(s)\}$ are equivalent to classic multivariate PCAs of the A and B coefficients matrices, respectively. In this case $\Psi = \Pi = I$ so that $\beta = V^X\nu(V^Y)'$. This is the case of the orthonormal wavelet approximation of sample curves considered in this paper.

3.4 Wavelet approximation of sample curves

In practice, basis coefficients of predictor and response sample curves need to be estimated from discrete time observations $\{x_{wk} : k = 1,\ldots,K_w\}$ and $\{y_{wl} : l = 1,\ldots,L_w\}$ of each predictor and response sample curves $x_w(t)$ and $y_w(s)$ at a finite set of time points $(t_{wk} : k = 1,\ldots,K_w)$ and $(s_{wl} : l = 1,\ldots,L_w)$, respectively.

In order to estimate basis coefficients from irregularly distributed observations, we will consider that discrete time observations have been observed with some error

$$x_{wk} = x_w(t_{wk}) + \varepsilon_{wk} = \sum_{p=1}^{P} a_{wp}\vartheta_p(t_{wk}) + \varepsilon_{wk} \quad k = 1,\ldots,K_w$$

$$y_{wl} = y_w(s_{wl}) + \varepsilon_{wk} = \sum_{q=1}^{Q} b_{wq}\varphi_q(s_{wl}) + \varepsilon_{wk} \quad l = 1, \ldots, L_w.$$

There are many applications where sample paths are not smooth curves. Wavelets provide useful methods for analyzing data with intrinsically local properties, such as discontinuities and sharp spikes. They form orthonormal basis and enable multiresolution analysis by localizing a function in different phases of both time and frequency domains simultaneously, and thus offer some advantages over traditional Fourier expansions. In this paper we propose to estimate basis coefficients by orthogonal projection of each predictor and response sample curve on basis of wavelets on bounded intervals. For simplicity, we will summarize wavelet approximation for the functional response X.

Let ϕ and ψ be the scaling and wavelet functions for an orthogonal multiresolution analysis (MRA) of $L^2(\mathbb{R})$. It is known that there exist different wavelet families with the orthogonal and compact support restriction which can satisfy interesting properties from an approximating and a denoising point of view (Mallat, 1998).

In our case we should consider the adaptation of the wavelet analysis onto a bounded interval. For simplicity, one of the usual wavelet adaptation in the space $L^2[0,1]$ is considered in this work (Mallat, 1998). Among other consequences, the proposed methodology will be formulated in terms of the unit interval $[0,1]$. In fact, for any given $J \in \mathbb{N}$, let V_J be the subspace at level J of the considered MRA in $L^2[0,1]$ and let $\{\phi^*_{J,k} : k = 0, \ldots, 2^J - 1\}$ be the scaling orthonormal basis of V_J. The orthogonal approximation of each sample path of x, at resolution level J, can be formulated by

$$\mathcal{P}_J x_w(\tau) = \sum_{k=0}^{2^J-1} \lambda_{J,k}(w)\, \phi^*_{J,k}(\tau), \qquad \forall \tau \in [0,1], \tag{3.5}$$

where $\lambda_{J,k} = \int_0^1 X(\tau)\phi_{J,k}(\tau)\, d\tau$. In this way, the FPCA of $\mathcal{P}_J X$ can be viewed as an approximation to the FPCA of X.

A more sparsely decomposition of $\mathcal{P}_J X$ can be obtained as follows by applying the Discrete Wavelet Transform (DWT) to the coordinate vector $\lambda_J = (\lambda_{J,k})_k$:

$$\mathcal{P}_J x_w(\tau) = \lambda_{0,0}(w)\phi^*_{0,0}(\tau) + \sum_{j=0}^{J-1}\sum_{k=0}^{2^j-1} \gamma_{j,k}(w)\, \psi^*_{j,k}(\tau) \in V_0 + \sum_{j=0}^{J-1} W_j, \tag{3.6}$$

where $\{\phi^*_{0,0}\} \cup \bigcup_{j=0}^{J-1} \{\psi^*_{j,k}\}_{k=0}^{2^j-1}$ is an orthonormal basis.

Unfortunately, the discrete time observation assumption makes impossible to exactly compute the scaling and thus the wavelet coordinates for the expansions given in equations (3.5) and (3.6). Nevertheless, once the scaling coordinates $\lambda_{J,k}$ are approximated, the coordinates in equations (3.6) could be directly approximated by applying DWT to the approximated scaling coordinates.

Acknowledgements This research has been funded by project P06-FQM-01470 from *"Consejería de Innovación, Ciencia y Empresa. Junta de Andalucía, Spain"* and project MTM2007-63793 from *Dirección General de Investigación, Ministerio de Educación y Ciencia, Spain.*

References

[1] Aguilera, A.M., Ocaña, F.A. and Valderrama, M.J.: Forecasting with unequally spaced data by a functional principal component approach. Test. **8**(1), 233-254 (1999).

[2] Antoniadis, A. and Sapatinas, T.: Wavelet methods for continuous-time prediction using Hilbert-valued autoregressive processes. Journal of Multivariate Analysis. **87**(1), 133-158 (2003).

[3] Cardot, H., Ferraty, F. and Sarda, P.: Functional linear model. Statistics and Probability Letters. **45**, 11-22 (1999).

[4] Chiou, J-M., Müller, H-G. and Wang, J-L.: Functional response model. Statistica Sinica. **14**, 659-677 (2004).

[5] Daubechies, I.: Orthonormal bases of compactly supported wavelets. Communications on Pure and Applied Mathematics. **41**, 909-996 (1988).

[8] Ferraty, F. and View, P.: Nonparametric functional data analysis. Theory and practice. Springer-Verlag: New York (2006).

[7] Mallat, S.: A wavelet tour of signal processing. Academic Press: San Diego (1998).

[8] Ocaña, F.A., Aguilera, A.M. and Escabias, M.: Computational considerations in functional principal component analysis. Computational Statistics. **22**, 449-466 (2007).

[13] Ramsay, J. O. and Silverman, B. W.: Functional Data Analysis. Second edition, Springer-Verlag: New York (2005).

[10] Yao, F., Muller, H-G. and Wang, J-L.: Functional linear regression analysis for longitudinal data. The Annals of Statistics. **33**(6), 2873-2903 (2005).

Chapter 4
Functional Linear Regression with Functional Response: Application to Prediction of Electricity Consumption

Jaromír Antoch, Luboš Prchal, Maria Rosaria De Rosa and Pascal Sarda

Abstract Functional linear regression model linking observations of a functional response variable with measurements of an explanatory functional variable is considered. The slope function is estimated with a tensor product splines. Some computational issues are addressed by means of a simulation study. This model serves to analyze a real data set concerning electricity consumption in Sardinia. The interest lies in predicting either incoming weekend or incoming weekdays consumption curves if actual weekdays consumption is known.

4.1 Introduction

Our aim is to analyze the effect of a functional variable on a functional response by means of functional linear regression models. The application motivating this study concerns electricity consumption in Sardinia. The data set consists in 52 584 values of electricity consumption collected every hour

---

Jaromír Antoch
Charles University of Prague Department of Probability and Mathematical Statistics, Sokolovská 83, CZ – 186 75 Prague 8, Czech Republic, e-mail: jaromir.antoch@mff.cuni.cz

Luboš Prchal
Department of Probability and Mathematical Statistics, Sokolovská 83, CZ – 186 75 Prague 8, Czech Republic and Université Paul Sabatier, Institut de Mathématiques de Toulouse, UMR 5219, 118, route de Narbonne, F – 310 62 Toulouse Cedex, France, e-mail: lubos.prchal@mff.cuni.cz,prchal@cict.fr

Maria Rosaria De Rosa
ITIS Galileo Ferraris, Napoli, Italy, e-mail: marodero@libero.it

Pascal Sarda
Université Paul Sabatier, Institut de Mathématiques de Toulouse, UMR 5219, 118, route de Narbonne, F – 310 62 Toulouse Cedex, France, e-mail: sarda@cict.fr

within the period January 1, 2000, till December 31, 2005. The complete data series has been cut into 307 weeks for which the weekdays (Monday to Friday) and the weekends (Saturday and Sunday) have been separated, leading to the two sets of discretized curves of electricity consumption. The reason for this separation is substantial difference between weekdays and weekend consumptions. Our main purpose was to predict the consumption for the next weekdays and the consumption for the next weekend. In each case the (functional) predictor is the (discretized) curve of the present weekdays consumption.

We adopt a general framework for both situations, considering the data as observations of identically distributed random functional variables $\{X_i(s), Y_i(t), s \in I_1, t \in I_2\}$, $i = 1, \dots, n$, defined on the same probability space and taking values in some functional spaces. The most common is to consider the separable real Hilbert spaces $L^2(I_1)$ and $L^2(I_2)$ of square integrable functions defined on the compact intervals $I_1 \subset \mathbb{R}$ and $I_2 \subset \mathbb{R}$, which are equipped with the standard inner products. We focus on the functional linear relation of the type

$$Y_i(t) = \alpha(t) + \int_{I_1} X_i(s)\beta(s,t)\, ds + \varepsilon_i(t), \quad t \in I_2, \quad i = 1, \dots, n, \tag{4.1}$$

where $\alpha(t) \in L^2(I_2)$ and $\beta(s,t) \in L^2(I_1 \times I_2)$ are unknown functional parameters and $\varepsilon_1(t), \dots, \varepsilon_n(t)$ stand for a sample of i.i.d. centered random variables taking values in $L^2(I_2)$, $\varepsilon_i(t)$ and $X_i(s)$ being uncorrelated. For a generic interval I, the set $L^2(I)$ is equipped with its usual inner product $\langle \phi, \psi \rangle = \int_I \phi(t)\psi(t)dt$, $\phi, \psi \in L^2(I)$ and the associated norm $\|\phi\| = \langle \phi, \phi \rangle^{1/2}$.

In what follows we often omit arguments of the functional variables and parameters and simply write X_i, Y_i, ε_i and β instead of $\{X_i(s), s \in I_1\}$, $\{Y_i(t), t \in I_2\}$, $\{\varepsilon_i(t), t \in I_2\}$ and $\{\beta(s,t), s \in I_1, t \in I_2\}$, respectively. Notice that in model (4.1) X_i's represent a weekdays curves while Y_i's represent a weekend curves, or a weekday curve in which case $Y_i = X_{i+1}$, and the model (4.1) corresponds to an ARH(1) as defined in Bosq (2000).

Bivariate parameter $\beta(s,t)$ is estimated by means of a tensor product splines minimizing a penalized least squares criterion. Computational aspects, comments on "tuning" estimator parameters and some remarks on discretization and eventual curve presmoothing are discussed on the basis of a simulation study. Further, we switch to the problem of predicting the consumption by analyzing in a first step detrended data and adding to it in the second step trend adjustment.

General linear regression model (4.1) has been studied by several authors, e.g., by Ferraty and Vieu (2006), Müller and Wang (2003) or Ramsay and Silverman (2005). The case of a scalar response has been for the first time considered in Ramsay and Dalzell (1991).

4.2 Estimation procedure

Before defining our estimation procedure, return to the model (4.1) an discuss identifiability of the model, i.e. existence and uniqueness of the slope function β.

Let $\boldsymbol{B_j} = (B_{j1}, \dots B_{jd_j})', j = 1, 2$, denote the normalized B-splines basis of the spline space $\mathcal{S}_{q_j k_j}(I_j)$ of degree q_j defined on the interval I_j with $k_j - 1$ equidistant interior knots and $d_j = k_j + q_j$ being the dimension of $\mathcal{S}_{q_j k_j}(I_j)$, see Dierckx (1993). Our estimator $\widehat{\beta}$ of β is a bivariate spline with a tensor product representation defined as $\widehat{\beta}(s,t) = \boldsymbol{B_1}'(s)\widehat{\boldsymbol{\Theta}}\boldsymbol{B_2}(t)$, where the $d_1 \times d_2$ matrix $\widehat{\boldsymbol{\Theta}}$ satisfies

$$\widehat{\boldsymbol{\Theta}} = \underset{\boldsymbol{\Theta} \in \mathbb{R}^{d_1 \times d_2}}{\arg\min} \quad \frac{1}{n}\sum_{i=1}^{n} \left\| Y_i - \overline{Y} - \int_{I_i} \left(X_i(s) - \overline{X}(s)\right) \boldsymbol{B_1}'(s)\boldsymbol{\Theta}\boldsymbol{B_2}(\cdot)\, ds \right\|^2 + \quad (4.2)$$

$$\varrho\, \mathsf{Pen}(m, \boldsymbol{\Theta}),$$

with a penalty parameter $\varrho > 0$ and the penalty term given by

$$\mathsf{Pen}(m, \boldsymbol{\Theta}) = \sum_{m_1=0}^{m} \frac{m!}{m_1!(m-m_1)!} \int_{I_2}\int_{I_1} \left[\frac{\partial^m}{\partial s^{m_1}\partial t^{m-m_1}} \boldsymbol{B_1}'(s)\boldsymbol{\Theta}\boldsymbol{B_2}(t)\right]^2 dsdt.$$

An explicit solution of (4.2) is derived using Kronecker product notation. Alternatively one can approximate this solution by a simpler matrix version $\widetilde{\boldsymbol{\Theta}}$ if one replaces $\mathsf{Pen}(m, \boldsymbol{\Theta})$ in the minimization task (4.2) with

$$\widetilde{\mathsf{Pen}}(m, \boldsymbol{\Theta}) = \int_{I_2}\int_{I_1} \left\{ \left[\boldsymbol{B}_1^{(m)\prime}\boldsymbol{\Theta}\boldsymbol{B}_2^{(0)}\right]^2 + \left[\boldsymbol{B}_1^{(0)\prime}\boldsymbol{\Theta}\boldsymbol{B}_2^{(m)}\right]^2 \right\} ds\, dt.$$

Hence, the approximating estimator of the functional parameter $\beta(s,t)$ is defined as $\widetilde{\beta}(s,t) = \boldsymbol{B_1}'(s)\widetilde{\boldsymbol{\Theta}}\boldsymbol{B_2}(t)$. Numerical solution is performed with the use of an algorithm discussed by Benner, *et al.* (2002). The intercept parameter α can be estimated either by

$$\widehat{\alpha}(t) = \overline{Y}(t) - \int_{I_1} \widehat{\beta}(s,t)\overline{X}(s)\, ds, \qquad \forall t_2 \in I, \tag{4.3}$$

or approximated by $\widetilde{\alpha}(t)$ if $\widetilde{\beta}$ is used instead of $\widehat{\beta}$ in (4.3). This estimator was introduced and its symptotic behavior of $\widehat{\beta}$ in terms of error of prediction are studied in Prchal and Sarda (2007).

4.3 Computational aspects and simulations

Numerical calculation of the estimator $\widehat{\beta}$ and $\widehat{\alpha}$ requires proper choice of several parameters: order q_j of the splines, order of derivatives m, the numbers of the knots k_j and penalization parameter ρ. It is known that orders q_j and m do not play an important role compared to k_j and ρ. It is thus usual to take values equal to q_j=3 or 4 and $m = 2$. Besides, it has been stressed in similar context that a strategy to choose k_j and ρ is to fix the number of knots k_j reasonably large, and thus both to prevent oversmoothing and to control degree of smoothness of the estimator with the parameter ρ. In our simulation experiments we have used value of k between 15 and 30, while ρ was chosen to minimize the leave-one-out cross-validation criterion

$$\mathsf{cv}(\varrho) = \sum_{i=1}^{n} \int_{I_2} \left[Y_i(t) - \int_{I_1} \widehat{\beta}_i(s,t) X_i(s)\, ds \right]^2 dt, \tag{4.4}$$

where $\widehat{\beta}_i(s,t)$ is obtained from the data set with the i-the pair (X_i, Y_i) omitted. An alternative and computationally faster criterion $\widetilde{\mathsf{cv}}(\varrho)$ can be used when replacing $\widehat{\beta}_i$ with $\widetilde{\beta}_i$. From our experience the approximating criterion provides in many cases a value close to the one obtained by minimizing the criterion (4.4). It can also be used to provide a pivotal parameter for the search of the minimizer of (4.4).

In order to investigate behavior of our estimator for the data which are "under the control", a series of simulations was performed. More precisely, we considered a general regression context with i.i.d. pairs (X_i, Y_i), $i = 1, \ldots, n$, but we guess that the results essentially remains valid in the case of a weak dependence as is the case, for example, of the *ARH(1)* processes. Therefore, we have simulated independent Brownian motion trajectories $X_i(s)$, $i = 1, \ldots, n$, on $[0,1]$ discretized at p equidistant points t_j and studied two functional parameters $\beta_1(s,t) = 5\sin(2\pi s)\cos(2\pi t)$, and $\beta_2(s,t) = 20\exp\{-100(s-t)^2\}$. The error terms ε_i were drawn at each point t_j from a Gaussian white noise. Our simulation study shows that the approximating matrix solution is competitive with the exact estimator and, as concerns data fitting, behaves satisfactorily. If one primarily focuses on the functional parameter estimation, the exact solution should be preferred as it is more stable as concerns tuning parameters of the method. The matrix approach, however, can still be used throughout the cross-validation procedure at least as the pivot parameter, whose neighborhood is then seek throught by the exact method.

Surprisingly, in some situations a very small number of knots can be sufficient to obtain good estimators. As the matrix method behaves well and is fast, it is worth performing estimation for several knot setups – eventually a kind of cross-validation can be used for the knots as well. Interesting is also the case of errors-in-variables due to, e.g., not exact predictor registering, for which a presmoothing of the curves or functional total least squares might be involved, for details see Cardot *et al.* (2007).

4.4 Prediction of electricity consumption

Figure 4.1 represents the complete data sets of electricity consumption in Sardinia during the years 2000 − 2005. In order to eliminate (and estimate) increasing trend, we have performed a one-sided kernel smoother which is known to be an efficient way to eliminate the trend and at least a part of the seasonality in a nonparametric framework.

As said before, we have concentrated on week interval separating weekdays (Monday to Friday) from weekends (Saturday and Sunday) and estimated and predicted each part separately. More precisely, spline estimator $\widehat{\beta}$ or its approximation $\widetilde{\beta}$ were used to predict either weekdays or weekends consumption curves. Concerning the parameters, we have used cubic splines, i.e. $q_1 = q_2 = 4$, and a derivative order $m = 2$. The cross-validation criterion (4.4) was used to select the penalty parameter ϱ (with either $\widehat{\beta}$ or $\widetilde{\beta}$ as a basis estimator). The same criterion (only with the computationally faster estimator $\widetilde{\beta}$) has
been also used to select the number of knots. Different number of knot was considered for predictor and response basis which was performed in each case, that is for the weekdays prediction and for the weekends prediction.

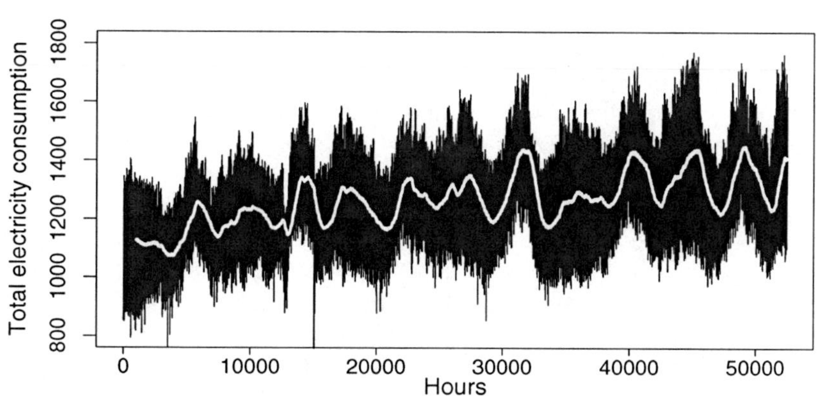

Fig. 4.1 Electricity consumption data with estimated trend.

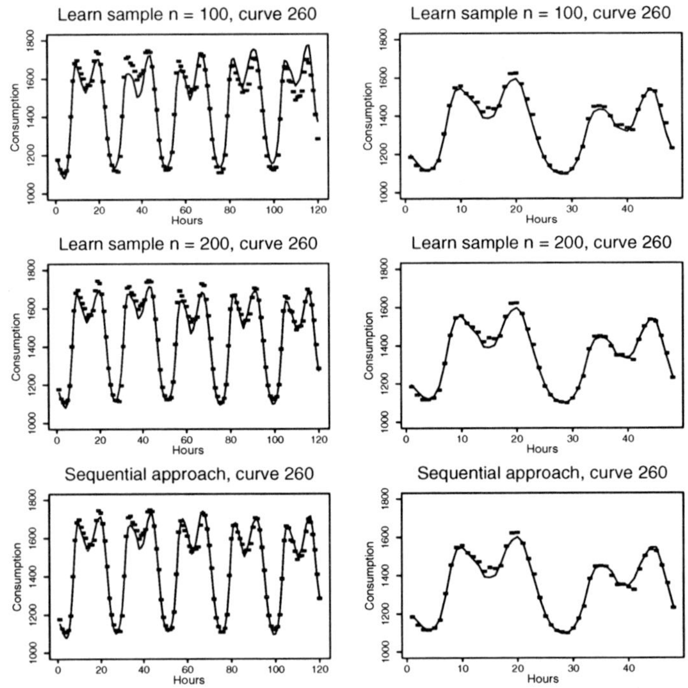

Fig. 4.2 Prediction of electricity data for the week 260 for both learn/test and sequential approaches. Dots stay for the observed data, while the solid lines present the prediction.

While for weekends prediction the number of knots is more or less the same for predictor basis and response basis, it turns out that for weekdays prediction considerably smaller number of knots for the predictor basis than the response one is sufficient despite the fact that both variables are of the same nature.

We have compared a *sequential approach* with a *learn-test* approach for detrended prediction. The trend adjustment was then performed via kernel estimation. Figure 4.2 shows a prediction of electricity data for the week 260 for both learn/test and sequential approaches as an example.

It appears that the functional approach is a competitive methodology to predict electricity consumption. Further, our experiments have confirmed satisfactory behavior of the approximating matrix estimator of the functional parameter that is computationally much simpler and faster than the exact vectorial one and provides fully competitive prediction results.

Finally, some open problems still remain. One of them is how to include in the estimation procedure the knowledge of special events such as festive days which may influence the prediction. Several solutions can be considered such as involving a longer consumption history as the predictor or replace the least squares approach with a more robust criterion.

Acknowledgements Authors would like to express their thanks to the members of the working group STAPH in Toulouse (`http://www.lsp.ups-tlse.fr/staph`) for fruitful discussions and valuable comments. The work of J. Antoch and L. Prchal is a part of the research project MSM 0021620839 of the MŠMT ČR, J. Antoch was partially supported by the grant GAČR 201/06/0186 and L. Prchal by the grant GAČR 201/05/H007.

References

[1] Benner, P., Quintana-Ortí, E.S. and Quintana-Ortí, G.: Numerical solution of discrete stable linear matrix equations on multicomputers. Parallel Algorithms Appl. **17**, 127 – 146 (2002).

[2] Bosq, D.: Linear Processes in Function Spaces. Lecture Notes in Statistics. **149**. Springer-Verlag, New-York (2000).

[3] Cardot, H., Crambes, C., Kneip, A. and Sarda, P.: Smoothing spline estimators in functional linear regression with errors-in-variables. Comput. Statist. Data Anal. **51**, 4832 – 4848 (2007).

[4] Dierckx, P.: Curve and Surface Fitting with Splines. Oxford University Press, Oxford (1993).

[8] Ferraty, F. and Vieu, P.: Nonparametric Functional Data analysis: Methods, Theory, Applications and Implementations. Springer-Verlag, London (2006).

[6] He, G., Müller, H.G. and Wang, J.L.: Extending correlation and regression from multivariate to functional data.: In Asymptotics in Statistics and Probability. Ed. Puri, M.L., VSP International Science Publishers, 301 – 315 (2003).

[7] Prchal, L. and Sarda, P.: Spline estimator for functional linear regression with functional response. Preprint. (2007).

[8] Ramsay, J.O. and Dalzell, C.J.: Some tools for functional data analysis (with discussion). J. R. Stat. Soc. Ser. B Stat. Methodol. **53**, 539 – 572 (1991).

[9] Ramsay, J.O. and Silverman, B.W.: Functional Data Analysis. 2nd edition. Springer, New York (2005).

Chapter 5
Asymptotic Normality of Robust Nonparametric Estimator for Functional Dependent Data

Mohammed Attouch, Ali Laksaci and Elias Ould-Saïd

Abstract We propose a family of robust nonparametric estimators for regression function based on kernel method. We establish the asymptotic normality of the estimator under the concentration properties on small balls of the probability measure of the functional explanatory dependent variables.

5.1 Introduction

The robust method used here belongs to the class of M-estimates introduced by Huber (1964). The literature on this estimation method is quite important when the data are vectors (see for instance Robinson (1984), Collomb and Härdle (1986), Boente and Fraiman (1989 and 1990), Fan *et al.* (1994) for previous results and Laïb and Ould-Saïd (2000), Boente and Rodriguez (2006) for recent advances and references).
In the functional case, Cadre (2001) studied the estimate of the median (without conditioning) of the distribution of a random variable taking its values in a Banach space. Cardot *et al.* (2005) used this robust approach to consider the linear model of regression quantile with the explanatory variable taking values in a Hilbert space. They established the L_2-convergence rate of the regression quantile estimators. Recently, Azzedine *et al.* (2006) obtained a rate of the almost complete convergence of the robust nonparametric regression estimator when the regressors are functional. The asymptotic normality of the

Mohammed Attouch
Univ. Djillali Liabès, B.P. 89, Sidi Bel Abbès 22000, Algérie, e-mail: attou_kadi@yahoo.fr

Ali Laksaci
Univ. Djillali Liabès, B.P. 89, Sidi Bel Abbès 22000, Algérie, e-mail: Laksaci@yahoo.fr

Elias Ould-Said
Univ. du Littoral Côte d'Opale, BP 699, 62228 Calais, France, e-mail: ouldsaid@lmpa.univ-littoral.fr

classical linear kernel estimators has been established by Masry (2005). Under the concentration properties on small balls of the probability measure of the underlying functional variable, Ezzahrioui and Ould-Saïd ((2005), (2006)) studied the asymptotic normality of the kernel estimator of the conditional mode and the conditional quantile in both iid and dependent cases, and Attouch *al.* (2007) obtained the asymptotic normality for regression function based in kernel method in the iid case. Among the wide literature concerning the functional case, we only refer to the good overviews in the parametric models given by Ramsay and Silverman (2002), (2005) and to the monograph of Ferraty and Vieu (2006) for the prediction problem in functional nonparametric statistics via the regression function, the conditional mode and the conditional quantile estimation by the kernel method.
In this work, we establish asymptotic normality of the kernel M-estimate under less restrictive conditions related to some regularity of the model, the topology structure of the functional data space. Our results are applied to derive asymptotic normality of the predictor estimate, to build confidence curve.

5.2 Model and estimation

Let $(X_n, Y_n)_{n\geq 1}$ distributed as (X, Y) which is a random pair valued in $\mathcal{F}\times\mathbb{R}$, where $\mathcal{F}$ is a semi-metric space, $d(.,.)$ denoting the semi-metric. For any x in $\mathcal{F}$, we consider ψ_x a real-valued Borel function satisfying some regularity conditions to be stated below. The nonparametric model studied in this paper, denoted by θ_x, is implicitly defined as a zero with respect to (w.r.t.) t of the following equation

$$\Psi(x,t) = \mathbb{E}\left[\psi_x(Y,t) \mid X = x\right] = 0. \tag{5.1}$$

We suppose that, for all $x \in \mathcal{F}$, θ_x exists and is the unique zero with respect to t of (5.1) (see, for instance Boente & Fraiman (1989) for this problem). The kernel estimate of $\Psi(x,t)$ is defined by

$$\widehat{\Psi}(x,t) = \frac{\sum_{i=1}^{n} K(h_n^{-1}d(x,X_i))\psi_x(Y_i,t)}{\sum_{i=1}^{n} K(h_n^{-1}d(x,X_i))}, \quad \forall t \in \mathbb{R}$$

where K is a kernel and h_n is a sequence of positive real numbers. A natural estimator of θ_x denoted by $\widehat{\theta_x}$, is a zero w.p.t. t of the

$$\widehat{\Psi}(x,t) = 0. \tag{5.2}$$

The robust method used in this paper is belongs in the class of M-estimates introduced by Huber (1964).

In the follows we suppose that $(X_n, Y_n)_{n\geq 1}$ are strongly mixing.

5.3 Hypothesis and results

In the following x will be a fixed point in $\mathcal{F}$, N_x will denote a fixed neighborhood of x and we set $\lambda_\gamma(u,t) = \mathbb{E}\left[\psi_x^\gamma(Y-t)|X=u\right]$ and $\Gamma_\gamma(u,t) = \mathbb{E}\left[(\psi_x')^\gamma(Y-t)|X=u\right]$, for $\gamma \in \{1,2\}$.
We need the following hypotheses:

(H1) $\mathbb{P}(X \in B(x,h)) = \phi_x(h) > 0$,

(H2) ψ_x is continuous differentiable function, monotone, bounded, w.r.t. the second component, and its derivative $\dfrac{\partial \psi_x(y,t)}{\partial t}$ is bounded and continuous at θ_x uniformly in y.

(H3) The function $\lambda_\gamma(\cdot,\cdot)$ satisfies the Lipschitz's condition w.r.t. the first one, that is: there exists a strictly positive constant b_γ such that:

$$\forall (u_1,u_2) \in N_x \times N_x,\ \forall t \in \mathbb{R},\ |\lambda_\gamma(u_1,t) - \lambda_\gamma(u_2,t)| \le C_1 d(u_1,u_2)^{b_\gamma}.$$

(H4) The function $\Gamma_\gamma(\cdot,\cdot)$ satisfies the Lipschitz's condition w.r.t. the first one, that is: there exists a strictly positive constant d_γ such that:

$$\forall (u_1,u_2) \in N_x \times N_x,\ \forall t \in \mathbb{R},\ |\Gamma_\gamma(u_1,t) - \Gamma_\gamma(u_2,t)| \le C_2 d(u_1,u_2)^{d_\gamma}.$$

(H5) The bandwidth h satisfies:

$$h \downarrow 0,\ \ \forall t \in [0,1]\ \lim_{h\to 0} \frac{\phi_x(th)}{\phi_x(h)} = \beta_x(t)\ \text{ and }\ n\phi(h) \to \infty\ \text{ as } n \to \infty.$$

(H6) The kernel K from $\mathbb{R}$ into $\mathbb{R}^+$ is a differentiable function supported on $[0,1]$. Its derivative K' exists and is such that there exist two constants C_3 and C_4 with $-\infty < C_3 < K'(t) < C_4 < 0$ for $0 \le t \le 1$.

(H7) $(X_i,Y_i)_{i\in\mathbb{N}}$ is an α-mixing sequence whose coefficients satisfy

$$\exists\, a > 0,\ \exists\, C > 0 :\ \forall n \in \mathbb{N}\ \alpha(n) \le Cn^{-a}.$$

(H8) $$0 < \sup_{i\neq j} \mathbb{P}\left((X_i,X_j) \in B(x,h) \times B(x,h)\right) = O\left(\frac{(\phi_x(h))^{(a+1)/a}}{n^{1/a}}\right).$$

(H9) $\lim_{n\to\infty} h = 0$ and $\exists \eta > 0$, such that, $Cn^{1-a+\eta} \le \phi_x(h) \le C'n^{\frac{1}{1-a}+\eta}$, with $a > 2$.

Theorem 5.1. *Assume that (H1)-(H9) hold, then $\widehat{\theta_x}$ exists and is unique with probability tending to 1, and for any $x \in \mathcal{A}$, we have*

$$\left(\frac{n\phi_x(h)}{\sigma^2(x,\theta_x)}\right)^{1/2} \left(\widehat{\theta_x} - \theta_x - B_n(x)\right) \xrightarrow{\mathcal{D}} \mathcal{N}(0,1) \text{ as } n \to \infty \tag{5.3}$$

with explicit expression of $\sigma^2(x,\theta_x)$ and $B_n(x)$.

5.4 Conditional confidence curve

To build confidence curve for the true value of θ given curve $X = x$, we use an asymptotic approximation provided by the following Corollary where $\sigma(x, \theta_x)$ is substituted by it estimate.

Corollary 5.1. *To remove the bias term $B_n(x)$ from the equation (5.3), we assume that $nh^{2b_1}\phi(h) \to 0$ on the bandwidth parameter h. Then, under the assumptions of Theorem 5.1, we have for any $x \in \mathcal{A}$*

$$\left(\frac{n\phi(h_n)}{\widehat{\sigma}_n^2(x, \widehat{\theta}_x)}\right)^{1/2} \left(\widehat{\theta}_x - \theta_x\right) \to \mathcal{N}(0,1) \quad as \quad n \to \infty.$$

From this Corollary we get the following $(1-\eta)$ confidence curve

$$\widehat{\theta}_x \pm t_{1-\eta} \times \left(\frac{\widehat{\sigma}_n^2(x, \widehat{\theta}_x)}{n\phi(h_n)}\right)^{1/2}$$

where $t_{1-\eta}$ denotes the $1-\eta$ quantile of the standard normal distribution.

References

[1] Benner, P., Quintana-Ortí, E.S. and Quintana-Ortí, G.: Numerical solution of discrete stable linear matrix equations on multicomputers. Parallel Algorithms Appl. **17**, 127 – 146 (2002).

[2] Bosq, D.: Linear Processes in Function Spaces. Lecture Notes in Statistics. **149**. Springer-Verlag, New-York (2000).

[3] Cardot, H., Crambes, C., Kneip, A. and Sarda, P.: Smoothing spline estimators in functional linear regression with errors-in-variables. Comput. Statist. Data Anal. **51**, 4832 – 4848 (2007).

[4] Dierckx, P.: Curve and Surface Fitting with Splines. Oxford University Press, Oxford (1993).

[8] Ferraty, F. and Vieu, P.: Nonparametric Functional Data analysis: Methods, Theory, Applications and Implementations. Springer-Verlag, London (2006).

[6] He, G., Müller, H.G. and Wang, J.L.: Extending correlation and regression from multivariate to functional data.: In Asymptotics in Statistics and Probability. Ed. Puri, M.L., VSP International Science Publishers, 301 – 315 (2003).

[7] Prchal, L. and Sarda, P.: Spline estimator for functional linear regression with functional response. Preprint. (2007).

[8] Ramsay, J.O. and Dalzell, C.J.: Some tools for functional data analysis (with discussion). J. R. Stat. Soc. Ser. B Stat. Methodol. **53**, 539 – 572 (1991).

[9] Ramsay, J.O. and Silverman, B.W.: Functional Data Analysis. 2nd edition. Springer, New York (2005).

Chapter 6
Measuring Dissimilarity Between Curves by Means of Their Granulometric Size Distributions

Guillermo Ayala, Martin Gaston, Teresa Leon and Fermín Mallor

Abstract The choice of a dissimilarity measure between curves is a key point for clustering functional data. In this paper we propose to obtain the granulometric distribution functions of the original curves and then calculate the dissimilarities between the new functional data. Good results have been obtained with two real examples.

6.1 Introduction

The issue of measuring the similarity between curves has been addressed by a number of authors in the literature. It is well known that the L_2 distance only provides a sensible criterion for clustering when the curves have approximately the same shape. Different semi-metrics between curves have been defined. Ferraty and Vieu (2003) proposed a semi-metric based on the derivatives of the curves and a semi-metric based on the functional principal component analysis. Other interesting approach can be found in Cerioli *et al.* (2005) where the original functional data are continuous trajectories evolving

---

Guillermo Ayala
Departamento de Estadística e I.O., Universidad de Valencia, Avda. Vicent Andres Estelles 1, 46100, Burjassot Spain e-mail: `guillermo.ayala@uv.es`

Martin Gaston
Departamento de Estadistica e I.O., Universidad Pública de Navarra. Campus de Arrosadía, 31006, Pamplona, Spain, e-mail: `martin.gaston@unavarra.es`

Teresa Leon
Departamento de Estadística e I.O., Universidad de Valencia , Avda. Vicent Andres Estelles 1, 46100, Burjassot Spain, e-mail: `teresa.leon@uv.es`

Fermín Mallor
Departamento de Estadistica e I.O., Universidad Pública de Navarra. Campus de Arrosadía, 31006, Pamplona, Spain, e-mail: `mallor@unavarra.es`

over time. These authors propose a new dissimilarity measure based on their sequence of local extrema (maxima or minima). Some statistical methods usually work with transformations of the original data. Registration of functional data is a basic transformation for curves Ramsay and Silverman (1997) in which the arguments of the different functional data are modified. The idea underlying curve registration is that two functions can differ because of two sources of variation: the usual amplitude variation and phase variation. A transformation of the time scale permits the pointwise comparison of the two functions. The package FDA (2007) (Ramsay *et al.* (2007)) includes curve registration (function *registerfd*). Dynamic time warping technique for aligning curves is studied in Morlini (2005) as an essential preliminary in many applications before classification.

In this paper, we propose a new preprocessing technique which takes into account the shape of the curves using tools from Mathematical Morphology. This technique, originally developed at sixties by Matheron and Serra at the Ecole des Mines in Paris, describes an image (or a function in our case) using set operations (extended to functions). A general reference about Mathematical Morphology is Soille (2003). In particular, our approach is heavily based on granulometries, essentially, they are size distributions associated with the original functions and have been used as shape-size descriptors in binary images and as texture descriptors for gray-level images. Our proposal is to obtain the granulometric size distribution of the original curves as a preliminary step and then different dissimilarity measures between the granulometric cumulative distribution functions will be used to group the original functional data. Different empirical studies will be given showing the better performance of the clustering techniques applied to this transformed functional data instead of the original ones. Section 2 contains some basic definitions and preliminaries. Finally, Section 3 shows the methods and the classification results for two datasets: radar waveforms and vertical forces exerted on the ground by both feet during the sit-to-stand movement.

6.2 Basic concepts

Although the main idea in Functional Data Analysis (FDA) is to take into account the continuous feature of the data, they are collected as n observed digitized curves $\{\chi_i(t_j); j = 1, \ldots p\}_{i=1,\ldots,n}$, where the observation points $\{t_j\}_{j=1}^{p}$ are usually equidistant. We use the indices $\{1, \ldots p\}$ to index our sets: $\chi(t_j) = \chi(j)$. Morphological operators extract relevant structures of the set under study by probing it with another set of known shape called structuring element (SE from now on) in the 1-D case it is usually a discrete interval. For the sake of simplicity, most of the definitions in this section will be given for the particular case of 1-D images, i.e. discrete functions.

The *erosion* of a function f by a SE B is the function $\varepsilon_B(f)$ defined at $x \in \{1, \dots p\}$ as $[\varepsilon_B(f)](x) = min_{b \in B} f(x+b)$. The *dilation* of a function f by a SE B, $\delta_B(f)$, at $x \in \{1, \dots p\}$ is defined as $[\delta_B(f)](x) = max_{b \in B} f(x+b)$. The ***structural*** *opening* or simply the *opening* of f by a symmetric SE B, $\gamma_B(f)$ is defined as $\gamma_B(f) = \delta_B(\varepsilon_B(f)) = \delta_B \varepsilon_B(f)$. The subgraph of the opened function is equivalent to the union of the translations of the SE when it fits the subgraph of the original function.

The *granulometry* of a given set G is a formalization of the intuitive concept of size distribution over the family $\mathcal{A}$ of subsets of G. It was defined by Matheron in Matheron (1975). This granulometric theory was extended to functions in Dougherty (1992).

From now on, let $B = (-1, 0, 1)$ and λ a homothetic parameter. A family of openings $\{\gamma_{\mu_i B}\}$ where $\mu_i \in \{1, 2, ..., p\}$ is a granulometry if for all $\mu_i, \mu_j \in \{1, 2, ..., p\}$ and for all function f, $\mu_i \leq \mu_j \Rightarrow \gamma_{\mu_i B}(f) \geq \gamma_{\mu_j B}(f)$. In this paper we will consider the following granulometric size distribution:

$$F_f(\lambda) = 1 - \frac{m(\lambda)}{m(0)} \text{ for } \lambda \geq 0, \tag{6.1}$$

where $m(\lambda) = \int \gamma_{\lambda B}(f)$.

According to this formula each granulometric curve is a function of the structuring element size which plots the area under the opened curve versus the area under the original curve, then they reflect the shape-size of the original curves. Many applications of granulometries and their associated size distributions have been published. Some examples appear in Ayala(2001), Sabourin (1997).

6.3 Methods and experimental results

Following Ferraty and Vieu (2006), the statistical modeling for treating curve data consists in looking at them as a sample of independent realizations $\{X_1, \dots, X_n\}$ of some functional variables distributed like X and taking values in some abstract infinite dimensional space (E, d), where d is a measure of similarity between curves (perhaps a metric or a semi-metric).

It can be easily proved that for $X_h(t) = X(t + h)$ the original X and the translated X_h will have the same granulometric size distribution i.e. any dissimilarity measure based on granulometries will be myope to translations of the domains.

This is a very useful property for some real data as the dataset of 472 radar waveforms obtained with the satellite Topex/Poseidon (http://www.lsp.ups-tlse.fr/staph/npfda/). This dataset was previously analyzed in Dabo *et al.* (2007) and our results can be compared with those given there. Figure 6.1 displays two curves together with their corresponding granulometries. We have used different metrics and semi-metrics to compute dissimilarities between the granulometric curves and also different clustering methods to group them. As an example, we will show the results obtained by using the semi-metric based on the functional principal component analysis selecting the first two components, and the agglomerative clustering method *agnes* which is fully described in Kaufman (1990). The average distance has been used for different groups. Similarly to *et al.* (2007), Figure 6.2 shows a random selection of curves from each one of the five clusters. In our opinion the curves in every cluster are quite homogenous. Other choices provide good results, in particular we have also performed a partitioning around medoids (PAM) Kaufman (1990) with five groups using the distance L_2 to obtain the dissimilarity matrix. The visual inspection of the curves and the contingency table of the counts at each combination of the groups 11, 12, 21, 22 and 23 obtained in Dabo *et al.* (2007) and those obtained from the PAM method (see Table 6.1) suggest that the second classification is better for the curves in the Groups 21, 22 and 23, although we should say that probably our "c2" is too big.

Table 6.1 Comparison between classification obtained by Dabo et al. and PAM

	G11	G12	G21	G22	G23
c1	1	23	61	9	28
c2	143	76	16	3	2
c3	0	0	0	0	40
c4	0	1	7	6	39
c5	0	0	0	0	17

The average silhouette widths are 0.64 and 0.66 respectively i.e. in both cases we get a clear cluster structure. Let us recall (see Kaufman (1990)) that the silhouette width for the i-th observation, $s(i)$, is defined as $s(i) = (b(i) - a(i))/max\{a(i), b(i)\}$ where $a(i)$ is the average dissimilarity between the observation i and all other points of the cluster to which i belongs meanwhile $b(i)$ is the mean dissimilarity between i and its neighboring cluster i.e. the nearest one to which i does not belong. Clearly, observations with a large $s(i)$ are very well clustered, $s(i)$ around 0 means that the observation is between two clusters, and observations with a negative $s(i)$ are probably placed in the wrong cluster. The average silhouette width is obtained as: $\sum_{i=1}^{n} \frac{s(i)}{n}$, this quantity ranges in $(-1, 1)$ and it has been used both to evaluate the quality of a classification and to estimate the *correct* number of clusters: the partition with the maximum average silhouette width is taken as the optimal partition. It is commonly accepted that average silhouette width values greater than 0.50 indicate that a *reasonable* classification has been achieved.

Our second dataset is a sample of vertical forces exerted on the ground by both feet during the sit-to-stand movement. Two groups were chosen, 59 healthy volunteers and 44 back-pain-patients. For each subject, we had five trials of the experiment. The five original curves were smoothed, registered and their mean was calculated (10). The mean curve evaluated at a regular grid is the datum per patient. We have performed a PAM clustering using the semi-metrics based on PCA. The average silhouette width for two groups is 0.6. Their sizes are 63 and 40 respectively. The first one includes 56 healthy volunteers and the second one includes 37 patients. Our approach has provided a good classification in this particular example.

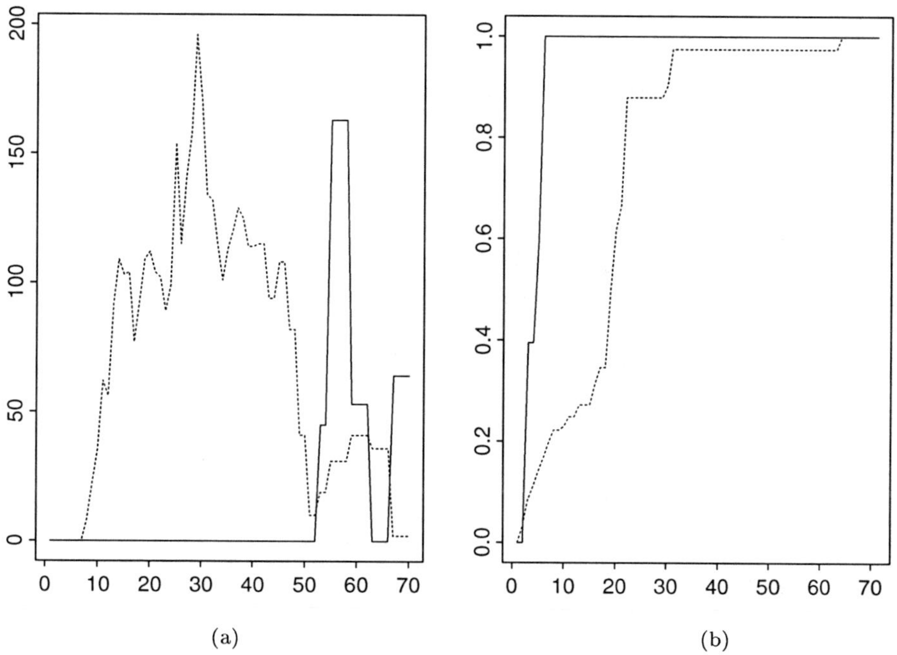

Fig. 6.1 Two satellite curves, (a), with their corresponding granulometric distribution functions, (b)

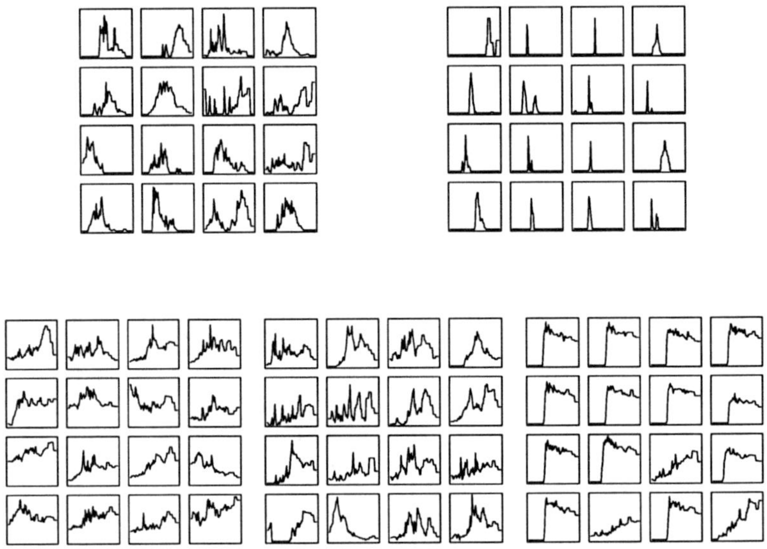

Fig. 6.2 Classification using agglomerative nesting and the average distance

Acknowledgements The authors are indebted to the Spanish Ministry of Education and Science for financing this research with grant TIN2006-10134.

References

[1] Ayala, G. and Domingo, J.: Spatial size distributions. applications to shape and texture analysis, IEEE Transactions on Pattern Analysis and Machine Intelligence. **23** (12), 1430–1442 (2001).

[2] Cerioli, A., Laurini, F. and Corbellini, A.: Functional cluster analysis of financial time series, in: New developments in Classification and Data Analysis. Proceedings of the meeting of the CLADAG of the Italian Statistics Society.Univ. Bologna. (2003), Springer. (2005).

[3] Dabo-Niang, S., F. Ferraty, Vieu, P.: On the using of modal curves for radar waveforms classification, Computational Statistics and Data Analysis. **51**, 4878–4890 (1998).

[4] Dougherty, E.: Euclidean gray-scale granulometries: Representation and umbra inducement, Journal of Mathematical Imaging and Vision **1** (1), 7–21 (1992).

[5] Ferraty, F. and Vieu, P.: Curves discrimination: a nonparametric functional approach, Computational Statistics and Data Analysis. **44**, 161–173 (2003).

[6] Ferraty, F. and Vieu, P.: Nonparametric Functional Data Analysis. Theory and Practice, Springer. (2006).

[7] Kaufman, L. and Rousseeuw, P.: Finding Groups in Data: An Introduction to Cluster Analysis, Wiley, New York. (1990).

[8] Matheron, G.: Random sets and Integral Geometry, Wiley, London. (1975).

[9] Morlini, I.: On the dynamic time warping for computing the dissimilarity between curves, in: New developments in Classification and Data Analysis. Proceedings of the meeting of the CLADAG of the Italian Statistics Society.Univ. Bologna. (2003), Springer. (2005).

[10] Page, A., Ayala, G., Leon, M. T., Peydro, M. F. and Prat, J. M.: Normalizing temporal patterns to analyze sit-to-stand movements by using registration of functional data, Journal of Biomechanics **39** (13) 2526–2534 (2006).

[11] Ramsay, J., Silverman, B.: Functional Data Analysis, Springer. (1997).

[12] Ramsay, J. O., Wickham, H. and Graves, S.: fda: Functional Data Analysis, r package version 1.2.2 (2007).
http://www.functionaldata.org

[13] Sabourin, R., Genest, G., Pretêux, F.: Off-line signature verification by local granulometric size distributions, IEEE Transactions on Pattern Analysis and Machine Intelligence **19** (9) 976–988 (1997).

[14] Soille, P.: Morphological Image Analysis. Principles and Applications, 2nd ed., Springer-Verlag. (2003).

Chapter 7
Supervised Classification for Functional Data: A Theoretical Remark and Some Numerical Comparisons

Amparo Baíllo and Antonio Cuevas

Abstract The nearest neighbors (k-NN) method is a simple, easy to motivate procedure for supervised classification with functional data. We first consider a recent result by Cerou and Guyader (2006) which provides a sufficient condition to ensure the consistency of the k-NN method. We give some concrete examples in which such condition is fulfilled. Secondly, we show the results of a comparative study, performed via simulations and some real-data examples, involving the k-NN procedure (as a "benchmark choice") together with other some recently proposed methods for functional classification.

7.1 Introduction

Supervised classification is a major topic in the emerging field of functional data analysis (see Ramsay and Silverman 2005, Ferraty and Vieu 2006 for recent monographies). In functional classification the aim is to predict the class or label Y of an observation X taking values in a separable metric space $(\mathcal{F}, d)$. For simplicity we will assume that the only possible values of Y are 0 or 1.

Classification of a new observation x from X is carried out by constructing a mapping $g : \mathcal{F} \longrightarrow \{0, 1\}$, called a classifier, which maps x into its predicted label and whose probability of error is given by $P\{g(X) \neq Y\}$. It is well known (see, e.g., Devroye, Györfi and Lugosi 1996) that the Bayes classifier

Amparo Baíllo
Departamento de Análisis Económico: Economía Cuantitativa, Universidad Autónoma de Madrid, 28049 Madrid, Spain, e-mail: amparo.baillo@uam.es

Antonio Cuevas
Departamento de Matemáticas, Universidad Autónoma de Madrid, 28049 Madrid, Spain, e-mail: antonio.cuevas@uam.es

$$g^*(x) = 1_{\{\eta(x) \geq 1/2\}}$$

attains the lowest probability of error, where $\eta(x) = P\{Y = 1|X = x\}$ denotes the regression function of Y on X. However, since the Bayes classifier depends on the unknown distribution of (X, Y), the aim in supervised classification is actually to construct a reasonable classifier g_n based on a training sample $\mathcal{X}_n = \{(X_i, Y_i)\}_{i=1,\ldots,n}$ of i.i.d. copies of (X, Y).

One first, obvious way of dealing with the classification problem is to reduce the dimension of the data from infinite to finite and then to apply a multivariate classification technique, such as the linear or the nearest neighbour discriminant rules. Some dimension reduction techniques that have been used in the literature of functional data are filtering (see, e.g., Biau, Bunea and Wegkamp 2005), partial least squares (see Preda, Saporta and Lévéder 2007) or principal components (see Ramsay and Silverman 2005, Müller 2005).

There are other classification techniques for functional data which do not rely on dimension reduction. For instance, it is possible to construct a classifier $g_n(x) = 1_{\{\eta_n(x) \geq 1/2\}}$ by thresholding an estimator η_n of the regression function η. Cérou and Guyader (2005) consider the regression estimator

$$\eta_n(x) = \frac{1}{k} \sum_{i=1}^{n} 1_{\{X_i \in k(x)\}} Y_i \tag{7.1}$$

where "$X_i \in k(x)$" means that "X_i is one of the k nearest neighbours of x (with respect to the metric d)". By thresholding the regression function given by (7.1) we get the k-nearest neighbour classifier

$$g_n(x) = 1_{\{\eta_n(x) \geq 1/2\}}, \tag{7.2}$$

which consists simply of taking a majority vote over the Y_i's such that the corresponding X_i's are in the subset of the k nearest neighbours of x. In Preda, Saporta and Lévéder (2007) a regression estimator η_n for classification purposes is constructed using the theory of Reproducing Kernel Hilbert Spaces (RKHS) (see also Evgeniou *et al.* 2002, Wahba 2002).

Other popular data classification rules are kernel ones (see Devroye, Györfi and Lugosi 1996, Ferraty and Vieu 2006), where the classifier is given by

$$g_n(x) = \begin{cases} 0 \text{ if } \sum_{i=1}^{n} 1_{\{Y_i=0\}} K\left(\frac{d(x,X_i)}{h}\right) \geq \sum_{i=1}^{n} 1_{\{Y_i=1\}} K\left(\frac{d(x,X_i)}{h}\right), \\ 1 \text{ otherwise,} \end{cases}$$

and K is a non-increasing kernel function whose support is contained in $[0, \infty)$. If $K = 1_{[0,1]}$ we obtain the moving window rule, studied in the functional data context by Abraham, Biau and Cadre (2006).

Let us finally point out that Cuevas, Febrero and Fraiman (2007) have developed functional data classification tools based on depth notions. The observation x is classified into population 0 or 1 depending on the depth of x in the respective training sample.

Our presentation will have two parts. The first one is more theoretical and will be devoted to the study of some aspects related to the consistency of the nearest neighbour rule. The second part is practical and is motivated by the wish of determining, among several of the above described techniques, which of the them is at the same time simple, computationally low-cost and effective when applied to a wide range of real data.

7.2 Consistency of the nearest neighbour rule

The k-nearest neighbour classifier is said to be *weakly consistent* if

$$E(L_n) \to L^* \quad \text{as } n \to \infty,$$

where $L_n = P\{g_n(X) \neq Y|\mathcal{X}_n\}$ is the conditional probability of error of the classifier defined in (7.2). A classical result of Stone (1977) states that, if $(\mathcal{F}, d) = (\mathbb{R}^m, \|\cdot\|)$, where $1 \leq m < \infty$ and $\|\cdot\|$ is the Euclidean norm, then the k-nearest neighbour classifier is universally weakly consistent. The term "universally" means that the result is independent of the distribution of (X, Y). However, when X is infinite dimensional, in order to obtain consistency results, it seems to be necessary to place assumptions on the regularity of the regression function η with respect to P_X. More concretely, Cérou and Guyader (2005) have studied the following smoothness assumption on the regression function (see also Abraham, Biau and Cadre 2006).

(H1) Besicovitch condition: For every $\varepsilon > 0$

$$\lim_{\delta \to 0} \frac{1}{P_X(B_{X,\delta})} \int_{B_{X,\delta}} \eta(z) dP_X(z) = \eta(X) \quad \text{in probability},$$

where $B_{X,\delta} := \{z \in \mathcal{F} : d(X, z) \leq \delta\}$ is the closed ball with center X and radius δ.

Cérou and Guyader (2005) have proved that, if $(\mathcal{F}, d)$ is separable and if Besicovitch condition (H1) is fulfilled, then the nearest neighbour classifier is weakly consistent. A stronger condition which implies Besicovitch one is the following, also appearing in Cérou and Guyader (2005).

(H2) P_X-continuity: For every $\varepsilon > 0$

$$\lim_{\delta \to 0} \frac{1}{P_X(B_{X,\delta})} \int_{B_{X,\delta}} 1_{\{|\eta(z)-\eta(X)|>\varepsilon\}} dP_X(z) = 0 \quad \text{a.s.}$$

We will describe some families of distributions of (X, Y) under which condition (H2) holds.

7.3 Comparison of several classification techniques

We will compare the performance of several classification rules (nearest neighbour, PLS linear, Gaussian RKHS, depth-based and moving window) via some simulations and the analysis of different real data sets.

References

[1] Abraham, C., Biau, G. and Cadre, B.: On the kernel rule for function classification. Ann. Inst. Stat. Math. **58**, 619–633 (2006).

[2] Biau, G., Bunea, F. and Wegkamp, M. H.: Functional classification in Hilbert spaces. IEEE Transactions on Information Theory. **51**, 2163–2172 (2005).

[3] Cérou, F. and Guyader, A.: Nearest neighbor classification in infinite dimension. ESAIM: Probability and Statistics. **10**, 340–355 (2006).

[4] Cuevas, A., Febrero, M. and Fraiman, R.: Robust estimation and classification for functional data via projection-based depth notions. Computational Statistics. **22**, 481–496 (2007).

[5] Devroye, L., Györfi, L. and Lugosi, G.: A Probabilistic Theory of Pattern Recognition. Springer. New York. (1996).

[6] Evgeniou , T., Poggio, T. Pontil, M. and Verri, A.: Regularization and statistical learning theory for data analysis. Computational Statistics and Data Analysis. **38**, 421–432 (2002).

[7] Ferraty, F. and Vieu, P.: Nonparametric Functional Data Analysis. Springer, New York. (2006).

[8] Liu, Y. and Rayens, W.: PLS and dimension reduction for classification. Computational Statistics. **22**, 189–208 (2007).

[9] Müller, H.-G.: Functional modelling and classification of longitudinal data. Scandinavian Journal of Statistics. **32**, 223–240 (2005).

[10] Preda, C., Saporta, G. and Lévéder, C.: PLS classification of functional data. Computational Statistics. **22**, 223–235 (2007).

[11] Ramsay, J. O. and Silverman, B.: Functional Data Analysis. Second edition. Springer-Verlag. New York. (2005).

[12] Stone, C. J.: Consistent nonparametric regression. Ann. Statist. 5, 595–645 (1977).

[13] Wahba, G.: Soft and hard classification by reproducing kernel Hilbert space methods. Proceedings of the National Academy of Sciences. **99**, 16524–16530 (2002).

Chapter 8
Local Linear Regression for Functional Predictor and Scalar Response

Amparo Baíllo and Aurea Grané

Abstract The aim of this work is to introduce a new nonparametric regression technique in the context of functional covariate and scalar response. We propose a local linear regression estimator and study its asymptotic behaviour. Its finite-sample performance is compared with a Nadayara-Watson type kernel regression estimator via a Monte Carlo study and the analysis of two real data sets.

8.1 Introduction

There is nowadays a large number of fields where functional data are collected: environmetrics, medicine, finance, pattern recognition, ... This has led to the extension of finite dimensional statistical techniques to the infinite dimensional data setting. A classical statistical problem is that of regression: studying the relationship between two observed variables with the aim to predict the value of the response variable when a new value of the auxiliary one is observed.

In this work we consider the regression problem with functional auxiliary variable X taking values in $L^2[0,1]$ and scalar response Y. A sample of random elements (X_i, Y_i), $1 \leq i \leq n$, is observed, where the X_i are independent and identically distributed as X and only recorded on an equispaced grid $t_0, t_1, \ldots, t_N$ of $[0,1]$ whose internodal space is $w = 1/N$. It is assumed that the response variable Y has been generated as

---

Amparo Baíllo
Departamento de Análisis Económico: Economía Cuantitativa, Universidad Autónoma de Madrid, 28049 Madrid, Spain, e-mail: amparo.baillo@uam.es

Aurea Grané
Departamento de Estadística, Universidad Carlos III de Madrid, 28903 Getafe (Madrid), Spain, e-mail: agrane@est-econ.uc3m.es

$$Y_i = m(X_i) + \varepsilon_i, \quad i = 1, \ldots, n \tag{8.1}$$

and that the errors ε_i are independent, with zero mean and finite variance σ_ε^2, and are also independent from any of the X_j.

In the context of regression with functional data a common assumption is that $m(x)$ is a linear function of x. The linear model has been studied in a large number of works: see, e.g., Cardot, Ferraty and Sarda (2003), Ramsay and Silverman (2005), Cai and Hall (2006) and Hall and Horowitz (2007). Extensions of this model have been considered, for instance, by James (2002), Ferré and Yao (2003), Cardot and Sarda (2005) or Müller and Stadtmüller (2005). However, when dealing with functional data, it is difficult to gain an intuition on whether the linear model is adequate at all or which is the parametric model that would best fit the data, since graphical techniques are of scarce use here.

Here we are interested in estimating the regression function m in a non-parametric fashion. This problem has already been considered, for instance, by Ferraty and Vieu (2006), who study a kernel estimator of Nadaraya-Watson type

$$\hat{m}_K(x) := \frac{\sum_{i=1}^n Y_i K_h(\|X_i - x\|)}{\sum_{i=1}^n K_h(\|X_i - x\|)}, \tag{8.2}$$

where $K_h(\cdot) := h^{-1}K(\cdot/h)$, $h = h_n$ is a positive smoothing parameter and $\|\cdot\|$ denotes the $L^2[0,1]$ norm. From now on K is assumed to be an asymmetrical decreasing kernel function. Observe that the estimator $\hat{m}_K(x)$ is the value of a minimizing the weighted squared error

$$\text{WSE}_0(x) = \sum_{i=1}^n (Y_i - a)^2 K_h(\|X_i - x\|).$$

Thus the kernel estimator given by (8.2) is locally approximating m by a constant (a zero-degree polynomial). However, in the context of nonparametric regression with finite-dimensional auxiliary variables, local polynomial smoothing has become the "golden standard" (see Fan 1992, Fan and Marron 1993, Wand and Jones 1995). Local polynomial smoothing at a point x fits a polynomial to the pairs (X_i, Y_i) for those X_i falling in a neighbourhood of x determined by a smoothing parameter h. In particular, the local linear regression estimator locally fits a polynomial of degree one. Here we plan to extend the ideas of local linear smoothing to the functional data setting, giving a first answer to the *open question 5* in Ferraty and Vieu (2006): "How can the local polynomial ideas be adapted to infinite dimensional settings?"

8.2 Local linear smoothing for functional data

Local polynomial smoothing is based on the assumption that the regression function m is smooth enough to be locally well approximated by a polynomial. Thus from now on we will assume that m is differentiable in a neighbourhood of x and, consequently, for every z in this neighbourhood we may approximate $m(z)$ by a polynomial of degree 1, that is, $m(z) \simeq a + \langle b, z - x \rangle$, where $a = m(x)$, $b = b(x) \in L^2[0,1]$ and $\langle \, , \rangle$ denotes the $L^2[0,1]$ inner product. Then the weighted squared error

$$\mathrm{WSE}(x) := \sum_{i=1}^{n} (Y_i - m(X_i))^2 \, K_h(\|X_i - x\|)$$

may be approximated by

$$\mathrm{WSE}_1(x) = \sum_{i=1}^{n} (Y_i - (a + \langle b, X_i - x \rangle))^2 \, K_h(\|X_i - x\|). \qquad (8.3)$$

A first naive answer to the question posed by Ferraty and Vieu 2006 would be to find the values $\hat{a}$ and $\hat{b}$ optimizing (8.3). Then we would take $\hat{m}_{LL}(x) = \hat{a}$ as the local linear estimator of $m(x)$, the regression function at x (see Fan 1992). However, the minimization of WSE_1 may be achieved by a "wiggly" $\hat{b}$ that forces $\hat{m}_{LL}(x)$ to adapt to all the data points in a neighbourhood of x. This is usually overcome by reducing the dimension of parameter b via an intermediate step of smoothing or regularization.

Here we expand b and X_i using the Fourier trigonometric basis $\{\phi_j\}_{j \geq 1}$ of $L^2[0,1]$ (see Ramsay and Silverman 2005)

$$b = \sum_{j=1}^{\infty} b_j \phi_j \qquad \text{and} \qquad X_i - x = \sum_{j=1}^{\infty} c_{ij} \phi_j \qquad (8.4)$$

with $b_j = \langle b, \phi_j \rangle$ and $c_{ij} = \langle X_i - x, \phi_j \rangle$. Then

$$\mathrm{WSE}_1 = \sum_{i=1}^{n} \left(Y_i - \left(a + \sum_{j=1}^{\infty} b_j c_{ij} \right) \right)^2 K_h(\|X_i - x\|).$$

The regularization step consists in truncating the series at a certain cut-off J. Thus we will minimize the following approximation to WSE_1

$$\mathrm{AWSE}_1 := \sum_{i=1}^{n} \left(Y_i - \left(a + \sum_{j=1}^{J} b_j c_{ij} \right) \right)^2 K_h(\|X_i - x\|). \qquad (8.5)$$

In matrix notation, if $\bar{Y} = (Y_1, \ldots, Y_n)'$,
$\mathbf{W} = \mathrm{diag}(K_h(X_1 - x), \ldots, K_h(X_n - x))$ and

$$\mathbf{C} = \begin{pmatrix} 1 & c_{11} & \ldots & c_{1J} \\ 1 & c_{21} & \ldots & c_{2J} \\ \vdots & & & \vdots \\ 1 & c_{n1} & \ldots & c_{nJ} \end{pmatrix}$$

and assuming that $\mathbf{C}'\mathbf{W}\mathbf{C}$ is a nonsingular matrix, the values of a and b_j, for $j = 1, \ldots, J$, minimizing AWSE_1, are

$$\begin{pmatrix} \hat{a} \\ \hat{b}_1 \\ \vdots \\ \hat{b}_J \end{pmatrix} = (\mathbf{C}'\mathbf{W}\mathbf{C})^{-1}\mathbf{C}'\mathbf{W}\bar{Y}.$$

Finally, our proposal for the local linear estimator of $m(x)$ is

$$\hat{m}_{LL}(x) = \hat{a} = \beta_1'(\mathbf{C}'\mathbf{W}\mathbf{C})^{-1}\mathbf{C}'\mathbf{W}\bar{Y}, \tag{8.6}$$

where β_1 is the $(J+1) \times 1$ vector having 1 in the first entry and 0's in the rest.

We refer to Barrientos-Marín (2007), chapter 3, for a simplified version of this approach. This author substitutes the linear functional given by $\langle b, X_i - x \rangle$ in expression (8.3) by a linear function $b\,\beta(X_i, x)$, where $b \in \mathbb{R}$ and β is an operator taking values in $\mathbb{R}$.

8.3 Performance of the estimator $\hat{m}_{LL}$

We will show a theoretical result, proved in Baíllo and Grané (2007), on the asymptotic behaviour of the local linear estimator introduced in Section 2. More concretely, we will give conditions under which the mean squared error

$$E((\hat{m}_{LL}(x) - m(x))^2 | X_1, \ldots, X_n)$$

converges to 0 as $n \to \infty$ and $J \to \infty$. Under an additional assumption on the fractal order of X, we will obtain convergence rates to zero of the mean squared error. This rates agree with the asymptotic results for the kernel estimator appearing in Ferraty and Vieu (2006), p. 208, in the sense that the more concentrated X is around x (as measured by the so-called small ball probability), the faster the local linear estimator will converge to the true regression function.

We will also compare the finite-sample performance of $\hat{m}_{LL}$ with the kernel estimator $\hat{m}_K$ via a Monte Carlo study and the analysis of real data sets. It will be seen that, in all the scenarios considered, the local linear regression estimator performs better than the kernel one, in the sense that the mean squared prediction error is lower.

References

[1] Baíllo, A. and Grané, A.: Local linear regression for functional predictor and scalar response. Working Paper 07-61, Statistics and Econometric Series 15, Universidad Carlos III de Madrid. (2007).

[2] Barrientos-Marín, J.: Some practical problems of recent nonparametric procedures: testing, estimation, and application. PhD Thesis, Departamento de Fundamentos del Análisis Economico, Universidad de Alicante, Spain. (2007).

[3] Cai, T. T. and Hall, P.: Prediction in functional linear regression. Ann. Statist. **34**, 2159–2179. (2006).

[4] Cardot, H., Ferraty, F. and Sarda, P.: Spline estimators for the functional linear model. Statistica. Sinica. **13**, 571–591. (2003).

[5] Cardot, H. and Sarda, P.: Estimation in generalized linear models for functional data via penalized likelihood. J. Multivariate Anal. **92**, 24–41 (2005).

[6] Fan, J. (1992). Design-adaptive nonparametric regression. J. Amer. Statist. Assoc. **87**, 998–1004. (2005).

[7] Fan, J. and Marron, J. S.: Local Regression: Automatic Kernel Carpentry: Comment. Statistical. Science. **8**, 129–134.(1993).

[8] Ferraty, F. and Vieu, P.: Nonparametric Functional Data Analysis. Springer, New York. (2006).

[9] Ferré, L. and Yao, A. F.: Functional sliced inverse regression analysis. Statistics. **37**, 475–488. (2003).

[10] Hall, P. and Horowitz, J. L.: Methodology and convergence rates for functional linear regression. Ann. Statist. **35**, 70–91. (2007).

[11] James, G. M.: Generalized linear models with functional predictors. J. R. Stat. Soc. Ser. B **64**, 411–432. (2002).

[12] Müller, H.-G. and Stadtmüller, U.: Generalized functional linear models. Ann. Statist. **33**, 774–805. (2005).

[13] Ramsay, J. O. and Silverman, B.: *Functional Data Analysis.* Second edition. Springer-Verlag, New York. (2005).

[14] Wand, M. P. and Jones, M. C.: Kernel Smoothing. Chapman and Hall. (1995).

Chapter 9
Spatio-temporal Functional Regression on Paleoecological Data

Avner Bar-Hen, Liliane Bel and Rachid Cheddadi

Abstract The aim of this presentation is modeling the relationship between genetic diversity (represented by a positive number) and curves of temperature and precipitation. Our model links the genetic measure to the climate curves through a functional regression. The interaction in climate variables is assumed to be bilinear and the methodology accounts for the spatial dependence among the observations.

9.1 Introduction

Influence of climate on biodiversity is an important ecological question. Various theory try to link climate change to allelic richness and therefore to predict the impact of global warning on genetic diversity.

The aim of this presentation is to modelize the relationship between genetic diversity in the European beech forests(represented by a positive number) and curves of temperature and precipitation reconstructed from the past.

Our model links the genetic measure to the climate curves through a linear functional regression. The interaction in climate variables is assumed to be bilinear and the methodology accounts for the spatial dependence among the observations.

Avner Bar-Hen
Laboratoire MAP5 (CNRS-UMR 8145), 45, rue des Saints-Pères, 75270 Paris, France, e-mail: avner@math-info.univ-paris5.fr

Liliane Bel
AgroParisTech, 16 rue Claude Bernard, F-75231 Paris cedex 05, France e-mail: Liliane.Bel@agroparistech.fr

Rachid Cheddadi
Institut des Sciences de l'Évolution, case postale 61 CNRS UMR 5554 34095 Montpellier, France, e-mail: cheddadi@isem.univ-montp2.fr

9.2 Data

Since a plant has a limited range of acceptable climate parameters, it is possible to reconstruct from pollen database climate variables. Temperature and precipitation were reconstructed from 216 locations from present to a variable date depending on available data. The pollen dataset was used to reconstruct climate variables, throughout Europe for the last 15 000 years of the Quaternary. Due to the methodology, each climate curve is sampled at irregular time for each location.

Genetic diversities were measured from variation at 12 polymorphic isozyme loci in the European beech (*Fagus sylvatica* L.) forests based on an extensive sample of 389 populations distributed throughout the species range. Based on these data, various index of diversity can be computed. They mainly characterise within or between population diversity.

Since the pollens were generally not collected in forest, genetic measure and climate curve are heterotopic. Temperature and precipitation curves are firstly estimated on a regular grid of time on sites where are collected the genetic measure. This is done by a spatio-temporal kriging assuming the covariance function is exponential and separable. Parameters are fitted empirically.

9.3 Functional regression

The functional linear regression model with functional or real response has been the focus of various investigations (see [5, 4, 1, 3]). We want to estimate the link between the real random response $d(s)$, the diversity at site s and $(\theta_1(t,s), \theta_2(t,s))_{t>0}$ the temperature and precipitation functions at site s. There are two points to consider for the modelization: (i) functional linear models need to be extended to incorporate interaction between climate functions; (ii) since the data are geo-referenced, observations cannot be considered as independent and we also need to extend functional modelization to spatial data.

We assume that the temperature and precipitation functions are square integrable random functions defined on some real compact set $[0, T]$. The very general model can be written as:

$$d(s) = f((\theta_1(t,s), \theta_2(t,s))_{T>t>0}) + \varepsilon_s$$

where f is an unknown functional from the space of the continuous functions from $\mathbb{R}^+$ to $\mathbb{R}$.

A linear model, with bilinear interaction can be written as

$$
\begin{aligned}
f(\theta_1,\theta_2) &= \mu + \int_{[0,T]} A(t)\theta_1(s,t)\mathrm{d}t + \int_{[0,T]} B(t)\theta_2(s,t)\mathrm{d}t + \\
&\iint_{[0,T]^2} C(t,u)\theta_1(t,s), \theta_2(u,s) du dt + \varepsilon_s \\
&= \mu + \langle A;\theta_1\rangle + \langle B;\theta_2\rangle + \langle C\theta_1;\theta_2\rangle + \varepsilon_s
\end{aligned}
$$

by the Riesz representation of linear and bilinear forms.
Expanding on the same orthonormal base $(e_i)_{i\in\mathbb{N}}$ we have

$$
\theta_1(s,t) = \sum_i \alpha_i(s)e_i(t) \quad \theta_2(s,t) = \sum_i \beta_i(s)e_i(t)
$$

$$
A(t) = \sum_i a_i(s)e_i(t) \quad B(t) = \sum_i b_i(s)e_i(t) \quad C(t,u) = \sum_{i,j} c_{ij}e_i(t)e_j(t)
$$

The question results in a linear regression on coefficients a_i, b_i, c_{ij}.

From a practical point of view the infinite sum is truncated. The effect of the truncature will be discussed. Different choices of the orthonormal basis will be presented.

Since the data are spatially located, our second extension concerns the correlation structure of the errors. Functional linear models generally consider independent observations. The data under study are a sample of regionalized variables, and the residuals exhibit spatial dependence. In order to estimate the regression coefficients by generalized least squares we proceed as usual in an iterative way: coefficients are estimated as if the observations were independent and then the spatial covariance is estimated on residuals. The coefficients are re-estimated with the fitted covariance.

It is important to quantify the predictive power of the proposed model. We use a leaving-one-out approach to quantify the quality of the model. For each observation $d(s_i)$ we compute the residual error between the observation and the prediction on the model based on all observation except $d(s_i)$. The mean average prediction error gives a global indication of the predictive power of the model. As usual, there is a trade-off between the quality of the modelization based on the observed data and the predictive power based on new observations. Parsimony is one important factor, which needs to be considered. The mean square average prediction statistic is used to choose the order of the decomposition of the main effects and the interaction. This also gives a tool to quantify the importance of the climate variable to explain the diversity.

To proceed with our data, we use a Fourier base of order 5. The results show a strong effect of the temperature function, and small effect of precipitation and interaction. When the change of climate just before the Holocene (9000 BP to present) was important the diversity is higher. This mostly concerns North and Western Europe (Figure 1).

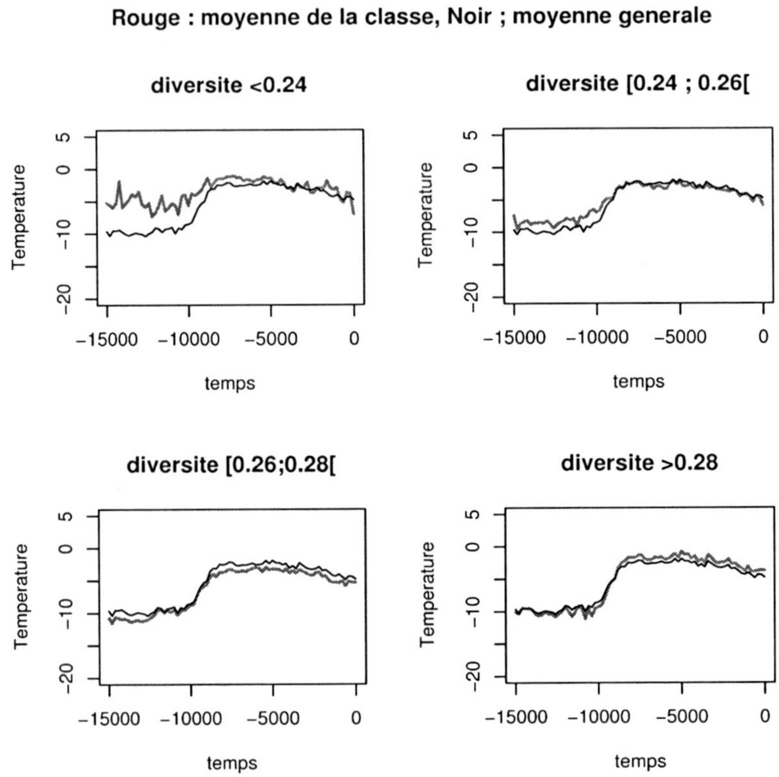

Fig. 9.1 Thin line: temperature average of the entire sample. Thick line: temperature average of sites with predicted diversity < 0.24, in $[0.24, 0.26[$, $[0.26, 0.28[$ and ≥ 0.28. Predicted diversity is low when the change of climate in -9000 is weak and becomes higher when the change of climate increases.

References

[1] Cardot, H., Ferraty, F. and Sarda, P.: Functional linear model. Statist. Probab. Lett. **45**, 11-22 (1999).

[2] Cressie N. and Huang. H.C.: Classes of nonseparable, spatio-temporal stationary covariance functions. J. Amer. Statist. Assoc. pp. 1330-134094 (1999).

[3] Fan, J. and Zhang, J.-T.: Two-step estimation of functional linear models with application to longitudinal data. J. R. Stat. Soc. Ser. B Stat. Methodol. **62**, 303-322 (2000).

[4] Faraway, J. J.: Regression analysis for a functional response. Technometrics. **39**, 254-261 (1997).

[5] Ramsay, J. O. and Silverman, B. W.: Functional Data Analysis. Springer, New-York. (1997).

[6] Rouhani S. and Myers D.E.: Problems in space-time kriging of geohydrological data. Math. Geol. **22**, 611-623 (1990).

Chapter 10
Local Linear Functional Regression Based on Weighted Distance-based Regression

Eva Boj, Pedro Delicado and Josep Fortiana

Abstract We consider the problem of nonparametrically predicting a scalar response variable y from a functional predictor χ. We have n observations (χ_i, y_i). We assign a weight $w_i = K\left(d(\chi, \chi_i)/h\right)$ to each χ_i, where d is a semi-metric, K is a kernel function and h is the bandwidth. Then we fit a Weighted (Linear) Distance-Based Regression, where the weights are as above and the distances are given by a possibly different semi-metric.

10.1 Introduction

Let $(\boldsymbol{\chi}, Y)$ be a random element where the first component $\boldsymbol{\chi}$ is a random element of a functional space (typically a function $\boldsymbol{\chi}$ from $[a, b] \subseteq \mathbb{R}$ to $\mathbb{R}$) and Y is a real random variable. We consider the problem of predicting the scalar response variable y from the functional predictor χ. We assume that we are given n i.i.d. observations $(\chi_i, y_i), i = 1, \ldots, n$, from $(\boldsymbol{\chi}, Y)$ as a training set. If we define the regression function $m(\chi) = E(Y|\boldsymbol{\chi} = \chi)$ a reasonable prediction of y would be an estimate of $m(\chi)$.

Ramsay and Silverman (2005) consider the linear functional regression model, where it is assumed that

$$m(\chi) = \alpha + \int_a^b \chi(t)\beta(t)dt, \text{ and then } y_i = m(\chi_i) + \varepsilon_i,$$

Eva Boj
Universitat de Barcelona, Barcelona, Spain, e-mail: `evaboj@ub.edu`

Pedro Delicado
Universitat Politècnica de Catalunya, Barcelona, Spain, e-mail: `pedro.delicado@upc.edu`

Josep Fortiana
Universitat de Barcelona, Barcelona, Spain, e-mail: `fortiana@ub.edu`

ε_i having zero expectation. The parameter β, $t \in [a,b]$, is a function and $\alpha \in \mathbb{R}$. They propose to estimate β and α by penalized least squares:

$$\min_{\alpha,\beta} \sum_{i=1}^{n} \left(y_i - \alpha - \int_T \chi_i(t)\beta(t)dt \right)^2 + \lambda \int_a^b (L(\beta)(t))^2 dt,$$

where $L(\beta)$ is a linear differential operator giving a penalty to avoid too much rough β functions and $\lambda > 0$ acts as a smoothing parameter.

Ferraty and Vieu (2006) say that this linear regression model is parametric because we need only a finite (and constant in n) number of functional elements to describe it. They consider a nonparametric functional regression model where only a few regularity assumptions are made on the regression function $m(\chi)$. They propose the following kernel estimator for $m(\chi)$:

$$\hat{m}_K(\chi) = \frac{\sum_{i=1}^{n} K(d(\chi,\chi_i)/h) y_i}{\sum_{i=1}^{n} K(d(\chi,\chi_i)/h)} = \sum_{i=1}^{n} w_i(\chi) y_i,$$

where $w_i(\chi) = K(d(\chi,\chi_i)/h) / \sum_{j=1}^{n} K(d(\chi,\chi_j)/h)$, K is a kernel function with support $[0,1]$, the bandwidth h is the smoothing parameter (depending on n), and $d(\cdot,\cdot)$ is a semi-metric in the functional space $\mathcal{F} = \{\chi : [a,b] \to \mathbb{R}\}$ to which the data χ_i belong. Examples of semi-metrics in $\mathcal{F}$ are L_2 distances between derivatives,

$$d_r^{deriv}(\chi,\gamma) = \left(\int_a^b \left(\chi^{(r)}(t) - \gamma^{(r)}(t) \right)^2 dt \right)^{1/2}.$$

Ferraty and Vieu (2006) prove that $\hat{m}_K(\chi)$ is a consistent estimator (in the sense of almost complete convergence) of $m(\chi)$ under regularity conditions on m, χ (involving small balls probability), Y and K. Ferraty, Mas, Vieu (2007) prove the mean squared convergence and asymptotic distribution of $\hat{m}_K(\chi)$.

The book of Ferraty and Vieu (2006) lists several interesting open problems concerning nonparametric functional regression. In particular, their *Open Question 5* face with including the local polynomial ideas in an infinite dimensional setting in order to extend the estimator $\hat{m}_K(\chi)$, that is a kind of Nadaraja-Watson regression estimator.

A first answer to this question is given in Baíllo and Grané (2007). They propose a natural extension of the finite dimensional local linear regression, by solving the problem

$$\min_{\alpha,\beta} \sum_{i=1}^{n} w_i(\chi) \left(y_i - \alpha - \int_T (\chi_i(t) - \chi(t))\beta(t)dt \right)^2,$$

where local weights $w_i(\chi) = K(||\chi - \chi_i||/h)/\sum_{j=1}^{n} K(||\chi - \chi_j||/h)$ are defined by means of L_2 distances (it is assumed that all the functions are in $L_2([a,b])$). Their estimator of $m(\chi)$ is $\hat{m}_{LL}(\chi) = \hat{\alpha}$. A closely related approach can be seen in Berlinet *et al.* (2007) and Barrientos (2007).

In this work we give an alternative response to the same open question. Our proposal rests on Distance-Based Regression (DBR) (Cuadras, 1989, Cuadras et al. 1990, Cuadras et al. 1996, Boj.et.al:2007), a prediction tool which can be applied to non-numerical explanatory variables while keeping compatibility with ordinary least squares regression (OLS), which appears as a particular case. We use WDBR, the weighted version of DBR, where each case (χ_i, y_i) has a weight $w_i > 0$. We assign a weight $w_i \propto K(d_1(\chi, \chi_i)/h)$ to observation i, where $d_1(\cdot, \cdot)$ is a semi-metric. Then we fit a WDBR, where the weights are as above and the distances between functions are given by a possibly different semi-metric $d_2(\cdot, \cdot)$.

10.2 Weighted distance-based regression (WDBR)

Let $\Omega = \{[1], \ldots, [n]\}$ be a set of n individuals randomly drawn from a population. Individual $[i]$ has weight $w_i \in (0,1)$. Let $\boldsymbol{w} = (w_1, \ldots, w_n)^T$ adding up to 1. For individual $[i]$ we have observed the value y_i of a continuous one-dimensional response, and we assume that the responses are w-centered (that is, $\boldsymbol{w}^T \cdot \boldsymbol{y} = 0$, where $\boldsymbol{y} = (y_1, \ldots, y_n)^T$). A distance function δ (being a metric or semi-metric) is defined between the elements of Ω. Let $\boldsymbol{\Delta} = (d_{i,j}^2)_{i=1..n, j=1..n}$ be the inter-individual squared distances matrix. The available information for the elements of the set Ω can be a mixture of quantitative and qualitative variables or, possibly, other nonstandard quantities, such as character strings or functions. The computation of distances d_{ij} is based on this information. The aim of the WDBR is to predict the response variable for a new individual $[n+1]$ from the same population, using $(d_{n+1,1}^2, \ldots, d_{n+1,n}^2)$, the vector of squared distances from $[n+1]$ to the remaining individuals, as the only available information.

WDBR operates as follows. We say that a $n \times q$ matrix $\bar{X}$, $q \leq n$, is a Euclidean configuration for $\boldsymbol{\Delta}$ if $\bar{X}$ verifies that the Euclidean distance between its rows i and j is equal to d_{ij}. It is assumed that such a configuration exists for $\boldsymbol{\Delta}$. A weighted version of Metric Multidimensional Scaling (see, e.g., Borg and Groenen, 2005 or Boj and Fortania, 2007) can be used to obtain $\bar{X}$ from $\boldsymbol{\Delta}$. Then a linear regression of $\boldsymbol{y}$ on $\bar{X}$ is estimated by Weighted Least Squares, giving a q-dimensional estimated regression coefficient $\hat{\beta}$. It can be proven (Boj and Fortania, 2007) that $\hat{\boldsymbol{y}} = \bar{X}\hat{\beta}$ is an intrinsic quantity, meaning that $\hat{\boldsymbol{y}}$ admits an alternative expression as a function of the distances $\boldsymbol{\Delta}$:

$$\hat{\boldsymbol{y}} = \mathbf{G}_w \cdot \left(\boldsymbol{D}_w^{1/2} \cdot \boldsymbol{F}_w^{+} \cdot \boldsymbol{D}_w^{1/2} \right) \cdot \boldsymbol{y}, \tag{10.1}$$

where $\boldsymbol{D}_w = \mathrm{Diag}(\boldsymbol{w})$, $\boldsymbol{J}_w = \boldsymbol{I}_n - \mathbf{1}_n \cdot \boldsymbol{w}^T$, $\mathbf{G}_w = -(1/2)\boldsymbol{J}_w \cdot \boldsymbol{\Delta} \cdot \boldsymbol{J}_w^T$, $\boldsymbol{F}_w = \boldsymbol{D}_w^{1/2} \cdot \mathbf{G}_w \cdot \boldsymbol{D}_w^{1/2}$, and $\boldsymbol{F}_w^+$ is the Moore-Penrose g-inverse of standardized inner-products matrix $\boldsymbol{F}_w$.

A new individual $[n+1]$ is represented as a q-vector $\mathbf{X}_{n+1}$ in the row space of $\bar{X}$ using the *Gower's add-a-point* formula (see Boj and Fortania, 2007 for the weighted version), giving the best q-dimensional approximation in the least squares sense to an exact Euclidean configuration of the whole set of $n+1$ individuals with their distances. It can be proven (Boj and Fortania, 2007) that $\hat{y}_{n+1} = \mathbf{X}_{n+1}\hat{\beta}$ can alternatively be expressed directly as a function of the distances:

$$\hat{y}_{n+1} = (1/2)\left(\boldsymbol{g}_w - \boldsymbol{d}_{[n+1]}\right) \cdot \boldsymbol{D}_w^{1/2} \cdot \boldsymbol{F}_w^+ \cdot \boldsymbol{D}_w^{1/2} \cdot \boldsymbol{y}, \tag{10.2}$$

where $\boldsymbol{g}_w$ is the row vector with the diagonal of $\mathbf{G}_w$.

Equations (10.1) and (10.2) are the core of WDBR. Observe that WDBR is a linear regression in the space where the Euclidean configuration $\bar{X}$ is included. In practice this configuration is not explicitly calculated. It is also remarkable that WDBR reproduces weighted least squares regression (WLS): if we start from a $n \times q$ matrix $\bar{X}$ of q continuous independent variables corresponding to n individuals (with weights given by $\boldsymbol{w}$) and we define $\boldsymbol{\Delta} = (d_{i,j}^2)$, d_{ij} being the Euclidean distance between rows i and j of $\bar{X}$, then $\hat{\boldsymbol{y}}_{WDBR} = \hat{\boldsymbol{y}}_{WLS}$ and $\hat{y}_{n+1,WDBR} = \hat{y}_{n+1,WLS}$, because $\bar{X}$ is trivially a Euclidean configuration for $\boldsymbol{\Delta}$. A particular example is when the i-th row of $\bar{X}$ is (x_i, x_i^2, x_i^3), $x_i \in \mathbb{R}$. Then doing the cubic weighted regression of y_i over x_i is equivalent to fitting WDBR with distances
$d(x_i, x_j) = \|(x_i, x_i^2, x_i^3) - (x_j, x_j^2, x_j^3)\|_2$.

10.3 Local linear distance-based regression

Let $(\chi_i; y_i)$, $i = 1, \ldots, n$, be a random sample of $(\boldsymbol{\chi}, Y)$, $Y \in \mathbb{R}$, $\boldsymbol{\chi} : [a,b] \to \mathbb{R}$. We want to estimate $m(\chi) = E(Y|\boldsymbol{\chi} = \chi)$ by a local linear regression around χ and we are doing that using WDBR. We consider the weights $w_i(\chi) = K(d_1(\chi, \chi_i)/h) / \sum_{j=1}^n K(d_1(\chi, \chi_j)/h)$, where d_1 is a semi-metric between functions. Let $\boldsymbol{\Delta}_2 = (d_2(\chi_i, \chi_j)^2)_{i=1..n, j=1..n}$ be the matrix of squared distances between functions defined from a possible different semi-metric d_2. We fit the WDBR using equation (10.1) from the elements $\boldsymbol{\Delta}_2$, $\boldsymbol{y} = (y_i)_{i=1..n}$, $\boldsymbol{w} = (w_i(\chi))_{i=1..n}$. We consider a new individual $[n+1]$ where the functional predictor is χ and we compute its squared distances to the other individuals χ_i: $\boldsymbol{d}_{2,[n+1]} = (d_2(\chi, \chi_1)^2, \cdots, d_2(\chi, \chi_n)^2)$. Then we use equation (10.2) to obtain the *local linear distance-based estimator* of $m(\chi)$:

$$\hat{m}_{LLDBR}(\chi) = \hat{y}_{n+1}.$$

Let us remark some important points. There are two semi-metrics involved in the local linear distance-based estimation: one of them, d_1, is used to compute the weight of observation χ_i around the function χ where the regression function is estimated, and the other, d_2, defines the distances between observations for computing the distance-based regression. The semi-metrics d_1 and d_2 can coincide or not. Observe that the local linear distance-based estimator of $m(\chi)$ is really a local linear estimator in the space where the semi-metric d_2 is a Euclidean distance.

Assume that d_1 and d_2 coincide and that they are the L_2 distance in $L_2([a,b])$, that is, $d_1 = d_2 = d_0^{deriv}$. Then the local linear distance-based estimator $\hat{m}_{LLDBR}(\chi)$ coincide with the local linear estimator $\hat{m}_{LL}(\chi)$ proposed in Baíllo and Grané (2007). Assume now that $d_2(\chi, \gamma) = 0$ for all functions χ and γ. Then the local linear distance-based estimator $\hat{m}_{LLDBR}(\chi)$ fits locally a constant around χ and then it coincides with the kernel estimator $\hat{m}_K(\chi)$ introduced by Ferraty and Vieu (2006).

Let K be the uniform kernel and assume that $h > \max_{i,j}(d_1(\chi_i, \chi_j))$. Then a (global) distance-based regression is fitted, that is a linear regression fit in the space where the semi-metric d_2 is a Euclidean distance.

The local linear distance-based estimation is also valid for predictors that are no functional data. For instance, it is valid for multivariate continuous data ($x_i \in \mathbb{R}^p$), mixed data (multivariate x_i with some components being continuous and other being qualitative), textual data or any other kind of data for which we are able to compute distances between individuals. Consider, for instance, that $x_i \in \mathbb{R}$, $d_1(x_i, x_j) = |x_i - x_j|$, $d_2(x_i, x_j) = \|(x_i, x_i^2, x_i^3) - (x_j, x_j^2, x_j^3)\|$. Then the estimator $\hat{m}_{LLDBR}(x)$ coincide with fitting a local cubic polynomial regression (see the end of Section 2).

Our proposal for estimating $m(\chi)$ non-parametrically by local linear distance-based regression is very flexible, including as particular cases the local polynomial regression for real predictor variables. So we consider that this proposal is a satisfactory answer to *Open question 5* in Ferraty and Vieu (2006).

10.4 A real data example: Spectrometric Data

We consider the *Spectrometric Data* described in Chapter 2 of Ferraty and Vieu (2006). This dataset includes 215 individuals, each of them being a sample of chopped meat. For each individual the function χ_i, relating absorbance versus wavelength, has been recorded for 100 values of wavelength in the range 850-1050 nm. An additional response variable is observed: y_i, the sample fat content obtained by analytical chemical processing. Given that obtaining a spectrometric curve is less expensive that determining the fat content by chemical analysis, it is important to predict the fat content y from the spectrometric curve χ.

Following Section 7.2 in Ferraty and Vieu (2006) we divide the sample in a training sample (the first 160 cases) and a test sample (the last 55 cases). The performance of different functional prediction methods is measured by the empirical mean square prediction error in the sample test: $MSPE = (1/55)\sum_{i=161}^{215}(\hat{y}_i - y_i)^2$.

Ferraty and Vieu (2006) use three functional predictors for this data set: nonparametric estimators of conditional expectation (functional kernel estimator as $\hat{m}_K(\chi)$), conditional mode and conditional median. The implementation of these estimators allows a variable bandwidth h based on k-nearest neighbours, where k is locally selected by cross-validation. The authors recommend to use the semi-metric based on the second order derivatives (d_2^{deriv}). We have used the R routines accompanying Ferraty and Vieu (2006) (the script `npfda-specpredRS.txt` to be specific) to recreate the results included in the book. The numbers we have obtained are shown in Table 10.1 with the label FV2006.

In order to have results that we can directly compare with our proposals, we have computed the functional kernel estimators with fixed bandwidth selected by cross-validation and based on the semi-metrics d_r^{deriv}, $r = 1, 2, 3$. We have used the R function from Ferraty and Vieu (2006) `funopare.kernel.cv`. The results are included in Table 10.1 with the label Kernel.FV.

We have implemented the local linear distance-based regression with automatic selection of the bandwidth by cross-validation. The usual way of implementing cross-validation has been modified as follows. Usually it is not possible to check the performance of a candidate bandwidth h being lower than $\max_i \min_j d_1(\chi_i, \chi_j) = \min_j d_1(\chi_{i^*}, \chi_j)$ because in this case there are not enough data in the ball centered at χ_{i^*} with radius h to fit the distance-based regression. So for an observation χ_i having less than 3 neighbours at distance h, we enlarge h to h_i allowing to include 3 neighbours in the ball centered at χ_{i^*} with radius h_i. So our implementation is with partially variable bandwidth.

An alternative implementation of functional kernel estimators is possible using local linear distance-based regression by selection $d_1 = d_r^{deriv}$, $r = 1, 2, 3$, and $d_2 \equiv 0$. The results are included in Table 10.1 with the label Kernel.LLDBR. The results do not coincide with those obtained using the function `funopare.kernel.cv` because the different way of bandwidth selection.

Finally we also show in Table 10.1 the results obtained by local linear distance-based regression for different combinations of distances d_1 and d_2, all of them using semi-metrics based on derivatives. First we fix d_2 equal to the L_2 distance between the original functions ($d_2 = d_0^{deriv}$) and use $d_1 = d_r^{deriv}$, $r = 1, 2, 3$. This way we do local linear regression in the space of the original functions for different semi-metrics defining neighborhoods in this space. The case $d_1 = d_0^{deriv}$ and $d_2 = d_0^{deriv}$ corresponds to the local linear estimator proposed by Baíllo and Grané (2007). The case $d_1 = d_2^{deriv}$ and $d_2 = d_0^{deriv}$ represents an improvement on the kernel method (see the row with label Kernel.LLDBR d_2^{deriv}) because now a local linear regression is fitted instead computing a local average. The best fitting is obtained when using $d_1 = d_2^{deriv}$ and $d_2 = d_2^{deriv}$: local linear regression in the space of second derivatives. This choice of distances d_1 and d_2 is also the most natural one taken into account the recommendations of Section 7.2 in Ferraty and Vieu (2006).

We conclude that local linear distance-based regression is a very flexible tool with good results in practice.

Functional predictor	MSPE
FV2006 Cond. Expect.	1.92
FV2006 Cond. Mode.	2.94
FV2006 Cond. Median.	4.84
Kernel.FV d_0^{deriv}	139.36
Kernel.FV d_1^{deriv}	11.93
Kernel.FV d_2^{deriv}	5.37

Functional predictor	MSPE
Kernel.LLDBR d_0^{deriv}	52.08
Kernel.LLDBR d_1^{deriv}	6.85
Kernel.LLDBR d_2^{deriv}	3.52
$d_1 = d_0^{deriv}$, $d_2 = d_0^{deriv}$	7.94
$d_1 = d_1^{deriv}$, $d_2 = d_0^{deriv}$	2.12
$d_1 = d_2^{deriv}$, $d_2 = d_0^{deriv}$	1.43
$d_1 = d_1^{deriv}$, $d_2 = d_1^{deriv}$	2.91
$d_1 = d_2^{deriv}$, $d_2 = d_2^{deriv}$	1.03

Table 10.1 Mean square prediction error (MSPE) for different functional predictors.

References

[1] Baíllo, A. and Grané, A.: Local linear regression for functional predictor and scalar response, Univ. Carlos III de Madrid, Statistics and Econometric Series, 07-61 (2007).

[2] Barrientos-Marin, Jorge .: Some Practical Problems of Recent Nonparametric Procedures: Testing, Estimation, and Application. Univ. de Alicante (2007).

[3] Berlinet, A., Elamine, A. and Mas, A.: Local linear regression for functional data", ArXiv e-prints. 0710.5218, **710**, http://adsabs.harvard.edu/abs/2007arXiv0710.5218B (2007).

[4] E. Boj and Claramunt, M. M. and J. Fortiana.: Selection of Predictors in Distance-Based Regression. Comm. in Statistics. Simulation and Computation. **36**, 87–98 (2007).

[5] Boj, E. and Fortiana, J.: Weighted Distance-Based Regression, Unpublished (2007).

[6] Borg, Ingwer and Groenen, Patrick .: Modern Multidimensional Scaling: Theory and Applications (2nd ed), Springer-Verlag, New York (2005).

[7] Cuadras, C. M. and Arenas, C.: A distance based regression model for prediction with mixed data. Comm. in Statistics A. Theory and Methods, **19**, 2261-2279 (1990).

[8] Cuadras Carles M. and Josep Fortiana.: Distance-Based Multivariate Two Sample Tests IMUB. Institut de Matemàtica de la Universitat de Barcelona. **334** (2003).

[9] Cuadras, C. M.: Distance Analysis in discrimination and classification using both continuous and categorical variables. dodge:1989, 459–473 (1989).

[10] Cuadras, C. M., Arenas, C. and Fortiana, J.: Some computational aspects of a Distance-Based model for Prediction. Comm. in Statistics. Simulation and Computation **25**, 593–609 (1996).

[20] Ferraty F. and Vieu P.: Nonparametric modelling for functional data. Springer-Verlag, New York. (2006).

[12] Ferraty, F., Mas, A. and Vieu, P.: Nonparametric regression on functional data: Inference and practical aspects. Australian and New Zeland J. Stats. **49** (3), 267–286 (2007).

[13] Muñoz-Maldonado, Y., Staniswalis, J.G. and Irwin, L.N. and Byers, D.: A similarity analysis of curves. Canadian Journal of Statistics. **30**, 373-381 (2002).

[14] Ramsay, J. and Silverman, B.: Functional Data Analysis (Second Edition) Spinger-Verlag, New York. (2005).

Chapter 11
Singular Value Decomposition of Large Random Matrices (for Two-Way Classification of Microarrays)

Marianna Bolla, Katalin Friedl and András Krámli

Abstract Asymptotic behavior of the SVD of blown up matrices exposed to Wigner-noise is investigated. It is proved that such an $m \times n$ matrix almost surely has a constant number of large singular values (of order $\sqrt{mn}$), while the rest of the singular values are of order $\sqrt{m+n}$, as $m, n \to \infty$. An algorithm, applicable to two-way classification of microarrays, is also given that finds the underlying block structure.

11.1 Introduction

In this paper the theory of random matrices – with sizes tending to infinity – is applied and developed to find linear structure in large real-world data sets like internet or microarray measurements. Because of the large sizes, classical statistical methods cannot be used immediately.

In *Bolla (2005)*, large symmetric blown up matrices burdened with a so-called symmetric Wigner-noise were investigated (see Definitions 2.1, 2.3). It was proved that such an $n \times n$ matrix has some protruding eigenvalues (of order n), while the majority of the eigenvalues is at most of order $\sqrt{n}$ with probability tending to 1, as $n \to \infty$. These provide a useful tool to recognize linear structure in large symmetric real matrices, such as weight matrices of

Marianna Bolla
Institute of Mathematics, Budapest University of Technology and Economics, Budapest, e-mail: marib@math.bme.hu

Katalin Friedl
Department of Computer Science, Budapest University of Technology and Economics, Budapest, e-mail: friedl@cs.bme.hu

András Krámli
Bolyai Institute, University of Szeged, Budapest, e-mail: kramli@informatika.ilab.sztaki.hu

random graphs on a large number of vertices produced by communication, social, or cellular networks. Our goal is to generalize these results for the SVD of large rectangular random matrices and to apply them to the contingency table matrix formed by categorical variables in order to perform two-way clustering of these variables.

11.2 Singular values of a noisy matrix

Definition 11.1. The $m \times n$ real matrix $\mathbf{W}$ is a *Wigner-noise* if its entries w_{ij} $(1 \leq i \leq m,\ 1 \leq j \leq n)$ are independent random variables, $\mathbb{E}(w_{ij}) = 0$ and the w_{ij}'s are uniformly bounded (i.e., there is a constant $K > 0$ such that $|w_{ij}| \leq K$).

The name is originated from the seminal paper *Wigner (1958)* on the distribution of eigenvalues of random matrices. The term Wigner-noise in statistics was first used in *Bolla (2005)*.

According to a generalization of a theorem of *Füredi and Komlós (1981)* to rectangular matrices, the following result is valid for $\mathbf{W}$.

Lemma 11.1. *The maximum singular value of the Wigner-noise* $\mathbf{W}$ *is at most of order* $\sqrt{m+n}$ *with probability tending to 1, as* $n, m \to \infty$.

Definition 11.2. The $m \times n$ real matrix $\mathbf{B}$ is a *blown up matrix*, if there is an $a \times b$ so-called *pattern matrix* $\mathbf{P}$ with entries $0 \leq p_{ij} \leq 1$, further there are positive integers $m_1, \dots, m_a$ with $\sum_{i=1}^{a} m_i = m$ and $n_1, \dots, n_b$ with $\sum_{i=1}^{b} n_i = n$, respectively, such that the matrix $\mathbf{B}$ can be divided into $a \times b$ blocks, the block (i, j) being an $m_i \times n_j$ matrix with entries all equal to p_{ij} $(1 \leq i \leq a,\ 1 \leq j \leq b)$.

Blown up structures are usual in graph theory and also sought for in microarray analysis where they are called chess-board patterns, cf. *Kluger et al (2003)*. Let us fix the matrix $\mathbf{P}$, blow it up to obtain the matrix $\mathbf{B}$, and let $\mathrm{A} = \mathbf{B} + \mathbf{W}$, where $\mathbf{W}$ is a Wigner-noise of appropriate size. We are interested in the properties of A when $m_1, \dots, m_a \to \infty$ and $n_1, \dots, n_b \to \infty$, roughly speaking, both at the same rate (in the sequel it will be called usual growth condition).

Proposition 11.1. *If the usual growth condition holds, then all the non-zero singular values of the* $m \times n$ *blown-up matrix* $\mathbf{B}$ *are of order* $\sqrt{mn}$.

Theorem 11.1. *Let* $\mathrm{A} = \mathbf{B} + \mathbf{W}$ *be an* $m \times n$ *random matrix, where* $\mathbf{B}$ *is a blown up matrix with positive singular values* $s_1, \dots, s_r$ *and* $\mathbf{W}$ *is a Wigner-noise of the same size. Then the matrix* A *almost surely has* r *singular values* $z_1, \dots, z_r$ *with* $|z_i - s_i| = \mathcal{O}(\sqrt{m+n})$, $i = 1, \dots, r$, *and for the other singular values* $z_j = \mathcal{O}(\sqrt{m+n})$, $j = r+1, \dots, \min\{m, n\}$ *hold almost surely, as* $m, n \to \infty$ *under the usual growth condition. (Here* $r \leq \min\{a, b\}$*.)*

Proof The statement follows from the analog of the Weyl's perturbation theorem for singular values of rectangular matrices. If $s_i(\mathrm{A})$ and $s_i(\mathbf{B})$ denote the ith singular values of the matrix in the argument in decreasing order then for the difference of the corresponding pairs $|s_i(\mathrm{A}) - s_i(\mathbf{B})| \le \max_i s_i(\mathbf{W}) = \|\mathbf{W}\|$, $i = 1, \dots, \min\{m, n\}$. Lemma 2.2 asserts that $\mathbf{W}$'s spectral norm $\|\mathbf{W}\| = s_1(\mathbf{W}) = \mathcal{O}(\sqrt{m+n})$ in probability (by large deviations, also almost surely), and this finishes the proof.

11.3 Classification via singular vector pairs

Let $\mathbf{Y}$ be the $m \times r$ matrix containing the left-hand side singular vectors $\mathbf{y}_1, \dots, \mathbf{y}_r$ of A in its columns. Similarly, let $\mathbf{X}$ be the $n \times r$ matrix containing the right-hand side singular vectors $\mathbf{x}_1, \dots, \mathbf{x}_r$ of A in its columns. We shall speak in terms of microarrays. Let the r-dimensional representatives of the genes be the row vectors of $\mathbf{Y}$: $\mathbf{y}^1, \dots, \mathbf{y}^m \in \mathbb{R}^r$, while the r-dimensional representatives of the conditions be the row vectors of $\mathbf{X}$: $\mathbf{x}^1, \dots, \mathbf{x}^n \in \mathbb{R}^r$. Let $S_a^2(\mathbf{Y})$ denote the a-variance of the genes' representatives in the clustering $A_1, \dots, A_a$:

$$S_a^2(\mathbf{Y}) = \sum_{i=1}^{a} \sum_{j \in A_i} \|\mathbf{y}^j - \bar{\mathbf{y}}^i\|^2, \quad \text{where} \quad \bar{\mathbf{y}}^i = \frac{1}{m_i} \sum_{j \in A_i} \mathbf{y}^j,$$

while $S_b^2(\mathbf{X})$ denotes the b-variance of the conditions' representatives in the clustering $B_1, \dots, B_b$:

$$S_b^2(\mathbf{X}) = \sum_{i=1}^{b} \sum_{j \in B_i} \|\mathbf{x}^j - \bar{\mathbf{x}}^i\|^2, \quad \text{where} \quad \bar{\mathbf{x}}^i = \frac{1}{n_i} \sum_{j \in B_i} \mathbf{x}^j.$$

Theorem 11.2. *With the above notation, for the a- and b-variances of the representation of the microarray* A *the relations*

$$S_a^2(\mathbf{Y}) = \mathcal{O}\left(\frac{m+n}{mn}\right) \quad \textit{and} \quad S_b^2(\mathbf{X}) = \mathcal{O}\left(\frac{m+n}{mn}\right)$$

hold almost surely, under the usual growth condition.

Proof idea The piecewise constant structure of singular vectors of $\mathbf{B}$ is used.

Hence, the addition of any kind of a Wigner-noise to a rectangular matrix that has a blown up structure $\mathbf{B}$ will not change the order of the protruding singular values, and the block structure of $\mathbf{B}$ can be reconstructed from the representatives of the row and column items of the noisy matrix A.

With an appropriate Wigner-noise, we can achieve that the matrix $\mathbf{B} + \mathbf{W}$ in its (i, j)-th block contains 1's with probability p_{ij}, and 0's otherwise. Thus, the noisy matrix A becomes a 0-1 random matrix of incidence relations between the genes and conditions.

11.4 Perturbation results for correspondence matrices

Sometimes the pattern matrix $\mathbf{P}$ is an $a \times b$ contingency table with entries that are nonnegative integers. Then the blown up matrix $\mathbf{B}$ can be regarded as a larger $(m \times n)$ contingency table that contains e.g., counts for two categorical variables with m and n different categories, respectively. For finding maximally correlated factors with respect to the marginal distributions of these two discrete variables, the technique of correspondence analysis is widely used, see *Benzécri et al. (1973)*. In case of a general pattern matrix $\mathbf{P}$ (with nonnegative real entries), the blown-up matrix $\mathbf{B}$ can also be regarded as a data matrix for two not independent categorical variables. As the categories may be measured in different units, a normalization is necessary. This normalization is made by dividing the entries of $\mathbf{B}$ by the square roots of the corresponding row and column sums. This transformation is identical to that of the correspondence analysis, and the transformed matrix remains the same when we multiply the initial matrix by a positive constant. Thus, it does not matter whether we started with a contingency or frequency table or just with a matrix with nonnegative entries. After performing correspondence transformation on $\mathbf{B}$, the resulting $\mathbf{B}_{corr}$ has entries in [0,1] and maximum singular value 1. It is proved that there is a significant gap between the k largest (where $k = \text{rank}\,(\mathbf{B}) = \text{rank}\,(\mathbf{P})$) and the other singular values of A_{corr}, the matrix obtained from the noisy matrix $\mathrm{A} = \mathbf{B} + \mathbf{W}$ by the correspondence transformation. This implies well two-way classification properties of the row and column categories (genes and expression levels).

11.5 Recognizing the structure

A construction is given, how a blown up structure behind a real-life matrix with a few protruding singular values and "well classifiable" corresponding singular vector pairs can be found.

Theorem 11.3. *Let $\mathrm{A}_{m\times n}$ be a sequence of $m \times n$ matrices, where m and n tend to infinity. Assume, that $\mathrm{A}_{m\times n}$ has exactly k singular values of order greater than $\sqrt{m+n}$ (k is fixed). If there are integers $a \geq k$ and $b \geq k$ such that the a- and b-variances of the row- and column-representatives are $\mathcal{O}(\frac{m+n}{mn})$, then there is a blown up matrix $\mathbf{B}_{m\times n}$ such that $\mathrm{A}_{m\times n} = \mathbf{B}_{m\times n} + \mathbf{E}_{m\times n}$, with $\|\mathbf{E}_{m\times n}\| = \mathcal{O}(\sqrt{m+n})$.*

In the proof we give an explicit construction for $\mathbf{B}_{m \times n}$ by means of metric classification methods. To find SVD of large rectangular matrices randomized algorithms are favored, e.g., *Frieze, Kannan (1999)*. They exploit the randomness of our data and provide good approximations of the underlying clusters only if originally there was a linear structure in our matrix.

References

[1] Benzecri, J. P. et al: L'Analyse dés Données. Tome 2. L'Analyse dés Correspondances. Dunod, Paris, (1973).
[2] Bolla, M.: Recognizing Linear Structure in Noisy Matrices, Linear Algebra and its Applications **402**, 228-244 (2005).
[3] Frieze, A., Kannan, R.: Quick Approximation to Matrices and Applications, Combinatorica **19**, (2) 175-220 (1999).
[4] Füredi, Z., Komlós, J.: The Eigenvalues of Random Symmetric Matrices, Combinatorica **1**, (3) 233-241 (1981).
[5] Kluger, Y., et al., Spectral Biclustering of Microarray Data: Coclustering Genes and Conditions, Genome Rerearch **13**, 703-716 (2003).
[6] Wigner, E. P.: On the Distribution of the Roots of Certain Symmetric Matrices, Ann. Math. **62**, 325-327 (1958).

Chapter 12
On Tensorial Products of Hilbertian Linear Processes

Denis Bosq

Abstract We study quadratic transformations for real and Hilbertian Linear Processes.

12.1 Introduction

Quadratic transforms of linear processes are of interest since they play an important part in estimation of autocovariance and prediction of these processes (cf Granger and Newbold (1976), Choi and Taniguchi (2002) among others).

In this work we study that kind of transformations for real and functional processes. For convenience we focus on autoregressive and moving average processes.

The main fact is conservation of these models by quadratic transformations. Properties of the resulting process are easy to derive in the autoregressive case when the moving average context appears to be more intricate.

12.2 The real case

In the real case tensorial product may be identified with usual product.

a) Transforming autoregressive processes

Let $(X_n, n \in \mathbf{z})$ be a real autoregressive process of order 1 (AR(1)) that satisfies the relation

$$X_n = \rho X_{n-1} + \varepsilon_n, \quad n \in \mathbf{z}$$

Denis Bosq
Name, Université Pierre et Marie Curie, Paris 6, France, e-mail: `bosq@ccr.jussieu.fr`

where $|\rho| < 1$ and $(\varepsilon_n,\ n \in \mathbf{z})$ is a (weak) white noise.

We suppose that

$$\text{(I)} - \qquad \mathrm{E}\,\varepsilon_n^4 = \mathrm{E}\,\varepsilon_0^4 < \infty,\ \mathrm{E}^{\mathcal{B}_{n-1}}(\varepsilon_n^2) = \sigma^2 > 0,\ \mathrm{E}^{\mathcal{B}_{n-1}}(\varepsilon_n) = 0,\ \ n \in \mathbf{z},$$

where $\mathcal{B}_{n-1} = \sigma(X_t,\ t \leq n-1)$ is the σ-algebra generated by X_t, $t \leq n-1$.

Then we have the following simple result :

Proposition 12.1.
Set

$$Z_n = X_n^2 - \frac{\sigma^2}{1-\rho^2},\quad n \in \mathbf{z},$$

then (Z_n) is an AR(1) such that

$$Z_n = \rho^2 Z_{n-1} + E_n,\quad n \in \mathbf{z}$$

where

$$E_n = (\varepsilon_n^2 - \sigma^2) + 2\rho X_{n-1}\varepsilon_n,\quad n \in \mathbf{z},$$

moreover (E_n) is a martingale difference adapted to $(\mathcal{B}_n)$.

Similarly, if (X_n) and (Y_n) are independent AR(1), (X_nY_n) is again an AR(1).

It is then possible to compute the best predictor of X_{n+1}^2 (resp. $X_{n+1}Y_{n+1}$) given $(X_t^2,\ t \leq n)$ (resp. $X_tY_t,\ t \leq n$) and to compare it with the best predictor given $(X_t,\ t \leq n)$ (resp. $X_tY_t,\ t \leq n$). We also compare the prediction errors.

b) Moving averages

We now consider the moving average of order 1 (MA(1)) defined by

$$X_n = \varepsilon_n + a\varepsilon_{n-1},\quad n \in \mathbf{z}$$

where $|a| < 1$ and (ε_n) satisfies (I) and the additional condition

$$\mathrm{E}\,(\varepsilon_{n-1}\varepsilon_n^3) = 0,\quad n \in \mathbf{z}.$$

Putting $Z_n = X_n^2 - \sigma^2(1+a^2)$, $n \in \mathbf{z}$, one obtains the following :

Proposition 12.2.
$(Z_n,\ n \in \mathbf{z})$ is a MA(1) defined by

$$Z_n = E_n + AE_{n-1}, \quad n \in \mathbf{z}$$

where $0 \leq A < 1$ and (E_n) is the innovation process given by

$$E_n = \sum_{j=0}^{\infty} (-A)^j X_{n-j}^2 - \sigma^2 \frac{1+a^2}{1+A}.$$

Contrary to the case of an AR(1) the relation

$$R = \rho^2 \tag{12.1}$$

(where $\rho = \mathrm{corr}(X_0, X_1)$ and $R = \mathrm{corr}(Z_0, Z_1)$) is not satisfied in general. Actually one has

$$R = \frac{1+\kappa_{(4)}}{1+\alpha\kappa_{(4)}} \rho^2$$

where $\kappa_{(4)} = \dfrac{\mathrm{E}\,\varepsilon_0^4}{2\sigma^4} - \dfrac{3}{2}$ is the 4th normed cumulant of ε_0 and $\alpha = (1+a^4)/(1+a^2)^2$. It follows that $0 \leq R < \dfrac{a^2}{1+a^4}$ and that (12.1) holds if and only if $\kappa_{(4)} = 0$, in particular if (X_n) is Gaussian. Note that (12.1) implies

$$\frac{A}{1+A^2} = \left(\frac{a}{1+a^2}\right)^2.$$

Similar results may be obtained for the product of two independent MA(1).

Finally we compare prediction of Z_{n+1} given $Z_t,\ t \leq n$ with prediction of Z_{n+1} given $X_t,\ t \leq n$.

12.3 The hilbertian case

Let H be a real separable Hilbert space and $(X_n,\ n \in \mathbf{z})$ a sequence of H-valued random variables. (X_n) is a (standard) **autogressive process of order 1** (ARH(1)) if it is stationary and such that

$$X_n = \rho(X_{n-1}) + \varepsilon_n, \quad n \in \mathbf{z} \tag{12.2}$$

where (ε_n) is a H-white noise and $\rho \in \mathcal{L}$ (the space of continuous linear operators from H to H, equipped with its usual norm $\|\cdot\| \cdot_{\mathcal{L}}$). If $\|\cdot\| \, \rho^{j_o}{}_{\mathcal{L}} < 1$ for some integer j_0, (12.2) has a unique solution, namely

$$X_n = \sum_{j=0}^{\infty} \rho^j(\varepsilon_{n-j}), \quad n \in \mathbf{z}.$$

Now let $\mathcal{S}$ be the space of Hilbert-Schmidt operators on H with its norm $\|\cdot\|\cdot_{\mathcal{S}}$, and consider the following assumptions

$$\text{(I)}' - \ \mathrm{E}\,\|\cdot\|\,\varepsilon_n \otimes \varepsilon_n{}^2_{\mathcal{S}} = \mathrm{E}\,\|\cdot\|\,\varepsilon_0 \otimes \varepsilon_0{}^2_{\mathcal{S}} < \infty, \ \mathrm{E}^{\mathcal{B}_{n-1}}(\varepsilon_n) = 0,$$
$$\mathrm{E}^{\mathcal{B}_{n-1}}(\varepsilon_n \otimes \varepsilon_n) = C_{\varepsilon_0},$$

where $C_{\varepsilon_0} = \mathrm{E}\,(\varepsilon_0 \otimes \varepsilon_0)$ and $\mathcal{B}_{n-1} = \sigma(X_t,\ t \leq n-1)$.
Let us set

$$Z_n = X_n \otimes X_n - C, \quad n \in \mathbf{z}$$

(where $C = \mathrm{E}\,(X_0 \otimes X_0)$) then :

Proposition 12.3.
(Z_n) is a $\mathcal{S}$-valued AR(1) process such that

$$Z_n = R(Z_{n-1}) + E_n, \quad n \in \mathbf{z}$$

where (E_n) is the $\mathcal{S}$-white noise given by

$$E_n = (X_{n-1} \otimes \varepsilon_n)\rho^* + \rho(\varepsilon_n \otimes X_{n-1}) + \varepsilon_n \otimes \varepsilon_n - C_{\varepsilon_0}, \quad n \in \mathbf{z}$$

and

$$R(s) = \rho s \rho^*, \quad s \in \mathcal{S}.$$

Moreover (E_n) is a $(\mathcal{B}_n)$ adapted martingale difference and

$$\|\cdot\|\, R^j{}^{(\mathcal{S})}_{\mathcal{L}} \leq \|\cdot\|\, \rho^j{}^2_{\mathcal{L}}, \quad j \geq 1.$$

Finally the case of a MAH(1) is more intricate. First set

$$\mathcal{G}_{\varepsilon_n} = \overline{\mathrm{e}}\left\{\ell(\varepsilon_n),\ \ell \in \mathcal{L}\right\}$$

where the closure is taken in the space $L^2_H(\Omega, \mathcal{A}, P)$, and let Π^{ε_n} be the orthogonal projection of ε_n. Then a **(non-standard) MAH(1)** associated with (ε_n) satisfies

$$X_n = \varepsilon_n + \Pi^{\varepsilon_n - 1}(X_n), \quad n \in \mathbf{z}. \tag{12.3}$$

Details concerning non-standard linear processes appear in Bosq (2007) and Bosq and Blanke (2007).

If a stationary process satisfies (12.3) we have :

Proposition 12.4.
$(X_n \otimes X_n - C)$ is a $\mathcal{S}$-valued non standard MA(1).

Some examples of applications are considered.

References

[1] Bosq, D.: General linear processes in Hilbert spaces and prediction. J. Statist. Plann. Inference 127(3), 879-94 (2007).

[2] Bosq, D., Blanke D.: Inference and Prediction in large dimensions. Wiley (and Dunod), Chichester, (2007).

[3] Choi, I.B., Taniguchi, M.: Prediction problems for square transformed stationary processes. Statist. Infer. Stoch. Process., 5, 1-22 (2002).

[4] Granger, C.W.J., Newhold, P.: Forecasting transformed series. J. Royal Statist. Soc., B, 38, 2, 189-203 (1976).

Chapter 13
Recent Results on Random and Spectral Measures with Some Applications in Statistics

Alain Boudou, Emmanuel Cabral and Yves Romain

Abstract In this talk we define and study first the convolution product of two spectral measures and secondly the tensor and convolution products of random measures. Then we propose some applications in stationary processes statistics.

13.1 Introduction and definitions

We begin by introducing three examples in the domain of stationary processes:
(i) An interpolation problem
Let $(X_n)_{n\in\mathbb{Z}}$ be a stationary series, how can we define all stationary series $(Y_n)_{n\in\mathbb{Z}}$ such as $Y_n = X_{nq}$?
(ii) A spatial process identification problem
Let $(X_n)_{n\in\mathbb{Z}}$ be a stationary series, how can we define all stationary series $(Y_{n,m})_{n,m\in\mathbb{Z}\times\mathbb{Z}}$ such as $Y_{nq,np} = X_n$?
(iii) An inverse Fourier transform problem

Alain Boudou
Name, Equipe Labo. de Stat. et Proba., Institut de Mathématiques UMR 5219
Université Paul Sabatier, 118 Route de Narbonne
F-31062 Toulouse cedex, France, e-mail: boudou@cict.fr

Emmanuel Cabral
Name, Equipe Labo. de Stat. et Proba., Institut de Mathématiques UMR 5219
Université Paul Sabatier, 118 Route de Narbonne
F-31062 Toulouse cedex, France, e-mail: cabral@cict.fr

Yves Romain
Equipe Labo. de Stat. et Proba., Institut de Mathématiques UMR 5219
Université Paul Sabatier, 118 Route de Narbonne
F-31062 Toulouse cedex, France, e-mail: romain@cict.fr

Let $(X_nY_n)_{n\in\mathbb{Z}}$ be a "multiplicative" series, how can we define the random measure whose Fourier transform is the considered series?

To resolve such kind of questions, we need to elaborate new tools on random and spectral measures. These results are presented in the following two and three sections. Before that we recall and give some definitions of random and spectral measures (cf. Azencott et *al.*, 1984; Birman et *al.*, 1996; Boudou et *al.*, 2002). A *random measure* (r.m.) defined on ξ (a σ-field of a set E) with values in a $\mathbb{C}$-Hilbert space H, is a mapping Z from ξ onto H such as:
(i) for all pairs (A, B) of disjoint elements of ξ, $Z(A \cup B) = Z(A) + Z(B)$ and
$< Z(A), Z(B) >= 0$;
(ii) for each decreasing sequence $(A_n)_{n\in\mathbb{N}}$ of elements of ξ converging to $\emptyset$, we have $\lim_n Z(A_n) = 0_H$.
Then it is easy to verify that the mapping $\mu_Z : A \in \xi \mapsto \|Z(A)\|^2 \in \mathbb{R}^+$ is a bounded measure.
The *stochastic integral* with respect to the r.m. Z can be defined as the unique isometry from $L^2(E, \xi, \mu_Z)$ onto $H_Z = \overline{vect}\{Z(A), A \in \xi\}$ which associates $Z(A)$ to $\mathbf{1}_A$, for all A in ξ.
When (E', ξ') is a second measurable space and $\mathcal{L}$ a measurable mapping from E into E' then the mapping $\mathcal{L}(Z) : A' \in \xi' \mapsto Z(\mathcal{L}^{-1}(A')) \in H$ is a r.m. called *the r.m. image of Z by $\mathcal{L}$*. It is easy to verify that $\mathcal{L}(\mu_Z) = \mu_{\mathcal{L}(Z)}$ and, if φ is an element of $L^2(E', \xi', \mu_{\mathcal{L}(Z)})$ then $\varphi \circ \mathcal{L}$ is a element of $L^2(E, \xi, \mu_Z)$ and $\int \varphi d\mathcal{L}(Z) = \int \varphi \circ \mathcal{L} dZ$.
When Z is a r.m. defined on $\mathcal{B}$ (resp. $\mathcal{B} \otimes \mathcal{B}$) the Borel σ-field of $\Pi = [-\pi, \pi[$ (resp. $\Pi \times \Pi$) we call *Fourier transform of Z* the series $(\int e^{i\cdot n} dZ)_{n\in\mathbb{Z}}$ (resp. $(\int e^{i(\cdot n + \cdot n')} dZ)_{(n,n')\in\mathbb{Z}^2})$ which is a stationary series, e.g.
$< \int e^{i\cdot n} dZ, \int e^{i\cdot m} dZ >=< \int e^{i\cdot(n-m)} dZ, \int e^{i\cdot 0} dZ >$ (resp.
$< \int e^{i(\cdot n + \cdot n')} dZ, \int e^{i(\cdot m + \cdot m')} dZ >=< \int e^{i(\cdot(n-m) + \cdot(n'-m'))} dZ, \int e^{i(\cdot 0 + \cdot 0)} dZ >)$.
Conversely, for each stationary series, we can associate a unique r.m. (called *the associated r.m.*) whose Fourier transform is the considered series.
A *spectral measure* (s.m.) *on ξ for H* is a mapping ε from ξ into $\mathcal{P}(H)$, the orthogonal projectors set of H, such as:
(i) $\varepsilon(E) = id_H$;
(ii) for all pairs (A, B) of disjoint elements of ξ, $\varepsilon(A \cup B = \varepsilon(A) + \varepsilon(B)$;
(iii) for each decreasing sequence $(A_n)_{n\in\mathbb{N}}$ of elements of ξ converging to $\emptyset$ and all X of H, $\lim_n \varepsilon(A_n)X = 0$.
Then we show that, for all pairs (A, B) of elements of ξ: $\varepsilon(A) \circ \varepsilon(B) = \varepsilon(A \cap B)$ and, for all X of H, the mapping $Z_\varepsilon^X : A \in \xi \mapsto \varepsilon(A)X \in H$ is a r.m. .
Naturally, with the previous notations, we verify that the mapping $\mathcal{L}(\varepsilon) : A' \in \xi' \mapsto \varepsilon(\mathcal{L}^{-1}(A')) \in \mathcal{P}(H)$ is a s.m. on ξ' for H called *the image s.m. of ε by $\mathcal{L}$*.

13.2 Convolution product of spectral measures

When U is a unitary operator of H (e.g. $U^*U = id_H = UU^*$), it is clear that, for all X of H, the series $(U_nX)_{n\in\mathbb{Z}}$ is stationary. If we denote by Z^X the associated r.m. defined on $\mathcal{B}$ with values in H, we can affirm that:
(i) for all A of $\mathcal{B}$, the mapping $\varepsilon(A) : X \in H \mapsto Z^X(A) \in H$ is an orthogonal projector;
(ii) the mapping $\varepsilon : A \in \mathcal{B} \mapsto \varepsilon(A) \in \mathcal{P}(H)$ is a s.m. on $\mathcal{B}$ for H called *the associated s.m. to U.*
It is clear that, for all X in H, $Z^X = Z_\varepsilon^X$ and then Z_ε^X is the associated r.m. to the stationary series $(U_nX)_{n\in\mathbb{Z}}$; and futhermore, $UX = \int e^{i\cdot 1} dZ_\varepsilon^X$. Conversely, given a s.m. ε on $\mathcal{B}$ for H, we can show that the mapping $X \in H \mapsto \int e^{i\cdot 1} dZ_\varepsilon^X \in H$ is a unitary operator whose associated s.m. is ε.
A unitary operator U with associated s.m. ε, may be written as a limit of Riemann's sum because of the equality

$$U = \lim_n \sum_{k=0}^{n-1} e^{i(-\pi + k\frac{2\pi}{n})} \varepsilon([-\pi + k\frac{2\pi}{n}, -\pi + (k+1)\frac{2\pi}{n}[);$$

in sense of the norm $\|A\| = sup\{\|Ax\|; \|x\| = 1\}$.
If U and V are two unitary operators of H with, respectively, associated s.m. ε and α, we show that $UV = VU$ if and only if, for all pairs (A, B) of elements of $\mathcal{B}$, the projectors $\varepsilon(A)$ and $\alpha(B)$ commute. In this case, the series $(U^nV^mX)_{(n,m)\in\mathbb{Z}\times\mathbb{Z}}$ is stationary. Denoting by $\mathfrak{z}^X$ its associated r.m., we can affirm that
- for all A of $\mathcal{B} \otimes \mathcal{B}$, the mapping $(\varepsilon \otimes \alpha)(A) : X \in H \mapsto \mathfrak{z}^X(A) \in H$ is an orthogonal projector;
- the mapping $\varepsilon \otimes \alpha : A \in \mathcal{B} \otimes \mathcal{B} \mapsto (\varepsilon \otimes \alpha)(A) \in \mathcal{P}(H)$ is a s.m. on $\mathcal{B} \otimes \mathcal{B}$ for H (called *the s.m. product of ε and α*).

We denote by P the measurable mapping $(\lambda_1, \lambda_2) \in \Pi \times \Pi \mapsto \lambda_1 \in \Pi$. The integration rules with respect to an image r.m. permit us to write

$$\int e^{i\cdot n} dP(\mathfrak{z}^X) = \int e^{i\cdot n} \circ P d\mathfrak{z}^X = \int e^{i(\cdot n + \cdot 0)} d\mathfrak{z}^X = U^nV^0X = U^nX;$$

so, by unicity of a r.m. associated to a stationary series, we have $P(\mathfrak{z}^X) = Z_\varepsilon^X$.
For all (A, X) of $\mathcal{B} \times H$, we have

$$\varepsilon(A)X = Z_\varepsilon^X(A) = P(\mathfrak{z}^X)A = \mathfrak{z}^X P^{-1}(A) = \varepsilon \otimes \alpha(A \times \Pi)X$$

and then, for all A of $\mathcal{B}$, $\varepsilon(A) = (\varepsilon \otimes \alpha)(A \times \Pi)$.
Similarly, for all B of $\mathcal{B}$, we show that $\alpha(B) = (\varepsilon \otimes \alpha)(\Pi \times B)$.
This last point permits us to affirm that $\varepsilon \otimes \alpha$ is the unique s.m. on $\mathcal{B} \otimes \mathcal{B}$ for H such as, for all pairs (A, B) of elements of $\mathcal{B}$, $(\varepsilon \otimes \alpha)(A \times B) = \varepsilon(A) \circ \alpha(B)$
.

Because of the topological group structure of Π, the mapping

$$S : (\lambda_1, \lambda_2) \in \Pi \times \Pi \mapsto \lambda_1 + \lambda_2 - 2\pi[\frac{\lambda_1 + \lambda_2 + \pi}{2\pi}] \in \Pi,$$

is continuous and also measurable. We may consider the s.m. on $\mathcal{B}$ for H, $S(\varepsilon \otimes \alpha)$, denoted by $\varepsilon * \alpha$ and called *the convolution product of the s.m.* ε *and* α.
For all X of H, one has:

$$\int e^{i\cdot 1} dZ^X_{\varepsilon * \alpha} = \int e^{i\cdot 1} dS(Z^X_{\varepsilon \otimes \alpha}) = \int e^{i\cdot 1} \circ S dZ^X_{\varepsilon \otimes \alpha} = \int e^{i(\cdot 1 + \cdot 1)} d\mathfrak{z}^X = UVX,$$

which implies that $\varepsilon * \alpha$ is the s.m. associated to the unitary operator UV.
The convolution product admits an identity element. The s.m. $\varepsilon_\Pi = \delta_0(\cdot) id_H$, where δ_0 is the Dirac measure concentrated n 0, is the s.m. associated to the unitary operator id_H, it commutes with all s.m. ε on $\mathcal{B}$ for H, and then, $\varepsilon_\Pi * \varepsilon = \varepsilon$.

We quote at last a distributivity property: if f is a continuous homomorphism defined from Π (resp. $\Pi \times \Pi$) into Π, if ε_1 and ε_2 two commuting s.m. on $\mathcal{B}$ (resp. $\mathcal{B} \otimes \mathcal{B}$) for H, then the s.m. $f\varepsilon_1$ and $f\varepsilon_2$ commute and $f(\varepsilon_1 * \varepsilon_2) = (f\varepsilon_1) * (f\varepsilon_2)$.
Let h be a (obviously continuous) homomorphism defined from $\mathbb{Z}$ into $\mathbb{Z}$ (resp. $\mathbb{Z} \times \mathbb{Z}$) and ${}^t h$ its transpose, that means ${}^t h$ is a homomorphism from Π, the dual space of $\mathbb{Z}$ (resp. $\Pi \times \Pi$ the dual space of $\mathbb{Z} \times \mathbb{Z}$) into Π such as $e^{i\lambda h(n)} = e^{i^t h(\lambda) n}$, for all (λ, n) of $\Pi \times \mathbb{Z}$ (resp. $e^{i<(\lambda_1,\lambda_2);h(n)>} = e^{i^t h(\lambda_1,\lambda_2) n}$, for all $((\lambda_1, \lambda_2), n)$ of $(\Pi \times \Pi) \times \mathbb{Z}$).
Then we show that if $(Y_n)_{n\in\mathbb{Z}}$ (resp. $(Y_{n,m})_{(n,m)\in\mathbb{Z}\times\mathbb{Z}}$) is a stationary series with associated r.m. Z_Y, ${}^t h Z_Y$ is the associated r.m. to the series $(Y_{h(n)})_{n\in\mathbb{Z}}$.
Also, given a stationary series $(X_n)_{n\in\mathbb{Z}}$ with associated r.m. Z_X, searching all series $(Y_n)_{n\in\mathbb{Z}}$ (resp. $(Y_{n,m})_{(n,m)\in\mathbb{Z}\times\mathbb{Z}}$) such as $(Y_{h(n)})_{n\in\mathbb{Z}} = (X_n)_{n\in\mathbb{Z}}$ consists of searching all r.m. Z_Y such as ${}^t h Z_Y = Z_X$.
To solve this equation (where the r.m. Z_Y is unknown) we consider a measurable mapping v from Π into Π (resp. $\Pi \times \Pi$) such as ${}^t h \circ v = id_\Pi$ and a s.m. ε_X on $\mathcal{B}$ for H such as $Z^{X_0}_{\varepsilon_X} = Z_X$.
Then we show that if ε' is a s.m. on $\mathcal{B}$ (resp. $\mathcal{B} \otimes \mathcal{B}$) for H which commutes with $v\varepsilon_X$ such as ${}^t h \varepsilon' = \varepsilon_\Pi$ then $Z^{X_0}_{\varepsilon' * v\varepsilon_X}$ is a r.m. solution and all solutions are of this form.
The direct part of this proposition is resulting from the properties of convolution product of s.m. quoted previously, e.g.:

$${}^t h(Z^{X_0}_{\varepsilon' * v\varepsilon_X}) = Z^{X_0}_{{}^t h(\varepsilon' * v\varepsilon_X)} = Z^{X_0}_{{}^t h\varepsilon' * {}^t h v\varepsilon_X} = Z^{X_0}_{\varepsilon_\Pi * \varepsilon_X} = Z^{X_0}_{\varepsilon_X} = Z_X.$$

If we choose as homomorphism h the mapping $n \in \mathbb{Z} \mapsto nq \in \mathbb{Z}$, we can then define all stationary series $(Y_n)_{n\in\mathbb{Z}}$ such as $Y_{nq} = X_n$, for all n of $\mathbb{Z}$, $(X_n)_{n\in\mathbb{Z}}$ being a given stationary series (cf. Boudou, 2003).

If we chooe as homomorphism h the mapping $n \in \mathbb{Z} \mapsto (nq, np) \in \mathbb{Z} \times \mathbb{Z}$, we can then define all stationary series $(Y_{n,m})_{n \in \mathbb{Z} \times \mathbb{Z}}$ such as $Y_{nq,mp} = X_n$, for all n of $\mathbb{Z}$, $(X_n)_{n \in \mathbb{Z}}$ being a given stationary series.

13.3 Tensor and convolution products of random measures

In this part, H_1 and H_2 are separable $\mathbb{C}$-Hilbert spaces. We use a complex extension of the functional tensor product (cf. Dauxois *et al.*, 1994) of A_1 and A_2, two bounded endomorphisms of H_1 and H_2 respectively, defined by the mapping

$$A_1 \overset{l}{\otimes} A_2 : K \in \sigma_2(H_1, H_2) \mapsto A_2 \circ K \circ A_1^* \in \sigma_2(H_1, H_2),$$

where $\sigma_2(H_1, H_2)$ is the set of all Hilbert-Schmidt operators from H_1 into H_2. It is easy to verify the following equalities:
$(A_1 \overset{l}{\otimes} A_2) \circ (A_1' \overset{l}{\otimes} A_2') = (A_1 \circ A_1') \overset{l}{\otimes} (A_2 \circ A_2')$ and $(A_1 \overset{l}{\otimes} A_2)^* = A_1^* \overset{l}{\otimes} A_2^*$.
We deduce that the functional tensor product of two projectors (resp. unitary operators) is a projector (resp. unitary operator). We can now establish that when ε_1 is a s.m. on $\mathcal{B}$ for H_1:
(i) for all A of $\mathcal{B}$, $E_1(A) = \varepsilon_1(A) \overset{l}{\otimes} id_{H_2}$ is a projector;
(ii) the mapping $A \in \mathcal{B} \mapsto E_1(A) \in \mathcal{P}(\sigma_2(H_1, H_2))$ is a s.m. on $\mathcal{B}$ for $\sigma_2(H_1, H_2)$ called *the right functional ampliation of* ε_1 *in regard to* H_2.
Similarly, if ε_2 is a s.m. on $\mathcal{B}$ for H_2, we can define *the left functional ampliation* E_2 *of* ε_2 *in regard to* H_1 by the equality $E_2(A) = id_{H_1} \overset{l}{\otimes} \varepsilon_2(A)$, for all A of $\mathcal{B}$.
It is clear that the ampliations E_1 and E_2 commute and we can consider the s.m. $E_1 \otimes E_2$ which is called *tensor product of* ε_1 *and* ε_2, and the *convolution s.m.* $E_1 * E_2$ (cf. Boudou and Romain, 2002).
If we denote by U_2 the unitary operator of H_2 with associated s.m. ε_2, and defined by

$$U_2 = \lim_n \sum_{k=0}^{n-1} e^{i(-\pi + k\frac{2\pi}{n})} \varepsilon_2([-\pi + k\frac{2\pi}{n}, -\pi + (k+1)\frac{2\pi}{n}[),$$

we deduce the equality

$$id_{H_1} \overset{l}{\otimes} U_2 = \lim_n \sum_{k=0}^{n-1} e^{i(-\pi + k\frac{2\pi}{n})} [id_{H_1} \overset{l}{\otimes} (\varepsilon_2([-\pi + k\frac{2\pi}{n}, -\pi + (k+1)\frac{2\pi}{n}[))]$$

which permits us to affirm that the unitary operator $id_{H_1} \overset{l}{\otimes} U_2$ admits E_2 for associated s.m.. If we denote by U_1 the unitary operator of H_1 with associated s.m. ε_1, we show similarly that $U_1^* \overset{l}{\otimes} id_{H_2}$ admits E_1 for associated s.m..
We can now affirm that $(U_1^* \overset{l}{\otimes} id_{H_2}) \circ (id_{H_1} \overset{l}{\otimes} U_2) = U_1^* \overset{l}{\otimes} U_2$ is the unitary operator with associated s.m. $E_1 * E_2$.
We consider now $(X_n^1)_{n\in\mathbb{Z}}$ (resp. $(X_n^2)_{n\in\mathbb{Z}}$) a stationary series of elements of H_1 (resp. H_2) with associated r.m. Z_1 (resp. Z_2) and let ε_1 (resp. ε_2) be the s.m. such as $(\varepsilon_1(A))(Z_1(\Pi)) = Z_1(A)$ (resp. $(\varepsilon_2(A))(Z_2(\Pi)) = Z_2(A)$), for all A of $\mathcal{B}$.
By the equality $(U_1^* \overset{l}{\otimes} U_2)^n(X_0^1 \otimes X_0^2) = X_{-n}^1 \otimes X_n^2$, we can deduce that $(X_{-n}^1 \otimes X_n^2)_{n\in\mathbb{Z}}$ is a stationary series of elements of $\sigma_2(H_1, H_2)$ with associated r.m. $Z_{E_1*E_2}^{X_0^1\otimes X_0^2}$. This r.m. is the image, by S, of $Z_{E_1\otimes E_2}^{X_0^1\otimes X_0^2}$ the r.m. which is the only r.m. defined on $\mathcal{B}\otimes\mathcal{B}$ with values in $\sigma_2(H_1, H_2)$ which associates $Z_1(A)\otimes Z_2(A)$ to $A_1 \times A_2$, for all pairs (A_1, A_2) of elements of $\mathcal{B}$. That is why we will call $Z_{E_1*E_2}^{X_0^1\otimes X_0^2}$ a *convolution product of r.m.* Z_1 *and* Z_2 and we will denote it by $Z_1 * Z_2$.
We suppose now that H_1 (resp. H_2) is $L^2(\Omega, \mathcal{B}_1, \mathbb{P})$ (resp. $L^2(\Omega, \mathcal{B}_2, \mathbb{P})$), where the sub σ-fields $\mathcal{B}_1$ and $\mathcal{B}_2$ are $\mathbb{P}$-independent. If $\mathcal{U}$ denotes the σ-field of Ω generated by the family $\{B_1 \cap B_2; (B_1, B_2) \in \mathcal{B}_1 \times \mathcal{B}_2\}$, we can affirm the existence of an isometry $\mathcal{I}$ between $\sigma_2(L^2(\mathcal{B}_1), L^2(\mathcal{B}_2))$ and $L^2(\mathcal{U})$ such that $\mathcal{I}(\bar{x}_1 \otimes x_2) = x_1 x_2$ for (x_1, x_2) in $L^2(\mathcal{B}_1)\times L^2(\mathcal{B}_2)$ (indeed $<\bar{x}\otimes y, \bar{x}'\otimes y'>=<x, x'><y, y'>= \int x\bar{x}'d\mathbb{P}\int y\bar{y}'d\mathbb{P} = \int xy\overline{x'y'}d\mathbb{P} =<xy, x'y'>$ for all pairs $((x, y), (x', y'))$ of elements of $L^2(\mathcal{B}_1) \times L^2(\mathcal{B}_2)$).
Endly we can verify that $\mathcal{I}\circ(Z_1 * Z_2)$ is a r.m. with values in $L^2(\mathcal{U})$ such that

$$\int e^{i\cdot n} d\mathcal{I}\circ(Z_1 * Z_2) = \mathcal{I}(\int e^{i\cdot n} dZ_1 * Z_2) = \mathcal{I}(X_{-n}^1 \otimes X_n^2) = \bar{X}_{-n}^1 X_n^2.$$

So, we may notice that $\mathcal{I}\circ(Z_1 * Z_2)$ is the r.m. associated to the stationary series $(\bar{X}_{-n}^1 X_n^2)_{n\in\mathbb{Z}}$. A multivariate version of this last result allow us some applications in frequency domain principal components analysis (cf. Boudou, Dauxois, 1994; Boudou, Romain, 2005). Futhermore, other extensions may be investigated as, for example, recent works for r.m. in Banach space (cf. Benchikh et *al.*, 2007)

References

[1] Azencott R., Dacunha-Castelle D.: Séries d'observations irrégulières. Modélisation et prévision. Techniques Stochastiques, Masson (1984).

[2] Birman M., Solomjak M.: Tensor product of a finite number of spectral measures is always a spectral measure. Integ. Eq. Oper. Th. **24**, 179-187 (1996).

[3] Benchikh T., Boudou A., Romain Y.: Mesures aléatoires opératorielle et banachique. Application aux séries stationnaires. C.R. Acad. Sci. Paris. SérieI, Math.**345**, 345-348 (2007).

[4] Boudou A.: Interpolation de processus stationnaires. C.R. Acad. Sci. Paris, SérieI, Math. **33**, 12, 1021-1024 (2003).

[5] Boudou A.: Groupe d'opérateurs unitaires déduit d'une mesure spectrale - une application. C.R. Acad. Sci. Paris. SerieI, Math.,**344**, 12, 791-794 (2007).

[6] Boudou A., Dauxois J.: Principal component analysis for a stationary random function defined on a locally compact abelian group. J. Multivariate Anal. **51**, no.1, 1-16 (1994).

[7] Boudou A., Romain Y.: On spectral and random measures associated to continuous and discrete time processes. Stat. Proba. Letters. **59**, 145-157 (2002).

[8] Boudou A., Romain Y.: Sur l'intégrale par rapport à une mesure aléatoire tensorielle et ses applications aux processus stationnaires multidimensionnels. Publi. Labo. Stat. Proba., **08-05**, 1-27, Univ. P. Sabatier, Toulouse. (2005).

[9] Dauxois J., Romain Y., Viguier S.: Tensor products and statistics. Lin. Alg. Appl. **210**, 59-88 (1994).

[10] Halmos, P.R. Introduction to Hilbert space and theory of spectral multiplicity. Reprint of the second (1957) edition. AMS Chelsea Publishing, Providence, RI (1998).

[11] Riesz F., Nagy B.SZ.: Leçons d'analyse fonctionnelle. Gauthiers-Villars, Paris (1968).

[12] Schaeffer H.H.: Banach lattices and positive operators. Springer-Verlag, Berlin Heidelberg New York (1974).

Chapter 14
Parameter Cascading for High Dimensional Models

David Campbell, Jiguo Cao, Giles Hooker and James Ramsay

Abstract This talk defines a general framework for parameter estimation that synthesizes a variety of common approaches and brings some important new advantages. The parameter cascade involves defining nuisance parameters as functions of structural parameters, and in turn defines structural parameters as functions of complexity parameters.

14.1 Introduction

High dimensional models often involve three classes of parameters. Nuisance parameters c are required to fit the data, are large in number, their number tends to depend on how much data is available, often define localized effects on the fit, and their values are seldom of direct interest. Structural parameters θ are the conventional kind; a small fixed number and their values are of interpretive importance. Above these are the complexity parameters γ that define the overall complexity of the solution.

This talk defines a general framework for parameter estimation that synthesizes a variety of common approaches and brings some important new advantages. The *parameter cascade* defines nuisance parameters as functions $c(\theta, \gamma)$ of structural and complexity parameters, and in turn defines structural

David Campbell
Dept. of Psychology 1205 Dr. Penfield Ave. Montreal, Quebec, Canada H3A 1B1

Jiguo Cao
Dept. of Psychology 1205 Dr. Penfield Ave. Montreal, Quebec, Canada H3A 1B1

Giles Hooker
Dept. of Psychology 1205 Dr. Penfield Ave. Montreal, Quebec, Canada H3A 1B1

James Ramsay
Dept. of Psychology 1205 Dr. Penfield Ave. Montreal, Quebec, Canada H3A 1B1, e-mail: ramsay@psych.mcgill.ca

parameters as functions $\theta(\gamma)$ of complexity parameters. These functional relationships are often defined by choosing three different optimization criteria corresponding to each level.

14.2 Inner optimization: nuisance parameters

It is common to define the lowest level or inner criterion $L(c|\theta,\gamma)$ as a regularized loss function with the penalty controlled by γ, as in

$$J(c|\theta,\gamma) = \sum_i^N [y_i - \beta' z_i - c'\phi(t_i)]^2 \\ +e^{-\gamma} c' [\int \| \frac{d^2\phi}{dt^2} - \alpha_0 \phi(t) - \alpha_1 \frac{d\phi}{dt} \|^2 dt] c$$

where $x(t) = c'\phi(t)$, z_i is a p-vector of covariate values, and $\phi(t)$ is a vector of K basis functions. There are three groups of parameters to estimate:

- The K coefficients in c defining the basis function expansion of $x(t)$.
- The $p+2$ model parameters α and β defining the data fitting model and the roughness penalty, respectively. For simplicity, we use θ to collect these two vectors together; $\theta = (\alpha', \beta')'$.
- The single smoothing parameter γ.

The regularization assures that $c(\theta,\gamma)$ is smooth in a specified sense, and effectively controls the degrees of freedom allocated to the nuisance parameters. But $c(\theta,\gamma)$ may also be defined explicitly, or by an algorithm whose result depends on θ and γ, as in kernel smoothing. This functional relationship between nuisance and other parameters is a generalization of the familiar *profiling* procedure often used in nonlinear regression, where the three optimization criteria are the same.

14.3 Middle optimization: structural parameters

The middle level optimization is usually an unregularized measure of fit, such as

$$H(\theta|\gamma) = \sum_i^N [y_i - \beta' z_i - c(\theta,\gamma)'\phi(t_i)]^2,$$

and the fact that the status of c as a parameter as been eliminated by replacing it by a function of the other two classes implicitly ensures regularization. Of course we need the derivative of $c(\theta,\gamma)$, and this is, by the *Implicit Function Theorem*,

$$\frac{dc}{d\theta} = -(\frac{\partial^2 F}{\partial \theta^2})^{-1}(\frac{\partial^2 F}{\partial \theta \partial c}).$$

14.4 Outer optimization: complexity parameters

Finally, the top level optimization is a measure of model complexity such as the generalized cross-validation measure of predictive complexity

$$G(\gamma) \sim \frac{\|[I - A(\gamma)]y\|^2}{\|[I - A(\gamma)\|^2},$$

where $A(\gamma)$ is the smoothing operator, is effectively a *Raleigh coefficient* showing the size of the residual vector $[I - A(\gamma)]y$ relative to the size of the *residual operator* $I - A(\gamma)$. The Implicit Function Theorem again gives us $\frac{d\theta}{d\gamma}$.

Estimation of confidence intervals and other inferential methods can proceed at this point by classical methods such as the delta method. The application will typically require further use of the Implicit Function Theorem to compute the required derivatives.

14.5 Parameter cascading precedents

This general framework can be seen to include a number of specific parameter estimation strategies in common use, such as the process of removing nuisance parameters by marginalizing a likelihood. Since the marginal likelihood

$$L^*(\theta|y) = \int L(\theta, c|y)p(c)dc$$

is a linear operation, it is necessarily the optimum of a functional quadratic optimization problem, and in fact minimizes

$$J(c|\theta, y) = \int [L(\theta, c|y) - L^*(\theta|y)]^2 e^{\ln p(c) + C} dc$$

for any constant C. We see here a *functional regression problem* in which function $L(\theta, c|y)$ is approximated by a marginal function $L^*(\theta|y)$ conditional on specific values of structural parameter θ and data y. What is missing in marginalization, however, is any counterpart of smoothing parameter γ that permits a continuum of regularization. But it seems perfectly feasible to remove this difficulty by appending a continuously controlled penalty to this definition of $J(c|\theta, y)$.

14.6 Parameter cascading advantages

The parameter cascade procedure brings important advantages to parameter estimation in the presence of nuisance parameters.

- Gradients and hessians at any level can be analytically computed using the Implicit Function Theorem.
- Interval estimation methods are readily at hand.
- Compared to marginalizing out the nuisance parameters employed in Bayesian approaches using MCMC, generalized profiling is
  - much faster,
  - much more stable,
  - much easier to program,
  - permits an adaptive control of the contribution of c to the fit,
  - requires no "tuning" by an MCMC expert, and
  - can be deployed to the user community much more conveniently.

References

[1] Cao, J., Ramsay, J. O.: Parameter cascades and profiling in functional data analysis. Computational Statistics. **22**, 335-351 (2007).
[2] Ramsay, J. O., Hooker, G., Cao, J. and Campbell, D.: Parameter estimation for differential equations: A generalized smoothing approach (with discussion). Journal of the Royal Statistical Society. Series B. **69**, 741-796 (2007).

Chapter 15
Advances in Human Protein Interactome Inference

Enrico Capobianco and Elisabetta Marras

Abstract Important cellular functions information can be obtained from decomposing Protein-Protein Interaction Networks (PPIN) into constituent groups (complexes, functional modules). Starting from well-covered model organisms (Yeast), our current efforts are shifting to a complex target organism (Homo Sapiens). It is through statistical techniques and machine learning algorithms that one can proceed with probabilistic steps: assigning unlabelled proteins (classification), inferring unknown functions (generalization), weighting interactions (scoring).

15.1 Introduction

Complex networks are among the most multidisciplinary areas currently investigated by research communities active in statistical mechanics, graph theory, statistics and probability, and involved in key application tasks in systems biology (gene regulatory identification and corresponding reverse engineering, protein-protein and protein-DNA interactions and protein signaling regulation), social studies (friends, web links, disease transmission, trade, opinion etc.), information and communication technologies (telephone, internet, etc.).

We focus on PPIN, whose continuing data release from large hightthroughput systems covering many organisms has generated a great opportunity for network applications, either directly to the experimental data or to the many tailored DB sources.

Enrico Capobianco
CRS4 Bioinformatics Laboratory, Technology Park of Sardinia, 09010 Pula (Cagliari) - Sardinia, Italy, e-mail: ecapob@crs4.it

Elisabetta Marras
CRS4 Bioinformatics Laboratory, Technology Park of Sardinia, 09010 Pula (Cagliari) - Sardinia, Italy, e-mail: lisa@crs4.it

In empirical studies, the sample space usually pertains to the proteins and their pairwise interactions. There are a finite number of validated interacting protein pairs (some experimentally observed, some computationally predicted), but also a much bigger set of non-interacting protein pairs (from which null models are built).

Other "omic" sources should be integrated to improve the accuracy of the global interaction map, but adding further complexity to the built-in features, i.e. high-dimensionality, multiresolution, noise, sparsity, etc.

A challenge is to shrink the interaction set to a calibrated putative interactome based on gold standard reference sets and scores computed from the likelihood of interactions.

A 'sparse representation' problem arises naturally in various different biological frameworks, especially where genomics, proteomics, metabolomics data sources refer to 'high throughput' technologies. With the term interactome it is indicated the whole set of molecular interactions in cells, and we are here interested in Protein-Protein Interaction Networks (PPIN).

A list of possible reasons inducing sparsity in PPIN includes:

- A small fraction of interacting pairs in the total set of potential protein pairs (1 in 600 possible pairs actually interact);
- Many false positive interacting pairs (also false negatives though), as 80000 interactions have been predicted in yeast by various high throughput methods, but only a few ($\sim$ 2400) are justified;
- Many missing values in biological datasets, with coverage going from $\sim 4\%$ in Y2H (Yeast Two-Hybrid), to $\sim 90\%$ for gene expression data, till 100% for sequence related features;
- Descriptive protein pairs features that are orders of magnitude less than the observed dimensionality, which suggests that the degrees of freedom of the problem depend on a small number of variables.

Common dimensionality reduction methods like principal component and factor analysis, and then variants such as independent component analysis and their kernelized versions (also combined with greedy techniques) have been employed with the aim to encapsulate information within a few salient dimensions.

This passage yields an embedding which emphasizes the role of a limited number of relevant features selected to support the hypotheses in the underlying biological system.

15.2 Methods

A probe (model organism, Yeast) is used to verify the power with which the employed methods reveal the connectivities in the protein space. The highly connected regions are identified as clusters, representing compartmentalized structures also referred as subnetworks, subgraphs, subproteomes.

Such structures provide a natural representation of possible physical interactions among groups of proteins. We expect these clusters to have a higher density of points than their surrounding regions. How to measure these densities depends on distance measures assigned to interacting proteins.

The connectivity degree and relevance of each cluster should then be validated (against complexes or functional modules, for instance). Clusters might be marginally interconnected, which suggests that threshold distances should be defined so to capture within- and between-clusters dependencies. This allows to reconstruct with a certain confidence the known network structures (i.e. target proteomes).

However, clusters can only be considered as proxies for determining the problems inherent to the identification and reconstruction tasks. More refined methodological solutions are currently considered. Assigning a probability function to each pairwise distance value and parameterizing each cluster allow to establish a likelihood function in a model framework that can be considered parametric or not, depending on the knowledge that we have of the parameters.

In turn, a parametric, semiparametric or nonparametric likelihood can be defined. If we consider a vector-valued parameter set consisting of features that justify and define the protein interactions, we might consider to specialize subsets of features in relation to each cluster, consider measurable and latent features, and endow the parameter set of a nuisance part in relation to unknown aspects of protein-protein relationships.

Inferring the global protein network structure in terms of dissected components calls for local manifold learning algorithms, where (Gaussian and non-Gaussian) mixture and/or latent variable models can perform quite efficiently, for instance. Then, through the approximation of their covariance structure by principal modes or eigenvectors, the goal of approaching the intrinsic manifold dimensionality can also be achieved.

15.3 Preliminary results

We have started to apply parametric mixture models, in particular Gaussian ones, to the problem under study. The usual practice is to run the Expectation Maximization (EM) algorithm as the optimization method that learns the parameters.

From the standpoint of our application domain, mixtures are relevant because they approximate densities of high dimensional data that lie on or near a low dimensional manifold. We expect that the relationships of protein-protein interactions (our data) and the intrinsic coordinates of the manifold (complexes, functional modules, etc.) could be locally linear and smoothly varying.

From a methodological standpoint, mixtures are interesting because they lead to flexible parametric and also non-parametric statistical inference, but in any case we can build through them a fully probabilistic model, valid also away from the training set, i.e, able to classify (proteins to groups), generalize (infer functions), predict (assign scores to proteins- clusters associations).

Mixtures of Gaussians are one of the possible choices we are currently considering. Extensions are under study, not only towards other parametric distributions but also for incorporating extra covariate information (thus leading to multiomic source integration), for examining multiresolution dynamics (by looking at possible time scale-dependent protein interaction effects), and for allowing flexibility in model selection as from the number of effectively needed components (sparsity versus redundancy aspects).

The EM algorithm is well-known to proceed by alternating till convergence an E-step and an M-step. The former step is employed to calculate an objective function Q (expectation of a complete data log-likelihood over the joint distribution of the unobservable data given the observed data) by using the current parameter estimates Ψ. The latter step is used to update the parameter estimates to Ψ' based on the optimization of the Ψ-dependent objective function Q.

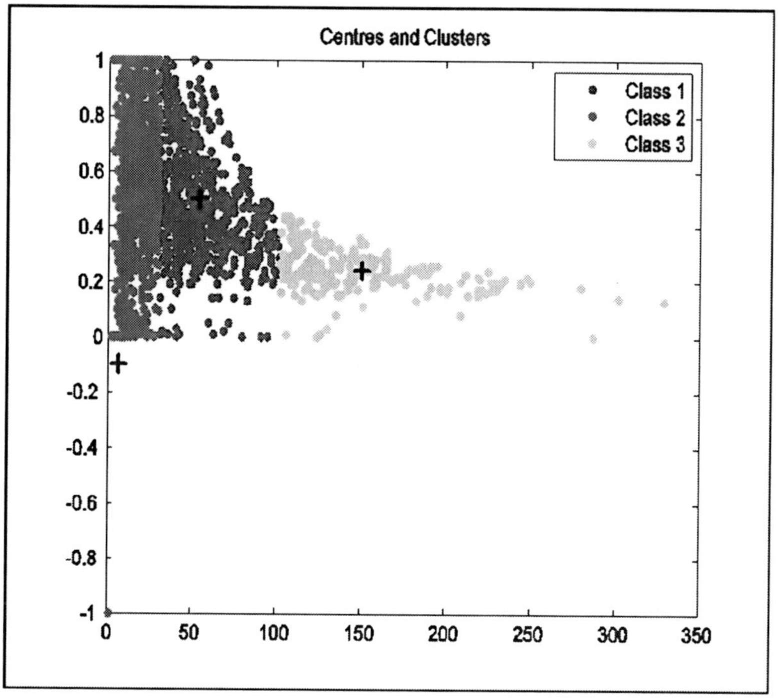

Fig. 15.1 Extracted groups

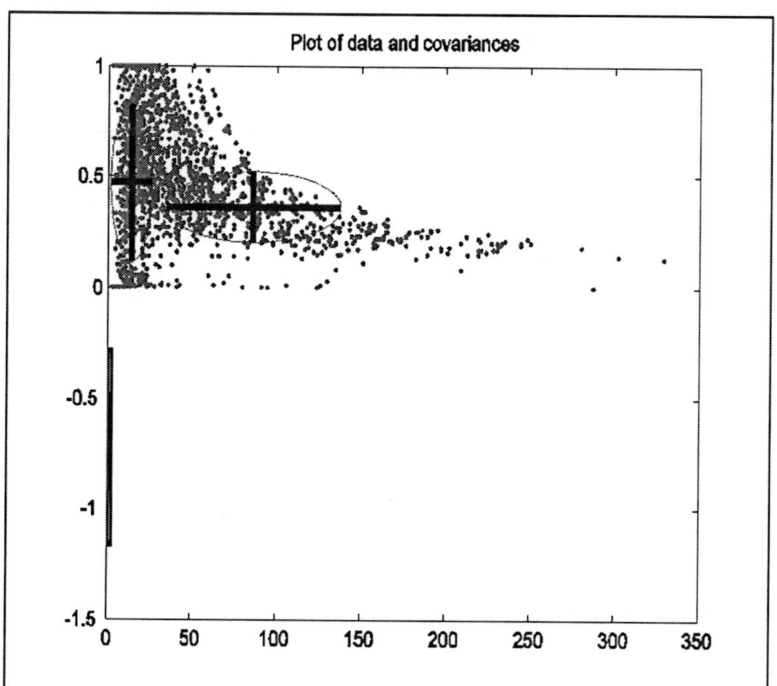

Fig. 15.2 Embedded variability.

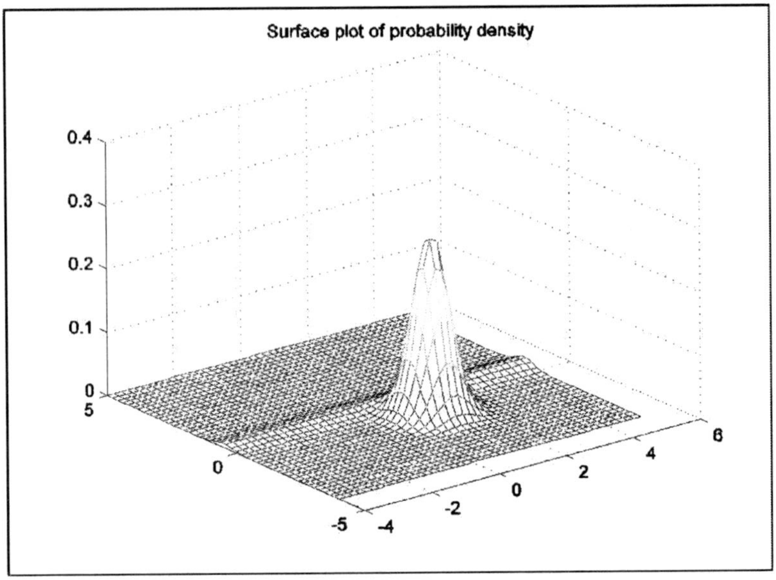

Fig. 15.3 Identified map.

The PPIN yeast data set (Bader, 2003) consists of 3632 nodes and 22500 interactions.

References

[1] small Bader et al.: Gaining Confidence in High-throughput protein interaction networks. Nature Biotechnology. **22**, 78-85 (2003).

[2] Uetz et al.: A comprehensive analysis of protein-protein interaction in Saccharomyces cerevisiae. Nature. **403**, 623-627 (2000).

[3] Gavin et al.: Proteome survey reveals modularity of the yeast cell machinery. Nature. **440**, 631-636 (2006).

[4] Krogan et al.: Global landscape of protein complexes in the yeast Saccharomyces cerevisiae. Nature. **440**, 637-643 (2006).

[5] Von Mering et al.: Comparative assessment of large-scale data sets of protein-protein interactions. Nature. **417**, 399-401 (2002).

[6] R. Jansen et al.: Bridging structural biology and genomics: assessing protein interaction data with known complexes. Science. **302**, 449-453 (2003).

[7] Bader et al.: A Bayesian Networks Approach for Predicting Protein-Protein Interactions from Genomic Data. Nature Biotechnology. **22** (1), 78-85 (2004).

[7] Dean et al.: Gaining confidence in high-throughput protein interaction networks. Mol Cel Proteom. **1** (5), 349-56 (2002).

[8] Xenarios et al.: Protein interactions: two methods for assessment of the reliability of high-throughput observations. NAR. **30** (1), 303-305 (2002).

Chapter 16
Functional Principal Components Analysis with Survey Data

Hervé Cardot, Mohamed Chaouch, Camelia Goga and Catherine Labruère

Abstract This work aims at performing Functional Principal Components Analysis (FPCA) thanks to Horvitz-Thompson estimators when the curves are collected with survey sampling techniques. Linearization approaches based on the influence function allow us to derive estimators of the asymptotic variance of the eigenelements of the FPCA. The method is illustrated with simulations which confirm the good properties of the linearization technique.

16.1 Introduction

Functional Data Analysis whose main purpose is to provide tools for describing and modeling sets of curves is a topic of growing interest in the statistical community. The books by Ramsay and Silverman (2002, 2005) propose an interesting description of the available procedures dealing with functional observations. These functional approaches have been proved useful in various domains such as chemometrics, economy, climatology, biology or remote sensing.

Hervé Cardot
Institut de Mathématiques de Bourgogne, Université de Bourgogne, 9 Avenue Alain Savary, BP 47870, 21078 DIJON Cedex, France, e-mail: herve.cardot@u-bourgogne.fr

Mohamed Chaouch
Institut de Mathématiques de Bourgogne, Université de Bourgogne, 9 Avenue Alain Savary, BP 47870, 21078 DIJON Cedex, France, e-mail: mohamed.chaouch@u-bourgogne.fr

Camelia Goga
Institut de Mathématiques de Bourgogne, Université de Bourgogne, 9 Avenue Alain Savary, BP 47870, 21078 DIJON Cedex, France, e-mail: camelia.goga@u-bourgogne.fr

Catherine Labruère
Institut de Mathématiques de Bourgogne, Université de Bourgogne, 9 Avenue Alain Savary, BP 47870, 21078 DIJON Cedex, France, e-mail: catherine.labruere@u-bourgogne.fr

The statistician generally wants, in a first step, to represent as well as possible a set of random curves in a small space in order to get a description of the functional data that allows interpretation. Functional principal components analysis (FPCA) gives a small dimension space which captures the main modes of variability of the data (see Ramsay and Silverman, 2002 for more details).

The way the data are collected is seldom taken into account in the literature and one generally supposes the data are independent realizations of a common functional distribution. However there are some cases for which this assumption is not fulfilled, for example when the realizations result from a sampling scheme. For instance, Dessertaine (2006) considers the estimation with time series procedures of a global demand for electricity at fine time scales with the observation of individual electricity consumption curves. More generally, there are now data (data streams) produced automatically by large numbers of distributed sensors which generate huge amounts of data that can be seen as functional. The use of sampling technique to collect them proposed for instance in Chiky and Hébrail (2007) seems to be a relevant approach in such a framework allowing a trade off between storage capacities and accuracy of the data.

We propose in this work to give estimators of the functional principal components analysis when the curves are collected with survey sampling strategies. Let us note that Skinner *et al.* (1986) have studied some properties of multivariate PCA in a survey framework. The functional framework is different since the eigenfunctions which exibit the main modes of variability of the data are also functions and can be naturally interpreted as modes of variability varying along time. In this new functional framework, we estimate the mean function and the covariance operator using the Horvitz-Thompson estimator. The eigenelements are estimated by diagonalization of the estimated covariance operator. In order to calculate and estimate the variance of the so-constructed estimators, we use the influence function linearization method introduced by Deville (1999).

This paper is organized as follows : Section 2 presents the functional principal components analysis in the setting of finite populations and defines then the Horvitz-Thompson estimator in the new functional framework. The generality of the influence function allows us to extend in section 3 the estimators proposed by Deville to our functional objects and to get asymptotic variances with the help of perturbation theory (Kato, 1966). Section 4 proposes a simulation study which shows the good behavior of our estimators for various sampling schemes as well as good approximations to their theoretical variances.

16.2 FPCA and sampling

16.2.1 FPCA in a finite population setting

Let us consider a finite population $U = \{1, \dots, k, \dots, N\}$ with size N not necessarily known

and a functional variable $\mathcal{Y}$ defined for each element k of the population U: $Y_k = (Y_k(t))_{t\in[0,1]}$ belongs to the separable Hilbert space $L^2[0,1]$ of square integrable functions defined on the closed interval $[0,1]$ equipped with the usual inner product $\langle .,. \rangle$ and the norm $\|.\|$. The mean function $\mu \in L^2[0,1]$, is defined by

$$\mu(t) = \frac{1}{N} \sum_{k\in U} Y_k(t), \quad t \in [0,1] \tag{16.1}$$

and the covariance operator Γ by

$$\Gamma = \frac{1}{N} \sum_{k\in U} (Y_k - \mu) \otimes (Y_k - \mu) \tag{16.2}$$

where the tensor product of two elements a and b of $L^2[0,1]$ is the rank one operator such that $a \otimes b(u) = \langle a, u \rangle b$ for all u in $L^2[0,1]$. The operator Γ is symmetric and non negative ($\langle \Gamma u, u \rangle \geq 0$). Its eigenvalues, sorted in decreasing order, $\lambda_1 \geq \lambda_2 \geq \dots \geq \lambda_N \geq 0$, satisfy

$$\Gamma v_j(t) \;=\; \lambda_j \; v_j(t), \quad t \in [0,1], \tag{16.3}$$

where the eigenfunctions v_j form an orthonormal system in $L^2[0,1]$, *i.e* $\langle v_j, v_{j'} \rangle = 1$ if $j = j'$ and zero else.

We can get now an expansion similar to the Karhunen-Loeve expansion or FPCA which allows to get the best approximation in a finite dimension space with dimension q to the curves of the population

$$Y_k(t) \approx \mu(t) + \sum_{j=1}^{q} \langle Y_k - \mu, v_j \rangle v_j(t), \quad t \in [0,1]$$

The eigenfunctions v_j indicate the main modes of variation along time t of the data around the mean μ and the explained variance of the projection onto each v_j is given by the eigenvalue

$$\lambda_j = \frac{1}{N} \sum_{k\in U} \langle Y_k - \mu, v_j \rangle^2 \; .$$

We aim at estimating the mean function μ and the covariance operator Γ in order to deduce estimators of the eigenelements (λ_j, v_j) when the data are obtained with survey sampling procedures.

16.2.2 The Horvitz-Thompson estimator

We consider a sample of n individuals s, *i.e.* a subset $s \subset U$, selected according to a probabilistic procedure $p(s)$ where p is a probability distribution on the set of 2^N subsets of U. We denote by $\pi_k = \Pr(k \in s)$ for all $k \in U$ the first order inclusion probabilities and by $\pi_{kl} = \Pr(k \;\&\; l \in s)$ for all $k, l \in U$ with $\pi_{kk} = \pi_k$, the second order inclusion probabilities. We suppose that $\pi_k > 0$ and $\pi_{kl} > 0$. We suppose also that π_k and π_{kl} are not depending on $t \in [0,1]$. We propose to estimate the mean function μ and the covariance operator Γ by replacing each total with the corresponding Horvitz-Thompson (HT) estimator (Horvitz and Thompson, 1952). We obtain

$$\widehat{\mu} = \frac{1}{\widehat{N}} \sum_{k \in s} \frac{Y_k}{\pi_k} \tag{16.4}$$

$$\widehat{\Gamma} = \frac{1}{\widehat{N}} \sum_{k \in s} \frac{Y_k \otimes Y_k}{\pi_k} - \widehat{\mu} \otimes \widehat{\mu} \tag{16.5}$$

where the size N of the population is estimated by $\widehat{N} = \sum_{k \in s} \frac{1}{\pi_k}$ when it is not known. Then estimators of the eigenfunctions $\{\widehat{v}_j, j = 1, \ldots q\}$ and eigenvalues $\{\widehat{\lambda}_j, j = 1, \ldots q\}$ are obtained readily by diagonalisation (or spectral analysis) of the estimated covariance operator $\widehat{\Gamma}$. Let us note that the eigenelements of the covariance operator are not linear functions.

16.3 Linearization by influence function

We would like to calculate and estimate the variance of $\hat{\mu}$, $\widehat{v}_j$ and $\widehat{\lambda}_j$. The nonlinearity of these estimators and the functional nature of $\mathcal{Y}$ make the variance estimation issue difficult. For this reason, we adapt the influence function linearization technique introduced by Deville (1999) to the functional framework.

Let us consider the discrete measure M defined on $L^2[0,1]$ as follows
$M = \sum_U \delta_{Y_k}$
where δ_{Y_k} is the Dirac function taking value 1 if $\mathcal{Y} = Y_k$ and zero otherwise. Let us suppose that each parameter of interest can be written as a functional T of M. For example, $N(M) = \int dM$, $\mu(M) = \int \mathcal{Y} dM / \int dM$ and

$\Gamma(M) = \int (\mathcal{Y} - \mu(M)) \otimes (\mathcal{Y} - \mu(M))\, dM / \int dM$. The eigenelements given by (16.3) are implicit functionals T of M.
The measure M is estimated by the random measure $\widehat{M}$ defined as follows $\widehat{M} = \sum_U \frac{\delta_{Y_k}}{\pi_k} I_k$ with $I_k = 1_{\{k \in s\}}$. Then the estimators given by (16.4) and (16.5) are obtained by substitution of M by $\widehat{M}$, namely they are written as functionnals T of $\widehat{M}$.

16.3.1 Asymptotic properties

We give in this section the asymptotic properties of our estimators. In order to do that, one need that the population and sample sizes tend to infinity. We use the asymptotic framework introduced by Isaki & Fuller (1982). Let us suppose the following assumptions :

(A1) $\sup_{k \in U} \|Y_k\| \leq C < \infty$,

(A2) $\lim_{N \to \infty} \frac{n}{N} = \pi \in (0, 1)$,

(A3) $\min_{k \in U_N} \pi_k \geq \lambda > 0$, $\min_{k \neq l} \pi_{kl} \geq \lambda* > 0$ and $\overline{\lim}_{N \to \infty} n \max_{k \neq l} |\pi_{kl} - \pi_k \pi_l| < \infty$,

with λ and $\lambda*$ are two positive constant. We also suppose that the functional T giving the parameter of interest is an homogeneous functional of degree α, namely $T(rM) = r^\alpha T(M)$ and $\lim_{N \to \infty} N^{-\alpha} T(M) < \infty$. For example, μ and Γ are functionals of degree zero with respect to M. Let us note that the eigenelements of Γ are also functionals of degree zero with respect to M.

Let us also introduce the Hilbert-Schmidt norm, denoted by $\|\cdot\|_2$ for operators mapping $L^2[0,1]$ to $L^2[0,1]$.

We show in the next proposition that the our estimators are asymptotically design unbiased, $\lim_{N \to \infty} \left(E_p(T(\widehat{M})) - T(M) \right) = 0$, and consistent, namely for any fixed $\varepsilon > 0$ we have $\lim_{N \to \infty} P(|T(\widehat{M}) - T(M)| > \varepsilon) = 0$. Here, $E_p(\cdot)$ is the expectation with respect to $p(s)$.

Proposition 16.1. *Under hypotheses (A1), (A2) and (A3),*

$$E_p \|\mu - \widehat{\mu}\|^2 = O(n^{-1}), \quad E_p \left\| \Gamma - \widehat{\Gamma} \right\|_2^2 = O(n^{-1}).$$

If we suppose that the non null eigenvalues are distinct, we also have,

$$E_p \left(\sup_j \left| \lambda_j - \widehat{\lambda_j} \right| \right)^2 = O(n^{-1}), \quad E_p \|v_j - \widehat{v_j}\|^2 = O(n^{-1}) \quad \textit{for each fixed } j.$$

16.3.2 Variance approximation and estimation

Let define, when it exists, the influence function of a functional T at point $\mathcal{Y} \in L^2[0,1]$ say $IT(M, \mathcal{Y})$, as follows

$$IT(M, \mathcal{Y}) = \lim_{h \to 0} \frac{T(M + h\delta_{\mathcal{Y}}) - T(M)}{h}$$

where $\delta_{\mathcal{Y}}$ is the Dirac function at $\mathcal{Y}$.

Proposition 16.2. *Under assumption (A1), we get that the influence functions of μ and Γ exist and $I\mu(M, Y_k) = (Y_k - \mu)/N$ and $I\Gamma(M, Y_k) = \frac{1}{N}\left((Y_k - \mu) \otimes (Y_k - \mu) - \Gamma\right)$. If the non null eigenvalues of Γ are distinct then*

$$I\lambda_j(M, Y_k) = \frac{1}{N}\left(\langle Y_k - \mu, v_j\rangle^2 - \lambda_j\right)$$

$$Iv_j(M, Y_k) = \frac{1}{N}\left(\sum_{\ell \neq j} \frac{\langle Y_k - \mu, v_j\rangle \langle Y_k - \mu, v_\ell\rangle}{\lambda_j - \lambda_\ell} v_\ell\right).$$

In order to obtain the asymptotic variance of $T(\widehat{M})$ for T given by (16.1), (16.2) and (16.3), we write the first-order von Mises expansion of our functional in $\widehat{M}/N$ "near" M/N and use the fact that T is of degree 0 and $IT\left(M/N, Y_k\right) = N \cdot IT\left(M, Y_k\right)$,

$$T(\widehat{M}) = T(M) + \sum_{k \in U} IT\left(M, Y_k\right)\left(\frac{I_k}{\pi_k} - 1\right) + R_T\left(\frac{\widehat{M}}{N}, \frac{M}{N}\right).$$

Proposition 16.3. *Suppose the hypotheses (A1), (A2) and (A3) are fulfilled. Consider the functional T giving the parameters of interest defined in (16.1), (16.2), (16.3). We suppose that the non null eigenvalues are distinct. Then $R_T\left(\frac{\widehat{M}}{N}, \frac{M}{N}\right) = o_p(n^{-1/2})$ and the asymptotic variance of $T(\widehat{M})$ is equal to $V_p[\sum_{k \in s} IT\left(M, Y_k\right)\frac{I_k}{\pi_k}] = \sum_U \sum_U (\pi_{kl} - \pi_k \pi_l) \frac{IT(M, Y_k)}{\pi_k} \frac{IT(M, Y_l)}{\pi_l}$.*

One can remark that the asymptotic variance given by the above result is not known. We propose to estimate it by the HT variance estimator with $IT\left(M, Y_k\right)$ replaced by its HT estimator. We obtain

$$\widehat{V}_p(\widehat{\mu}) = \frac{1}{\widehat{N}^2}\sum_{k\in s}\sum_{\ell\in s}\frac{1}{\pi_{k\ell}}\frac{\Delta_{k\ell}}{\pi_k\pi_\ell}(Y_k-\widehat{\mu})\otimes(Y_\ell-\widehat{\mu})$$

$$\widehat{V}_p\left(\widehat{\lambda}_j\right) = \frac{1}{\widehat{N}^2}\sum_{k\in s}\sum_{\ell\in s}\frac{1}{\pi_{k\ell}}\frac{\Delta_{k\ell}}{\pi_k\pi_\ell}\left(\langle Y_k-\widehat{\mu},\widehat{v}_j\rangle^2-\widehat{\lambda}_j\right)\left(\langle Y_\ell-\widehat{\mu},\widehat{v}_j\rangle^2-\widehat{\lambda}_j\right)$$

$$\widehat{V}_p\left(\widehat{v}_j\right) = \sum_{k\in s}\sum_{\ell\in s}\frac{1}{\pi_{k\ell}}\frac{\Delta_{k\ell}}{\pi_k\pi_\ell}\widehat{Iv}_j(M,Y_s)\otimes\widehat{Iv}_j(M,Y_\ell)$$

where $\Delta_{k\ell} = \pi_{kl} - \pi_k\pi_l$ and $\widehat{Iv}_j(M,Y_\ell) = \frac{1}{\widehat{N}}\left(\sum_{\ell\neq j}\frac{\langle Y_k-\widehat{\mu},\widehat{v}_j\rangle\langle Y_k-\widehat{\mu},\widehat{v}_\ell\rangle}{\widehat{\lambda}_j-\widehat{\lambda}_\ell}\widehat{v}_\ell\right)$. Cardot *et al.* (2007) show that under the assumptions (A1)-(A3), these estimators are asymptotically design unbiased and consistent.

16.4 A Simulation study

In our simulations all functional variables are discretized in $p = 100$ equispaced points in the interval $[0,1]$. We consider a random variable Y distributed as brownian motion on $[0,1]$. We make $N = 10000$ replications of Y and construct then two strata U_1 and U_2 with different variances and with sizes $N_1 = 7000$ and $N_2 = 3000$. Our population U is the union of the two strata. Then we estimate the eigenelements of the covariance operator for two different sampling designs (Simple Random Sampling Without Replacement (SRSWR) and stratified) and two different sample sizes $n = 100$ and $n = 1000$. To evaluate our estimation procedures we make 500 replications of the previous experiment. Then estimation errors for the first eigenvalue and the first eigenvector are evaluated by considering the following loss criterions $\frac{\lambda_1-\widehat{\lambda}_1}{\lambda_1}$ and $\frac{||v_1-\widehat{v}_1||}{v_1}$, with $||.||$ is the Euclidiean norm. Linear approximation by influence function gives reasonable estimation of the variance for small size samples and accurates estimations as far as n gets large enough ($n = 1000$). We also note that the variance of the estimators given by stratified sampling turns out to be smaller than those by SRSWR.

References

[1] Cardot, H, Chaouch, M, Goga, C. and Labruère, C.: Functional Principal Components Analysis with Survey Data. Preprint (2007).

[2] Chiky, R, Hébrail, G.: Generic tool for summarizing distributed data streams. Preprint (2007).

[3] Dauxois, J., Pousse, A., and Romain, Y.: Asymptotic theory for the principal component analysis of a random vector function: some applications to statistical inference. J. Multivariate Anal. **12**, 136-154 (1982).

[4] Dessertaine A.: Sondage et séries temporelles: une application pour la prévision de la consommation electrique. 38èmes Journées de Statistique. Clamart. (Juin 2006).

[5] Deville, J.C.: Variance estimation for complex statistics and estimators: linearization and residual techniques. Survey Methodology. **25**, 193-203 (1999).
[6] Horvitz, D.G. and Thompson, D.J.: A generalization of sampling without replacement from a finite universe. J. Am. Statist. Ass. **47**, 663-685 (1952).
[7] Isaki, C.T. and Fuller, W.A.: Survey design under the regression superpopulation model. J. Am. Statist. Ass. **77**, 89-96 (1982).
[8] Kato, T.: Perturbation theory for linear operators. Springer Verlag, Berlin (1966).
[9] Ramsay, J. O. and Silverman, B.W.: Functional Data Analysis. Springer-Verlag, 2nd ed. (2005).
[10] Skinner, C.J, Holmes, D.J, Smith, T.M.F. : The Effect of Sample Design on Principal Components Analysis. J. Am. Statist. Ass. **81**, 789-798 (1986).

Chapter 17
Functional Clustering of Longitudinal Data

Jeng-Min Chiou and Pai-Ling Li

Abstract This study considers two clustering criteria to achieve different goals of grouping similar curves. These criteria are based on the minimal L^2 distance and the maximal functional correlation defined in this study, respectively. Each cluster centers on a subspace spanned by the cluster mean and covariance eigenfunctions of the underlying random functions. Clusters can thus be identified by the subspace projection of curves.

17.1 Introduction

This study discusses functional data clustering. Individual observations are viewed as realizations of random functions, and random functions are assumed to follow a stochastic process possibly coupled with random scale effects. The stochastic process comprises a mixture of cluster sub-processes. The proposed functional clustering method, k-centers projected functional clustering (FC), accounts for both the means and modes of variation differentials between clusters by predicting cluster membership with a reclassification step. This step comprises nonparametric mean and covariance updating schemes to estimate cluster structures. These functional structures help predict cluster membership of each curve based on the varied nonparametric random effect models of the truncated Karhunen-Loève expansion. This study discusses two clustering criteria to achieve two different goals of clustering for curve similarity. These two criteria are based on the minimal L^2 distance and the maximal functional correlation defined in this study. The former

Jeng-Min Chiou
Institute of Statistical Science, Academia Sinica, Taiwan, e-mail: jmchiou@stat.sinica.edu.tw

Pai-Ling Li
Department of Statistics, Tamkang University, Taiwan, e-mail: plli@stat.tku.edu.tw

considers curve similarity through the L^2 distance, and the latter mainly is concerned with shape similarity, regardless of shifted means and scales. The following section briefly describes clustering procedures for both criteria.

17.2 The methods

Suppose that n random functions or curves, $Y_1, \ldots, Y_n$, are independently sampled from a mixture of stochastic processes Y in $L^2(\mathcal{T})$.

The Y mixture process includes K sub-processes, and each sub-process corresponds to a cluster. The random cluster variable C for each individual cluster membership is randomly distributed among the clusters $\{1, \ldots, K\}$. Here, $L^2(\mathcal{T})$ is a Hilbert space of square integrable functions on a real interval $\mathcal{T} = [0, T]$. The inner product of two functions y_i and y_j in $L^2(\mathcal{T})$ defined by the integral operator $\langle y_i, y_j \rangle = \int y_i(t) y_j(t) \nu(t) dt$, and the norm $\|y_i\| = |\langle y_i, y_i \rangle|^{1/2}$. This study sets the weight function $\nu(t)$ as a uniform kernel on a compact support $\mathcal{T}$.

The L^2 clustering criterion

The mean $\mu^{(c)}$ and the covariance $\Gamma^{(c)}$ of the sub-process associated with cluster c are defined via conditioning such that $E(Y(t) \mid C = c) = \mu^{(c)}(t)$, $Cov(Y(s), Y(t) \mid C = c) = \Gamma^{(c)}(s,t)$, for $c \in \{1, \ldots, K\}$. Each of these sub-processes is assumed to possess a Karhunen-Loève expansion, with the corresponding eigenvalue-eigenfunction pairs $(\lambda_j^{(c)}, \phi_j^{(c)})$, such that $\langle \Gamma^{(c)}(\cdot, t), \phi_j^{(c)} \rangle = \lambda_j^{(c)} \phi_j^{(c)}(t)$, $t \in \mathcal{T}$. These eigenfunctions are orthonormal satisfying $\langle \phi_j^{(c)}, \phi_k^{(c)} \rangle = 1$ for $j = k$ and 0 otherwise.

The eigenvalues $\lambda_j^{(c)}$ are in non-increasing order, $\lambda_1^{(c)} \geq \lambda_2^{(c)} \geq \ldots$, with the property that $\sum_{j=1}^{\infty} \lambda_j^{(c)} < \infty$ for a L^2 stochastic process.

Consider the nonparametric random effect model $Y^{(c)}$ of Y, given the structure components with the mean $\mu^{(c)}$ and the covariance eigenfunctions $\phi_j^{(c)}$ for cluster c, such that

$Y^{(c)}(t) = \mu^{(c)}(t) + \sum_{j=1}^{\infty} \xi_j^{(c)}(Y) \phi_j^{(c)}(t)$,

where

$\xi_j^{(c)}(Y) = \langle Y - \mu^{(c)}, \phi_j^{(c)} \rangle$.

Although the expansion $Y^{(c)}$ is infinite dimensional, it is common in practical applications that a value M_c exists for a given functional data set such that the first leading M_c eigenfunctions can effectively span the process.

The value M_c must be chosen from the data and is always finite.

Choosing M_c leads to the truncated model,

$$\tilde{Y}^{(c)}(t) = \mu^{(c)}(t) + \sum_{j=1}^{M_c} \xi_j^{(c)}(Y) \phi_j^{(c)}(t). \tag{17.1}$$

If the cluster membership of Y actually belongs to cluster c, given an observed curve Y, then $\tilde{Y}^{(c)}$ is the truncated Karhunen-Loève expansion of Y. Otherwise, discrepancies exist between $\tilde{Y}^{(c)}$ and Y. According to this basic principle, the expression $\tilde{Y}^{(c)}$ serves as the basic model for predicting cluster memberships.

The L^2 distance between the curves serves as a reasonable distance measure in functional clustering among others. Given an observed curve $Y = y$ and the cluster structure components $\mu^{(c)}$ and $\phi_j^{(c)}$, the cluster membership is determined by the following criterion,

$$c_1^*(y) = \underset{c \in \{1,\ldots,K\}}{\arg\min} \|y - \tilde{y}^{(c)}\|^2, \tag{17.2}$$

where $\tilde{y}^{(c)}$ is the truncated Karhunen-Loève expansion obtained by (17.1). The clustering criterion (17.2) is used to reclassify the curves, given the (initial) clustering results. For the initial clustering, the classical multivariate k-means method clusters the marginal functional principal component scores, ignoring cluster membership attributes. Other multivariate clustering methods, such as hierarchical clustering methods and model-based approaches, can also perform initial clustering.

Criterion (17.2) suggests that each individual is associated with a cluster that centers on the corresponding mean and eigenfunctions via projection. This criterion is similar to k-means clustering, where the cluster centers are the multivariate sample means. In contrast, cluster centers in k-centers projected FC are stochastic structures consisting of cluster means and covariance eigenfunctions. These cluster centers are used to obtain the projection of a curve onto the functional principal component subspaces of individual clusters. This idea coincides with Bock (1987) as a functional version of a k-means type algorithm. For more details on this functional clustering method, please refer to Chiou and Li (2007).

The correlation criterion

The underlying shape patterns are often interesting for random functions accompanied with random scales. To cluster curves with shape similarities, consider the random function $Y_\theta(t) = \theta Y(t)$, where θ is a random scale effect. Let $Y_{\theta(c)}$ denote the random function of Y_θ in cluster c, $Y_{\theta(c)}(t) = \theta^{(c)} Y^{(c)}(t)$, such that the stochastic representation uses the structure components of cluster c with the random scale $\theta^{(c)}$, where $E\theta^{(c)} = 1$ and $\mathrm{var}(\theta^{(c)}) = \sigma^2_{\theta(c)}$. Assume that each sub-processes corresponds to a cluster structure comprising the underlying mean function $\mu^{(c)}$ and the orthonormal basis $\{\varphi_0^{(c)}, \varphi_1^{(c)}, \ldots\}$ of the stochastic expansion, setting $\varphi_0^{(c)} = 1$, for cluster $c = 1, \ldots, K$. Write $\mu^{(c)}(t) = \eta_0^{(c)} + \eta^{(c)}(t)$ where $\eta_0^{(c)} = \langle \mu^{(c)}, 1 \rangle$. These functions satisfy $\langle \eta^{(c)}, 1 \rangle = 0$, $\langle \varphi_r^{(c)}, \varphi_s^{(c)} \rangle = 1$ for $r = s$ and 0 otherwise.

Further, let $Y^{\mathcal{X}}_{(c)}(t) = Y_{\theta(c)}(t) - \langle Y_{\theta(c)}, 1\rangle$. This centering yields the random effects model, $Y^{\mathcal{X}}_{(c)}(t) = \theta^{(c)}\,\eta^{(c)}(t) + \sum_{r=1}^{\infty} \varepsilon_r^{(c)}\,\varphi_r^{(c)}(t)$, where $\varepsilon_r^{(c)} = \langle Y^{\mathcal{X}}_{(c)} - \theta^{(c)}\,\eta^{(c)},\, \varphi_r^{(c)}\rangle$. A truncated version is

$$\tilde{Y}^{\mathcal{X}}_{(c)}(t) = \theta^{(c)}\,\eta^{(c)}(t) + \sum_{r=1}^{M_c} \varepsilon_r^{(c)}\,\varphi_r^{(c)}(t), \tag{17.3}$$

where M_c is a properly chosen constant. Here, assume that the mean shape function $\eta^{(c)}$ does not belong to the space spanned by the M_c leading eigenfunctions for identifiability of $\theta^{(c)}$. Further, rescale $Y^{\mathcal{X}}_{(c)}$ and $\tilde{Y}^{\mathcal{X}}_{(c)}$ such that $Y^{\mathcal{Z}}_{(c)}(t) = Y^{\mathcal{X}}_{(c)}(t)/\|Y^{\mathcal{X}}_{(c)}\|$ and $\tilde{Y}^{\mathcal{Z}}_{(c)}(t) = \tilde{Y}^{\mathcal{X}}_{(c)}(t)/\|\tilde{Y}^{\mathcal{X}}_{(c)}\|$. To define functional correlation, apply the classical geometrical concept of an *angle* to the functional data setting (See p. 388 of Ramsay and Silverman, 2005). A functional correlation between two observed random functions y_i and y_j is thus defined as $\rho(y_i, y_j) = \langle y_i/\|y_i\|, y_j/\|y_j\|\rangle$. This functional correlation corresponds to the cosine function of an angle ϑ such that $\cos(\vartheta) = \rho(y_i, y_j)$, and thus $-1 \le \rho \le 1$. The larger the absolute value of ρ, the stronger the positive or negative association between the functions. This functional correlation serves as a similarity measure for the functional clustering of shape similarities.

In an attempt to group curves with similar shapes, the fixed and random intercepts and the random scales are treated as a nuisance. Only the cluster shape functions or structure components $\{\eta^{(c)}, \varphi_1^{(c)}, \varphi_2^{(c)}, \ldots\}$ are important in constructing the underlying shape. Curves with similar shapes are embedded in the cluster subspace spanned by the cluster structure components. Let y be a realization of the random function Y_θ and $y^{\mathcal{Z}} = y^{\mathcal{X}}/\|y^{\mathcal{X}}\|$. Further, let $y_{(c)}$ denote the function of y expanded by the structure components of cluster c, and let $\tilde{y}^{\mathcal{Z}}_{(c)} = \tilde{y}^{\mathcal{X}}_{(c)}/\|\tilde{y}^{\mathcal{X}}_{(c)}\|$, where $\tilde{y}^{\mathcal{X}}_{(c)}$ is defined as in (17.3). The best cluster membership of the function y is determined by maximizing the functional correlation between $y^{\mathcal{Z}}$ and $\tilde{y}^{\mathcal{Z}}_{(c)}$ such that

$$c_2^*(y) = \underset{c\in\{1,\ldots,K\}}{\arg\max}\ \rho(y, \tilde{y}_{(c)}) = \underset{c\in\{1,\ldots,K\}}{\arg\max}\ \langle y^{\mathcal{Z}}, \tilde{y}^{\mathcal{Z}}_{(c)}\rangle. \tag{17.4}$$

Note that maximizing the functional correlation $\langle y^{\mathcal{Z}}, \tilde{y}^{\mathcal{Z}}_{(c)}\rangle$ is equivalent to minimizing the L^2-distance $\|y^{\mathcal{Z}} - \tilde{y}^{\mathcal{Z}}_{(c)}\|^2$, observing that $\|y^{\mathcal{Z}} - \tilde{y}^{\mathcal{Z}}_{(c)}\|^2 = 2 - 2\langle y^{\mathcal{Z}}, \tilde{y}^{\mathcal{Z}}_{(c)}\rangle$. Thus the functional clustering criterion (17.4) can also be written as

$$c_2^*(y) = \arg\min c \in \{1, \ldots, K\}\|y^{\mathcal{Z}} - \tilde{y}^{\mathcal{Z}}_{(c)}\|^2. \tag{17.5}$$

The criterion in (17.5) is thus similar to the L^2 criterion (17.2). However, criterion (17.5) requires the additional standardization procedure to account for shape similarities, regardless of shifted means and scales.

In (17.3), estimating the random scale effects requires additional steps, which complicates the functional clustering procedure.

17.3 Discussion

The k-centers projected FC approach uses data-adaptive eigenbases for the random process expansion. These basis functions are determined through covariance functions. This approach has the advantage that the first few eigen-components chosen by functional principal component analysis maximize the percentage of total variation explained. In contrast to the proposed k-centers projected FC method, recent approaches based on clustering basis coefficients must choose the same basis functions for all clusters to use the fitted coefficients as proxies to be clustered. This may create some difficulties since proper basis functions must be chosen so that the fitted coefficients adequately reflect cluster differences. Tarpey and Kinateder (2003) raised this issue, and García-Escudero and Gordaliza (2005) discussed the relative merits of using different basis functions. In addition, most coefficient-based methods are designed for clustering according to mean functions. Unlike the k-centers projected FC approach, these methods do not consider differentiation in cluster covariance structures. In addition, the k-centers projected FC approach does not rely on any distributional assumptions, compared to most model-based clustering approaches, which require Gaussian model assumptions. As a by-product, using the k-centers projected FC method reveals the mean and covariance structures. This facilitates functional cluster analysis by providing a visual insight into clusters.

References

[1] Bock, H. H.: On the interface between cluster analysis, principal component analysis, and multidimensional scaling. In: H. Bozdogan, A. K. Gupta (eds): Multivariate statistical modeling and data analysis. D. Reidel, Dordrecht, 17-44 (1987).

[2] Chiou, J.-M. and Li, P.-L.: Functional clustering and identifying substructures of longitudinal data. J. R. Statist. Soc. B. **69**, 679-699 (2007).

[3] García-Escudero, L. A. and Gordaliza, A.: A proposal for robust curve clustering. J. Class. **22**, 185-201 (2005).

[4] Ramsay, J. O. and Silverman, B. W.: Functional Data Analysis. 2nd ed. Springer: New York. (2005).

[5] Tarpey, T. and Kinateder, K.K.J.: Clustering functional data. J. Class. **20**, 93-114 (2003).

Chapter 18
Robust Nonparametric Estimation for Functional Data

Christophe Crambes, Laurent Delsol and Ali Laksaci

Abstract It is well known that robust estimation provides an alternative approach to classical methods which is not unduly affected by the presence of outliers. Recently, these robust estimators have been considered for models with functional data. In this talk, we focus on asymptotic properties of a conditional nonparametric estimation of a real valued variable with a functional covariate. We present results dealing with convergence in probability, asymptotic normality and $\mathbb{L}^q$ errors.

18.1 Introduction

A common problem in statistics consists in trying to explain how a variable of interest Y is linked with a covariate X. This talk deals with this framework, where we assume that the variable to explain Y is real valued and the explanatory variable X takes values in a semi-metric functional space $(\mathcal{F}, d)$. This kind of variables, well-known as $f$unctional variables in literature allows to consider variables as functions (of time for instance), which is interesting since it is well adapted to the functional nature of the observations (see Ramsay and Silverman, 2002-2005). In this context, the most general model is the regression model when the covariate is functional, which writes

---

Christophe Crambes
Montpellier II, place Eugène Bataillon, 34095 Montpellier cedex, France, e-mail: `ccrambes@math.univ-montp2.fr`

Laurent Delsol
Université Toulouse III, 118 route de Narbonne, 31062 Toulouse cedex 9, France, e-mail: `delsol@cict.fr`

Ali Laksaci
Univ. Djillali Liabès, B.P. 89, Sidi Bel Abbès 22000, Algérie, e-mail: `alilak@yahoo.fr`

$$Y = r(X) + \varepsilon,$$

where r is an operator from $\mathcal{F}$ to $\mathbb{R}$ and ε is a random error variable. This model has already been studied from a nonparametric point of view (that is to say only with regularity assumptions on r). The book of Ferraty and Vieu (2006) gives an overview of the main results obtained for a kernel nonparametric estimator of r. However, this estimation of r seen as the conditional mean of Y given $X = x$ may be unadapted to some situations. For instance, the presence of outliers or considering heteroskedastic variables can lead to irrelevent results. Robust regression has been introduced to solve these problems. Since the first important results obtained in the sixties (see Huber (1964)), an important literature have been devoted to this domain (see for instance, Robinson, 1984, Collomb and Härdle, 1986, Boente and Fraiman, 1990, and Laïb and Ould-Saïd, 2000 for recent references). Concerning data of infinite dimension, the literature is relatively restricted (see Cadre, 2001, Cardot *et. al.*, 2004). Recently, Azzedine *et. al.* (2006) studied the almost complete convergence of robust estimators based on a kernel method. In the same context, Attouch *et. al.* (2007) studied the asymptotic normality of these estimators.

In this work, we propose to study robust estimators. We first recall the convergence in probability as well as an asymptotic normality result obtained in (1). Then, we give the asymptotic expressions of the dominant terms in $\mathbb{L}^p$ errors, extending the work of Delsol (2007). We finally apply robust estimation methods to problems of nonparametric statistics as for instance the prediction of time series.

18.2 Model

Let (X, Y) be a couple of random variables taking values in $\mathcal{F} \times \mathbb{R}$, where $\mathcal{F}$ is a semi-metric space, which semi-metric is denoted by d. For $x \in \mathcal{F}$, we consider a real measurable function ψ_x. The functional parameter studied in this work, denoted by θ_x, is the solution (with respect to t), assumed to be unique, of the following equation

$$\Psi(x,t) := \mathbb{E}\left[\psi_x\left(Y,t\right)|X = x\right] = 0. \tag{18.1}$$

In general, the function ψ_x is fixed by the statistician according to the situation he is confronted to. Some classic examples of ψ_x lead to the estimation of the conditional mean or conditional quantiles (see Ferraty and Vieu, 2006, Attouch *et. al.*, 2007). Now, given a sample $(X_i, Y_i)_{i=1,\dots,n}$ with the same law as (X, Y), a kernel estimator of $\Psi(x,t)$ is given by

$$\widehat{\Psi}(x,t) = \frac{\sum_{i=1}^n K\left(h^{-1}d(x,X_i)\right)\psi_x\left(Y_i,t\right)}{\sum_{i=1}^n K\left(h^{-1}d(x,X_i)\right)}, \qquad \forall t \in \mathbb{R}, \tag{18.2}$$

where K is a kernel and $h = h_n$ is a sequence of positive real numbers. Then, a natural estimator of θ_x is $\widehat{\theta}_n = \widehat{\theta}_n(x)$ given by

$$\widehat{\Psi}(x, \widehat{\theta}_n) = 0. \tag{18.3}$$

We can notice that, when $\psi_x(Y,t) = Y - t$, then $\widehat{\theta}_n$ is the estimator given in Ferraty and Vieu (2002) for the functional nonparametric regression. Let us also remark that, under the condition that $\sum_{i=1}^{n} K\left(h^{-1}d(x, X_i)\right)$ is not equal to zero, the definition of the estimator by (18.3) is equivalent to

$$\widehat{\rho}_n(x, \widehat{\theta}_n) := \sum_{i=1}^{n} K\left(h^{-1}d(x, X_i)\right) \psi_x\left(Y_i, \widehat{\theta}_n\right) = 0. \tag{18.4}$$

18.3 Asymptotic results

18.3.1 Convergence in probability and asymptotic normality

In this section we recall some results given in Attouch *et. al.*(2007) for independent and identically distributed couples $(X_i, Y_i)_{i=1,\ldots,n}$. Under some technical conditions but rather classic in this nonparametric context, Attouch *et. al.*(2007) obtain

$$\widehat{\theta}_n - \theta_x \xrightarrow[n\to+\infty]{\mathbb{P}} 0.$$

and

$$\left(\frac{nF(h_n)}{V_n(x)}\right)^{1/2} \left(\widehat{\theta}_n - \theta_x - B_n(x)\right) \xrightarrow[n\to+\infty]{\mathcal{L}} \mathcal{N}(0,1),$$

with explicit expressions for $V_n(x)$ and $B_n(x)$.

18.3.2 A uniform integrability result

We give a result of uniform integrability which is useful to get the convergence of the moments for $\widehat{\theta}_n$. Let $t \in \mathbb{R}$ be fixed. We give the result for independent and identically distributed couples (X_i, Y_i). We also can show, with stronger hypotheses, the same kind of result for arithmetically α-mixing couples. We set

$$F(h) = \mathbb{P}\left(d(X, x) \leq h\right),$$

known as the *small balls probabilities*, and we consider the following hypotheses.

$(H.1)$ There exist $p > 2$ and $C > 0$, such that, for X in an open neighbourhood of x, we have almost surely

$$\mathbb{E}\left[\left|\psi_x(Y, t)\right|^p |X\right] \leq C.$$

$(H.2)$ We assume that $\lim_{n \to +\infty} nF(h_n) = +\infty$.

$(H.3)$ K is supported on the compact $[0, 1]$, is bounded, and $K(1) > 0$.

Under the hypotheses $(H.1) - (H.3)$, for $0 \leq q < p$, the quantity

$$\left|\sqrt{nF(h_n)}\left(\Psi_n(x, t) - \mathbb{E}\left[\Psi_n(x, t)\right]\right)\right|^q,$$

is uniformly integrable, where $\Psi_n(x, t) = \dfrac{1}{nF(h_n)} \widehat{\rho}_n(x, t)$.

18.3.3 Moments convergence

We give the result for independent and identically distributed couples $(X_i, Y_i)_{i=1,\ldots,n}$. We also can show, with stronger hypotheses, the same kind of result for arithmetically α-mixing couples. We assume that ψ_x is $\mathcal{C}^1$ with respect to its second argument on a neighbor of θ_x. We note ζ_n the random variable (taking values between θ_x and $\hat{\theta}_n$) such that $\hat{\theta}_n - \theta_x = -\frac{\Psi_n(x,\theta_x)}{\frac{\partial \Psi_n}{\partial t}(x,\zeta_n)}$ and we define $B_n := -\frac{\mathbb{E}[\Psi_n(x,\theta_x)]}{\mathbb{E}\left[\frac{\partial \Psi_n}{\partial t}(x,\zeta_n)\right]}$. We assume that

$$Z_n := \sqrt{\frac{nF(h_n)}{V_n(x)}}\left(\widehat{\theta}_n - \theta_x - B_n(x)\right) \xrightarrow[n \to +\infty]{\mathcal{L}} W, \tag{18.5}$$

where W is a standard gaussian variable, and we have explicit expressions of $B_n(x)$ and $V_n(x)$. We suppose that assumptions $(H.1) - (H.3)$ are satisfied (with $t = \theta_x$), as well as some other technical conditions given below.

$(H.4)$ $t \mapsto \sup_y \left(\dfrac{\partial \psi_x}{\partial t}(y, t) - \dfrac{\partial \psi_x}{\partial t}(y, \theta_x)\right)$ is continuous in a neighborhood of θ_x.

$(H.5)$ There exists a constant N such that, almost surely in a neighborhood of x

$$\mathbb{E}\left[\left(\frac{\partial\psi_x}{\partial t}(Y_i,\zeta_n)-\frac{\partial\psi_x}{\partial t}(Y_i,\theta_x)\right)^2 \mid X_i\right]\leq N.$$

$(H.6)$ There exist some constants γ and δ such that

$$\mathbb{E}\left[\frac{\partial\psi_x}{\partial t}(Y,\theta_x)\mid X\right]1\!\!1_{\{d(X,x)\leq\delta\}}\geq\gamma 1\!\!1_{\{d(X,x)\leq\delta\}}.$$

$(H.7)$ $B_n(x)$ satisfies $\sqrt{nF(h_n)}B_n = O(1)$.

$(H.8)$ There exist $p' > 2$ and a constant $0 < C' < +\infty$ such that, for X in an open neighborhood of x, we have almost surely

$$\mathbb{E}\left[\left|\frac{\partial\psi_x}{\partial t}(Y,\zeta_n)\right|^{p'}|X\right]\leq C'.$$

$(H.9)$ There exist r and a constant $0 < M_0 < +\infty$ such that $\mathbb{E}\left[\left|\widehat{\theta}_n-\theta_x\right|^r\right]\leq M_0$.

Then, we have, for all $q < q'$ (we have an explicit definition of q', not given here)

$$\mathbb{E}\left[\left|\widehat{\theta}_n-\theta_x\right|^q\right]=\mathbb{E}\left[\left|B_n(x)+\sqrt{\frac{V_n(x)}{nF(h_n)}}W\right|^q\right]+o\left(\frac{1}{\sqrt{nF(h_n)}^q}\right).$$

More explicit asymptotic expressions of $\mathbb{L}^q$ errors can be obtained from the explicit expressions of $B_n(x)$ and $V_n(x)$ given in Attouch *et. al.*(2007) with the same approach as in Delsol (2007). These expressions may be usefull to choose the optimal bandwidth and give the first general $\mathbb{L}^q$ convergence rates results for robust estimators in models with functional data.

18.4 Application to time series prediction

In this example, we are interested in the application of robust statistics as a prediction tool. We use here a time series data similar to the one studied by Ferraty and Vieu (2006). It concerns the U.S. monthly petroleum consumption for electricity generation[1]. The objective of this study is to predict the total consumption one year given the curve the preceding year. The data are represented on figure 1.
In order to avoid the heteroskedasticity problem's, Ferraty and Vieu (2006) used transformed data with a logarithmic difference. However, we choose to study the prediction problem with the initial data, and we consider the

[1] data available at www.economagic.com

objective function $\psi_x(\cdot,\cdot) = \psi\left(\frac{\cdot - \cdot}{S(x)}\right)$ where $S(x) = Median|Y - med_x|$ is a robust measure of conditional scale, med_x is the conditional median of Y knowing $X = x$ and $\psi(t) = \frac{t}{\sqrt{1+t^2/2}}$. The choice of the smoothing parameter has an important influence, mainly in the balance between the bias and the variance of the estimator. Hence, we choose the parameter locally with the $\mathbb{L}^1$ cross validation on the number of nearest neighbors. The kernel is chosen to be quadratic. Another important parameter to fix is the semimetric d. For this example, we consider an entire family of semimetrics computed with the functional principal components analysis (see Besse *et al.*, 1997) with several dimensions q and choose by cross-validation the one that is the most fitted to data. We have plotted on figure 2 the result of the prediction during one year.

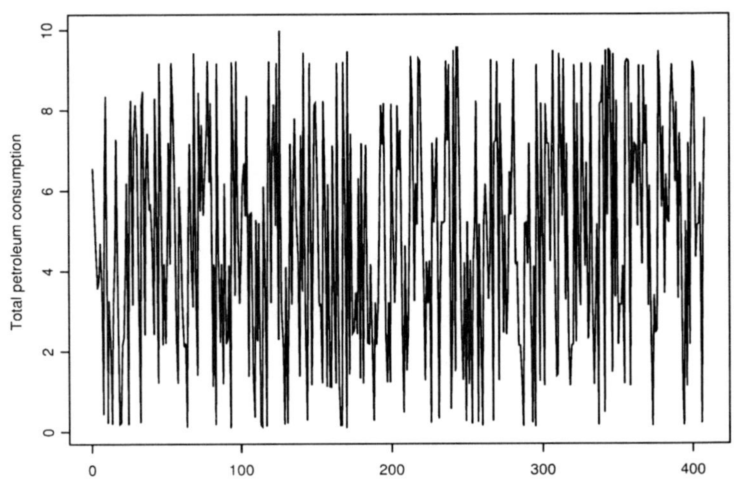

Fig. 18.1 Curves of energetic consumption.

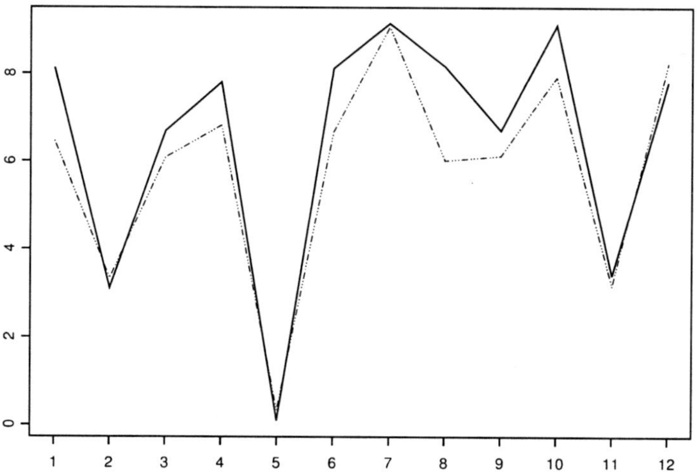

Fig. 18.2 Prediction of the energetic consumption during one year (real values: continuous line, predicted values: dashed line).

References

[1] M. Attouch, A. Laksaci. and E. Ould-Sa.:ïd. Asymptotic distribution of robust estimator for functional nonparametric models, Prépublication, LMPA No 314 Janvier (2007). Submitted.

[2] N. Azzeddine, A. Laksaci, E. Ould-Sa.:ïd. On the robust nonparametric regression estimation for functional regressor, Statist. and Probab. Lett. (2007), Accepted under minor revision.

[3] P. Besse, H. Cardot and F. Ferraty.: Simultaneous nonparametric regressions of unbalanced longitudinal data. Comput. Statist. Data Anal. **24**, 255-270 (1997).

[4] G. Boente, R. Fraiman.: Asymptotic distribution of robust estimators for nonparametric models from mixing processes. Ann. Statist. **18**, 891-906 (1990).

[5] B. Cadre.: Convergent estimators for the L_1-median of a Banach valued random variable. Statistics. **35**, 509-521 (2001).

[6] H. Cardot, C. Crambes, P. Sarda.: Quantiles regression when the covariates are functions. J. of Nonparam. Stat. **17**, 841-856 (2005).

[7] G. Collomb, W. Härdle.: Strong uniform convergence rates in robust nonparametric time series analysis and prediction: Kernel regression estimation from dependent observations. Stoch. Proc. Appl. **23**, 77-89 (1986).

[8] L. Delsol.: Régression non-paramétrique fonctionnelle: Expressions asymptotiques des moments. (2007). *Submitted.*

[9] F. Ferraty and P. Vieu.: The functional nonparametric model and application to spectrometric data. Comp. Statist. **17**, 545-564 (2002).

[10] F. Ferraty and P. Vieu.: Nonparametric functionnal data analysis. Springer-Verlag. New York. (2006).

[11] P.J. Huber.: Robust estimation of a location parameter. Ann. of the Math. Statist. **35**, 73-101 (1964).

[12] N. Laïb and E. Ould-Saïd.: A robust nonparametric estimation of the autoregression function under an ergodic hypothesis. Canad. J. Statist. **28**, 817-828 (2000).

[13] R. Robinson.: Robust Nonparametric Autoregression. Lecture Notes in Statistics, Springer-Verlag. New York. **26**, 247-255 (1984).

[14] J. Ramsay and B.W. Silverman.: Applied functional data analysis. Springer-Verlag, New York. (2002).

[15] J. Ramsay and B.W. Silverman.: Functional data analysis (Sec. Ed.). Springer-Verlag, New York. (2005).

Chapter 19
Estimation of the Functional Linear Regression with Smoothing Splines

Christophe Crambes, Alois Kneip and Pascal Sarda

Abstract We consider functional linear regression where a real variable Y depends on a functional variable X. The functional coefficient of the model is estimated by means of smoothing splines. We derive the rates of convergence with respect to the semi-norm induced by the covariance operator of X, which comes to evaluate the error of prediction. These rates, which essentially depend on the smoothness of the function parameter and on the structure of the predictor, are shown to be optimal over a large class of functions parameters and distributions of the predictor.

19.1 Introduction

In many fields of applications (climatology, teledetection, linguistics, ...), data come from the observation of continuous phenomenons of time or space. These data, known as *functional data* in the literature, are currently the subject of many works. For an overview of technics for the analysis of functional data, we can notably refer to the monographs Ramsay and Silverman (2002), Ramsay and Silverman (2005) and Ferraty and Vieu (2006).

We are interested here in the so-called *functional linear model*, where we want to explain the effects of a variable X on another variable Y. The variable

Christophe Crambes
Montpellier II, place Eugène Bataillon, 34095 Montpellier cedex, France, e-mail: `ccrambes@math.univ-montp2.fr`

Alois Kneip
Universität Bonn, Adenauerallee 24-26, 53113 Bonn, Germany, e-mail: `akneip@uni-bonn.de`

Pascal Sarda
IMT (UMR 5219), route de Narbonne, 31062 Toulouse cedex 1, France, e-mail: `Pascal.Sarda@math.ups-tlse.fr`

X (covariate) is a functional variable and is assumed to take its values in the space $L^2([0,1])$ of the functions $f : [0,1] \longrightarrow \mathbb{R}$ such that $\int_0^1 f(t)^2 dt$ is finite, while the variable Y (variable of interest) is a real-valued random variable. In this context, the functional linear regression writes, for a set of observations (X_i, Y_i), $i = 1, \ldots, n$ distributed as (X, Y),

$$Y_i = \alpha_0 + \langle \alpha, X_i \rangle + \varepsilon_i = \alpha_0 + \int_0^1 \alpha(t) X_i(t) dt + \varepsilon_i, \tag{19.1}$$

for $i = 1, \ldots, n$, where the parameter α_0 and the function $\alpha \in L^2([0,1])$ are unknown and ε_i is an error random variable satisfying $\mathbb{E}(\varepsilon_i) = 0$, $\mathbb{E}\left(\varepsilon_i^2\right) = \sigma_\varepsilon^2$ and $\mathbb{E}(\varepsilon_i X_i(t)) = 0$ for almost every t. Our goal is then to estimate the slope function α and the intercept α_0 from the observations (X_i, Y_i), $i = 1, \ldots, n$. This model has already been studied by several authors. The first works on this model can be found in Ramsay and Dalzell (1991). More recently in Cardot *et al.* (1999) and Cardot *et al.* (2003), two estimators of α have been proposed, the first one based on the functional principal component regression, and the other one based on regression splines. The estimator presented below is based on smoothing splines and has been introduced in Cardot *et al.* (2007) and Crambes *et al.* (2007).

19.2 Construction of the estimator

Consider the points $0 < t_1 < \ldots < t_p < 1$ where the curves X_i are observed. More precisely, we assume that $t_1 = 1/2p$ and $t_j - t_{j-1} = 1/p$ for $j = 2, \ldots, p$. Then, we consider the p-dimensional space of *natural splines* of degree $2m-1$ with knots $t_1, \ldots, t_p$, and a basis of this space $(b_1, \ldots, b_p)$ (see Eubank, 1988 for different candidates of bases). The estimator $\widehat{\alpha}$ of α is defined as the minimizer over all functions a in the Sobolev space $W^{m,2}([0,1])$ of

$$\frac{1}{n} \sum_{i=1}^{n} \left(Y_i - \overline{Y} - \frac{1}{p} \sum_{j=1}^{p} a(t_j)(X_i(t_j) - \overline{X}(t_j)) \right)^2 + \tag{19.2}$$
$$\rho \left(\frac{1}{p} \sum_{j=1}^{p} \pi_a^2(t_j) + \int_0^1 (a^{(m)}(t))^2 dt \right),$$

where m is a given positive integer and $\pi_a(t)$ is the best approximation, in a mean square sense, of a by a polynomial of degree $m-1$.

For every $\boldsymbol{a} = (a_1, \ldots, a_p)^\tau \in \mathbb{R}^p$, there exists a unique *spline interpolation* $s_{\boldsymbol{a}}$ explicitely defined from the vector $\boldsymbol{a}$ and the basis functions $b_1, \ldots, b_p$. Using the properties of natural splines, it is shown that $\widehat{\boldsymbol{\alpha}} = (\widehat{\alpha}(t_1), \ldots, \widehat{\alpha}(t_p))^\tau$ satisfies

$$\widehat{\alpha} = \frac{1}{np}\left(\frac{1}{np^2}\mathbf{X}^T\mathbf{X} + \frac{\rho}{p}\mathbf{A}_m\right)^{-1}\mathbf{X}^T\mathbf{Y}, \tag{19.3}$$

where $\mathbf{X}$ is the $n \times p$ matrix with general term $X_i(t_j)$, $\mathbf{Y}$ is the vector of length n with general term Y_i and $\mathbf{A}_m$ is an explicit $p \times p$ matrix linked with the derivatives of order m of the basis functions $b_1, \ldots, b_p$. Then, the estimator of α is defined by $\widehat{\alpha} = s_{\widehat{\boldsymbol{\alpha}}}$, while the estimation of α_0 is given by $\widehat{\alpha}_0 = \overline{Y} - \langle \widehat{\alpha}, \overline{X} \rangle$.

19.3 Convergence results

The convergence of the estimator $\widehat{\alpha}$ is analyzed with respect to the semi-norm induced by the covariance operator of X, $\Gamma := \mathbb{E}\left(\langle X - \mathbb{E}(X), . \rangle \left(X - \mathbb{E}(X)\right)\right)$. This semi-norm is defined by $\|u\|_{\Gamma}^2 = \mathbb{E}\left(\langle X - \mathbb{E}(X), u\rangle^2\right)$ and allows us to interpret our results in terms of prediction error. Indeed, for a new observation (X_{n+1}, Y_{n+1}) with X_{n+1} independent from $X_1, \ldots, X_n$, we predict Y_{n+1} by the quantity $\widehat{\alpha}_0 + \langle \widehat{\alpha}, X_{n+1}\rangle$ and we have

$$\mathbb{E}\left[\left((\widehat{\alpha}_0 + \langle \widehat{\alpha}, X_{n+1}\rangle) - (\alpha_0 + \langle \alpha, X_{n+1}\rangle)\right)^2 | \widehat{\alpha}_0, \widehat{\alpha}\right] = \|\widehat{\alpha} - \alpha\|_{\Gamma}^2 + O_P\left(n^{-1}\right). \tag{19.4}$$

The rates of convergence of our estimator depend essentially on regularity assumptions on the function α and on the curves X_i. More precisely, we make the following assumptions.

$(H.1)$ α is m times differentiable and $\alpha^{(m)} \in L^2([0,1])$.

$(H.2)$ There exists $0 < \kappa < 1$ such that, for every $\delta > 0$, there exists $0 < C_1 < +\infty$ satisfying

$$\mathbb{P}\left(|X(t) - X(s)| \leq C_1 |t-s|^{\kappa}, t, s \in [0,1]\right) \geq 1 - \delta.$$

$(H.3)$ There exist $0 < C_2 < +\infty$ and $q \in \mathbb{N}$ such that, for every $k \in \mathbb{N}$, there exists a sub-space $\mathcal{L}_k$ of $L^2([0,1])$ such that

$$\mathbb{E}\left(\inf_{f \in \mathcal{L}_k} \sup_{t \in [0,1]} |X(t) - f(t)|^2\right) \leq C_2 k^{-2q}.$$

$(H.4)$ There exists $C_3 > 0$ such that, for every r, s

$$Var\left(\frac{1}{n}\sum_{i=1}^{n}\langle X_i - \mathbb{E}(X), \zeta_r\rangle\langle X_i - \mathbb{E}(X), \zeta_s\rangle\right) \leq$$

$$\frac{C_3}{n}\mathbb{E}\left(\langle X - \mathbb{E}(X), \zeta_r\rangle^2\right)\mathbb{E}\left(\langle X - \mathbb{E}(X), \zeta_s\rangle^2\right),$$

where $(\zeta_r)_r$ are the eigenfunctions of Γ.

Under the previous assumptions, if $np^{-2\kappa} = O(1)$ and $\rho \sim n^{-(2m+2q+1)/(2m+2q+2)}$, we have

$$\|\widehat{\alpha} - \alpha\|_{\Gamma}^2 = O_P\left(n^{-(2m+2q+1)/(2m+2q+2)}\right). \tag{19.5}$$

Assumption $(H.2)$ allows to control the error resulting in the approximation of an integral by a discrete sum. It appears that $(H.3)$ is satisfied provided that the predictors X_i are smooth *i.e.* continuously differentiable at an order q_1 and $X_i^{(q_1)}$ being Lipschitz continuous with order r_1 and $q = [q_1 + r_1]$. On the other hand, $(H.3)$ may be satisfied for non smooth X_i as it is the case for Brownian motion. In any case, $(H.3)$ implies that the eigenvalues λ_r of Γ decrease rapidly in the sense that $\sum_{r=k+1}^{+\infty} \lambda_r = O(k^{-2r})$.

We show that the convergence rate (19.5) is optimal relatively to a certain class of functions α and curves X_i. These results are compared to the rates of convergence obtained in Cai and Hall (2006). These authors concentrate on the error $(\widehat{\alpha}_0 + \langle\widehat{\alpha}, x\rangle) - (\alpha_0 + \langle\alpha, x\rangle)$ for a fixed (non random) x, which is a major difference with our work.

References

[1] Cai, T.T. and Hall, P.: Prediction in functional linear regression. Annals of Statistics. **34**, 2159-2179 (2006).

[2] Cardot, H., Crambes, C., Kneip, A. and Sarda, P.: Smoothing splines estimators in functional linear regression with errors-in-variables. Computational Statistics and Data Analysis. special issue on functional data analysis, **51**, 4832-4848 (2007).

[3] Cardot, H., Ferraty, F. and Sarda, P.: Functional linear model. Statistics and Probability Letters. **45**, 11-22 (1999).

[4] Cardot, H., Ferraty, F. and Sarda, P.: Spline estimators for the functional linear model. Statistica Sinica. **13**, 571-591(2003).

[5] Crambes, C., Kneip, A. and Sarda, P.: Smoothing splines estimators for functional linear regression. To appear. (2007).

[6] Eubank, R.L.: Spline smoothing and nonparametric regression. Marcel Dekker. (1988).

[7] Ferraty, F. and Vieu, P.: Nonparametric functional data analysis: theory and practice. Springer, New York. (2006).

[8] Ramsay, J.O. and Dalzell, C.J.: Some tools for functional data analysis. Journal of the Royal Statistical Society, Series B. **53**, 539-572 (1991).

[9] Ramsay, J.O. and Silverman, B.W.: Applied functional data analysis. Springer, New York. (2002).

[10] Ramsay, J.O. and Silverman, B.W.: Functional data analysis (Second edition). Springer, New York. (2005).

[11] Utreras, F.: Natural spline functions, their associated eigenvalue problem. Numerische Mathematik. **42**, 107-117 (1983).

Chapter 20
A Random Functional Depth

Juan Cuesta-Albertos and Alicia Nieto-Reyes

Abstract We define an easy to compute and not computational demanding functional depth which gives results comparable to those obtained with more involved depths. When applied in finite-dimensional settings, it can be seen as an approximation to the Tukey depth as it uses a finite number of randomly chosen one-dimensional projections while the Tukey depth considers all-possible one-dimensional projections.

20.1 Introduction

Given a probability distribution P defined in a infinite-dimensional (or multidimensional) space $\mathcal{X}$, a depth tries to order the points in $\mathcal{X}$ from the "center (of P)" to the "outward (of P)". Obviously, this problem includes data sets if we consider P as the empirical distribution associated to the data set at hand.

Some functional [for instance, Fraiman and Muniz (2001) and López-Pintado and Romo (2006)] and several multidimensional depths [see Liu, Parelius and Singh (1999) and references there] have been proposed.

Here, we try to take advantage of the main result in Cuesta-Albertos, Fraiman and Ransford (2007) to introduce a new functional depth. Roughly speaking, this result establishes that a randomly chosen one-dimensional projection is enough to determine a distribution (see Section 3 for a more detailed description of this result). Thus, from a theoretical point of view, it should be possible to compute depths of points using only a randomly chosen one-dimensional projection.

Juan Cuesta-Albertos
Universidad de Cantabria Spain, e-mail: juan.cuesta@unican.es

Alicia Nieto-Reyes
Universidad de Cantabria Spain, e-mail: alicia.nieto@unican.es

This is the point of view chosen in Cuevas, Febrero and Fraiman (2007), where the authors propose to choose just a vector, v_1, at random and define the deepness of a given point x as its deepness in the one-dimensional subspace generated by v_1. However, there, they also propose to choose at random a finite number of vectors and take as depth of a given point x the mean of the one-dimensional depths obtained with each vector. They do that in order to get stability in the definition although, as stated, Theorem 4.1 in Cuesta-Albertos, Fraiman and Ransford (2007) provides the theoretical background for the first definition.

Here, we take a different point of view, inspired by the Tukey depth. If $x \in \mathbb{R}^p$, then, the Tukey depth of x with respect to P, $D_T(x, P)$, is the minimal probability which can be attained in the closed halfspaces containing x.

Let us see an equivalent definition. In the one-dimensional setting, it seems reasonable to order the points using the order induced by the function

$$x \to D_1(x, P) := \min\{P(-\infty, x], P[x, \infty)\}.$$

Given $v \in \mathbb{R}^p$, let Π_v be the projection of $\mathbb{R}^p$ on the one dimensional subspace generated by v. Thus, $P \circ \Pi_v^{-1}$ is the marginal of P on this subspace, and it is obvious that

$$D_T(x, P) = \inf\left\{D_1(\Pi_v(x), P \circ \Pi_v^{-1}) : v \in \mathbb{R}^p\right\}, \ x \in \mathbb{R}^p. \tag{20.1}$$

Our idea, here, is to compute the depths of points in separable Hilbert spaces as the infimum over a finite family of randomly chosen one-dimensional vectors, $v_1, ..., v_k$. That is, we replace in (20.1) the infimum over a nondenumerable number of vectors by the infimum over $v_1, ..., v_k$ and do the computations in a separable Hilbert space. This provides some stability to the definition of depth and, kept k low, we have an easily computable depth.

It worths to mention that, some other depths based on the consideration of all possible one-dimensional projections, but replacing $D_1(x, P)$ by some other function, have been proposed [see, for instance, Zuo (2003)]. We consider that what follows could be applied to all of them, but, we have chosen the Tukey depth to test it concretely.

Furthermore, it is well known that the most important drawback of the Tukey depth is the required computational time. This time is more or less reasonable if $p = 2$, but it becomes prohibitive even for $p = 8$ [see Mosler and Hoberg (2006)]. To reduce the time, in Zuo (2006), it is proposed to approximate their values using randomly selected projections. Thus, in some sense, Zuo (2006) can be considered as an antecedent of our depth.

In Section 2, we define the random functional depth and show some of its characteristics. Section 3 justifies the randomness in the definition of functional depth. Finally, in Section 4, we apply our depth to a functional classification problem.

In this contribution, we will be interested in probability distributions defined on a general separable Hilbert space which will be denoted by $\mathcal{X}$. All the random elements will be assumed to be defined on the same, rich enough, probability space (Ω, σ, P).

The computations have been carried out with MatLab. Computational codes are available from the authors upon request.

20.2 Functional depth

We begin with the definition of the random functional depth.

Definition 20.1. Let P be a probability distribution on $\mathcal{X}$. Let $x \in \mathcal{X}$, $k \in \mathbb{N}$ and let ν be an absolutely continuous distribution on $\mathcal{X}$. The random functional depth of x with respect to P based on a set $R = \{v_1, ..., v_k\}$ is

$$D_R(x, P) = \min\{D_1(\Pi_{v_i}(x), P \circ \Pi_{v_i}^{-1}) : v_i \in R \text{ for } i = 1, ..., k\}, \ x \in \mathcal{X},$$

where $v_1, ..., v_k$ are independent and identically distributed random vectors with distribution ν.

An interesting question is whether D_R satisfies the definition of depth given by Zuo and Serfling (2000). This definition includes four requirements. The first three (affine invariance, maximality at center and monotonicity relative to deepest point) are always fulfilled. However, the fourth one (vanishing at infinity) is not satisfied in general, but it holds if the dimension of $\mathcal{X}$ is finite.

Furthermore, we have that the random functional depth can be consistently estimated.

Theorem 20.1. *Let $v_1, ..., v_k \in \mathcal{X}$. Let P be a probability distribution on $\mathcal{X}$, and let $\{P_n\}$ be a sequence of empirical distributions computed on a random sample taken from P which is independent of the vectors $v_1, ..., v_k$.*

Then, conditionally on $R = \{v_1, ..., v_k\}$, we have that

$$\sup_{x \in \mathcal{X}} |D_R(x, P_n) - D_R(x, P)| \to 0, \ \text{almost surely } [P].$$

The proof is based on the real case of the Glivenko-Cantelli Theorem.

For a broader exposition of the random functional depth, see Cuesta-Albertos and Nieto-Reyes (2008a) and (2008b).

We close this section noticing that there exists the possibility of extending the results in this contribution to Banach spaces due to the generalization of Theorem 4.1 in Cuesta-Albertos, Fraiman and Ransford (2007) which appears in Cuevas and Fraiman (2007).

20.3 Randomness

Obviously, $D_R(x, P)$ is a random variable. It may seem a bit strange to take a random quantity to measure the depth of a point, which is inherently not-random. We have two reasons to take this point of view.

Firstly, Theorem 4.1 in Cuesta-Albertos, Fraiman and Ransford (2007) shows that if P and Q are probability distributions on $\mathcal{X}$, ν is an absolutely continuous distribution on $\mathcal{X}$ and

$$\nu\{v \in \mathcal{X} : P \circ \Pi_v^{-1} = Q \circ \Pi_v^{-1}\} > 0,$$

then $P = Q$. In other words, if we have two different distributions and randomly choose a marginal of them, those marginals are almost surely different. In fact, it is also required that at least one of the distributions is determined by its moments, but this is not too important for the time being. According to this result, one randomly chosen projection is enough to distinguish between two distributions on $\mathcal{X}$. Since the depths determine one-dimensional distributions, a depth computed on just one random projection allows to distinguish between two distributions.

Secondly, let us consider the case in which $\mathcal{X}$ is finite dimensional. If the support of ν is $\mathcal{X}$, and, for every k, $R_k \subset R_{k+1}$, then

$$D_{R_k}(x, P) \geq D_{R_{k+1}}(x, P) \to D_T(x, P), \quad \text{a.s.} \tag{20.2}$$

Therefore, if we choose a large enough k, the effect of the randomness in D_{R_k} will be negligible. Of course, the question of interest here is to learn how large k must be, because values of k that are too large would make this definition useless.

We propose to choose the right k depending on the problem. For instance, with bootstrap in test problems or cross-validation in supervised classification problems (see Section 4). Furthermore, to have an idea about the range of possible values, it could be a possibility to choose k based on the comparison between D_T and D_{R_k} in several multidimensional cases. However, the long computation times required to obtain D_T make those comparisons unpractical. Instead of this, we have decided to choose a situation in which the deepness of the points are clearly defined and can be easily computed with a different depth.

If P is an elliptical distribution with parameters μ and Σ, every depth should be a monotone function of the Mahalanobis depth, where, given $x \in \mathcal{X}$ ($= \mathbb{R}^p$ in this case) this depth is

$$D_M(x, P) := \frac{1}{1 + (x - \mu)^t \Sigma^{-1} (x - \mu)}.$$

Thus, from (20.2), the larger the k, the larger the resemblance between $D_{R_k}(\cdot, P)$ and a monotone function of $D_M(\cdot, P)$.

For getting an idea about the optimal number of projections required we will present some simulations, where the selection has been based on the Spearman correlation coefficient between the random functional and the Mahalanobis depths. Note that depths only try to rank points according to their closeness to the center of P. Thus, it is reasonable to measure the resemblance between $D_{R_k}(\cdot, P)$ and $D_M(\cdot, P)$ looking only at the ranks of the points, which is equivalent to employing the Spearman correlation coefficient.

Those simulations have been carried out for different dimensions, sample sizes and distributions with independent and dependent marginals. Moreover, since, in practice, we do not know P, and we only have a random sample of P we have also introduced some simulations in which μ and Σ are estimated. We have obtained no obvious differences between them.

20.4 Analysis of a real data set

In this section we try to classify by sex some growth curves. The data are very well known and have been taken from the file growth.zip, downloaded from the URL ftp://ego.psych.mcgill.ca/pub/ramsay/FDAfuns/Matlab.

We have applied a leave-one-out cross-validation (CV) procedure to those curves. In order to classify a given curve, we have employed the procedures based on depths proposed in López-Pintado and Romo (2006) and Cuevas, Febrero and Fraiman (2007) and, also, some combinations of them.

For applying the previous procedures, we have to compute random functional depths. Thus, two problems has to be fixed. The first one consists in choosing the distribution to be employed to select the projections. The second one refers to fixing the number of one dimensional projections to be considered.

First problem is included in a more general and important one: which is the optimal distribution to choose the projections in a given problem? Of course, the answer should depend on the problem at hand and it is in progress. However, we have done some preliminary steps and, here we choose the optimal distribution by CV in a parametric family which includes the standard Brownian Motion. The results are encouraging since the difference between employing the standard Brownian Motion or a distribution in the family by CV is a reduction of about 10% in the rate of error.

The number of projections is also chosen by CV between $\{1, 3, 5, ..., 99\}$.

There are some procedures to compare with. Some of them are based on functional depths [see López-Pintado and Romo (2006)], some on the k-nearest neighbors procedure [see, for instance, Biau, Bunea and Wegcamp (2005)], and some others on nonparametric regression techniques [see Abraham, Biau and Cadre (2006), Baíllo and Grané (2007) and Ferraty and Vieu (2003)].

In fact, one of the employed classification procedures classifies the curve in the group whose deepest curve is closest to it. Obviously, the deepest curves can be replaced by another representative curve like the median or the modal curves [see Ferraty and Vieu (2006)].

Moreover, since every functional data, at the end, belong to the discrete word, in practice they are discrete and, then procedures like Random Forests [see Breiman (2001)] can be also considered.

References

[1] Abraham, C., Biau, G. and Cadre, B.: On the kernel rule for function classification. Ann. Inst. Statist. Math. **58**, 619-633 (2006).

[2] Baíllo, A. and Grané A.: Local Linear regression for functional predictor and scalar response. Preprint. (2007)

[3] Biau, G., Bunea, F. and Wegcamp, M.H.: Functional Classification in Hilbert Spaces. IEEE Transact. Informat. Theo.**51**, 2163-2172 (2005).

[4] Breiman, L.: Random Forests. Machine Learning, 45, 5-32 (2001).

[5] Cuesta-Albertos, J. A., Fraiman, R. and Ransford, T.: A sharp form of the Cramér-Wold theorem. J. Theoret. Probab. **20** 201-209 (2007).

[6] Cuesta-Albertos, J. A. and Nieto-Reyes, A.: The Random Tukey Depth. Preprint. (2008a).

[7] Cuesta-Albertos, J. A. and Nieto-Reyes, A.: The Tukey and the random Tukey depths characterize discrete distributions. Journal of Multivariate Analysis, doi:10.1016/j.jmva.2008.02.017. (2008b).

[8] Cuevas, A., Febrero, M. and Fraiman, R.: Robust estimation and classification for functional data via projection-based depth notions. To appear in Computation. Statist. (2007).

[9] Cuevas, A. and Fraiman, R.: On depth measures and dual statistics: A methodology for dealing with general data. Preprint. (2007).

[10] Ferraty, F. and Vieu, P.: Curves discrimination: a nonparametric functional approach. Computat. Statist. Data Anal.**44**, 161-173 (2003).

[11] Ferraty, F. and Vieu, P.: Nonparametric Functional Data Analysis. Springer Series in Statistics (2006).

[12] Fraiman, R. and Muniz, G.: Trimmed means for functional data. Sociedad Estadística e Investigación Operatica. Test. **10**(2), 419-440 (2001).

[13] Liu, R.Y., Parelius, J.M. and Singh, K.: Multivariate analysis by data depth: descriptive statistics, graphics and inference. The Annals of Statisticas. **27**(3), 783-858 (1999).

[14] López-Pintado, S. and Romo, J.: Detph-based classification for functional data. DIMACS Series. **72**, 103-119 (2006).

[15] Mosler, K. and Hoberg, R.: Data analysis and classification with the zonoid depth. DIMACS Series. **72**, 49-59 (2006).

[16] Zuo, Y. and Serfling, R.: General notions of statistical depth function. Ann. Statist. **28**(2), 461-482 (2000).

[17] Zuo, Y.: Projection-based depth functions and associated medians. Ann. Statist. **31**(5), 1460-1490 (2003).

[18] Zuo, Y.: Multidimensional trimming based on projection depth. Ann. Statist. **34** (5), 2211-2251 (2006).

Chapter 21
Parametric Families of Probability Distributions for Functional Data Using Quasi-Arithmetic Means with Archimedean Generators

Etienne Cuvelier and Monique Noirhomme-Fraiture

Abstract Parametric probability distributions are central tools for probabilistic modeling in data mining, and they lack in functional data analysis (FDA). In this paper we propose to build this kind of distribution using jointly Quasi-arithmetic means and generators of Archimedean copulas. We also define a density adapted to the infinite dimension of the space of functional data. We use these concepts in supervised classification.

21.1 QAMML distributions

Let $(\Omega, \mathcal{A}, P)$ a probability space and $\mathcal{D}$ a closed real interval. A *functional random variable (frv)* is any function from $\mathcal{D} \times \Omega \to \mathbb{R}$ such for any $t \in \mathcal{D}, X(t,.)$ is a real random variable on $(\Omega, \mathcal{A}, P)$. Let $L^2(\mathcal{D})$ be the space of square integrable functions (with respect to Lebesgues measure) $u(t)$ defined on $\mathcal{D}$.

If $f, g \in L^2(\mathcal{D})$, then the pointwise order between f and g on $\mathcal{D}$ is defined as follows :

$$\forall t \in \mathcal{D}, f(t) \leq g(t) \iff f \leq_{\mathcal{D}} g. \tag{21.1}$$

It is easy to see that the pointwise order is a partial order over $L^2(\mathcal{D})$, and not a total order. We define the *functional cumulative distribution function (fcdf)* of a *frv* $\underline{X}$ on $L^2(\mathcal{D})$ computed at $u \in L^2(\mathcal{D})$ by :

$$F_{\underline{X},\mathcal{D}}(u) = P[\underline{X} \leq_{\mathcal{D}} u]. \tag{21.2}$$

Etienne Cuvelier
Facultés Universitaires Notre-Dame de la Paix Faculté d'Informatique 21, rue grandgagnage 5000 Namur, Belgique, e-mail: ecu@info.fundp.ac.be

Monique Noirhomme-Fraiture
Facultés Universitaires Notre-Dame de la Paix Faculté d'Informatique 21, rue grandgagnage 5000 Namur, Belgique, e-mail: mno@info.fundp.ac.be

To compute the above probability, let us remark that, it is easy to compute the probability distribution of the value of $X(t)$ for a specific value of t, and this for any $t \in \mathcal{D}$. Then we define respectively the *surface of distributions* and the *surface of densities* as follow :

$$G : \mathcal{D} \times \mathbb{R} \to [0,1] : (t,y) \mapsto P[X(t) \leq y] \tag{21.3}$$

$$g : \mathcal{D} \times \mathbb{R} \to [0,1] : (t,y) \mapsto \frac{\partial}{\partial t} G(t,y) \tag{21.4}$$

We can use various methods for determining suitable g and G for a chosen value of $\underline{X}$. Thus for example, if $\underline{X}$ is a Gaussian process with mean value $\mu(t)$ and standard deviation $\sigma(t)$, then, for any $(t,y) \in \mathcal{D} \times \mathbb{R}$, we have :

In the following we will always use the function G with a function u of $L^2(\mathcal{D})$, so, for the ease of the notations, we will write : $G[t;u] = G[t, u(t)]$. We will use the same notation for g. In what follows we define our parametric families of probability distributions.

Let $\underline{X}$ be a frv, $u \in L^2(\mathcal{D})$ and G its *Surface of Distributions*. Let also ϕ be a continuous strictly decreasing function from $[0,1]$ to $[0,\infty]$ such that $\phi(0) = \infty$, $\phi(1) = 0$, where $\psi = \phi^{-1}$ must be completely monotonic on $[0,\infty[$ i.e. $(-1)^k \frac{d^k}{dt^k}\psi(t) \geq 0$ for all t in $[0,\infty[$ and for all k. We define the *Quasi-Arithmetic Mean of Margins Limit (QAMML)* distribution of $\underline{X}$ by :

$$F_{\underline{X},\mathcal{D}}(u) = \psi\left[\frac{1}{|\mathcal{D}|}\int_{\mathcal{D}} \phi\left(G[t;u]\right) dt\right]. \tag{21.5}$$

The function ϕ is called the *QAMML* generator. In fact the expression (21.5) can be seen as the limiting (or continuous) case of two other expressions. The first expression, which is obvious and gives its name to (21.5), use a quasi-arithmetic mean M :

$$F_{\underline{X},\mathcal{D}}(u) = \lim_{n\to\infty} M\left\{G[t_1;u], \ldots, G[t_n;u]\right\} \tag{21.6}$$

where $\{t_1, \ldots, t_n\} \subset \mathcal{D}$ is a subset of points in $\mathcal{D}$, preferably equidistant. In the discrete case, a quasi-arithmetic mean is a function $M : [a,b]^n \to [a,b]$ defined as follows:

$$M(x_1, \ldots, x_n) = \psi\left(\frac{1}{n}\sum_{i=1}^{n} \phi(x_i)\right) \tag{21.7}$$

where ϕ is a continuous strictly monotonic real function and $\psi = \phi^{-1}$.

The second limiting case links the *QAMML* distributions to the classical approximation : $P[\underline{X} \leq_{\mathcal{D}} u] = H(u(t_1), \ldots, u(t_n))$, using the archimedean copulas:

$$F_{\underline{X},\mathcal{D}}(u) = \lim_{n\to\infty} \psi\left[\sum_{i=1}^{n} \phi\left(G^*[t_i;u]\right)\right] \tag{21.8}$$

where * is the following transformation, applied to margins:

$$G^*(x) = \psi\left(\frac{1}{n}\phi\left(G(x)\right)\right). \tag{21.9}$$

Let us remind that a copula is a multivariate cumulative distribution function defined on the n-dimensional unit cube $[0, 1]^n$ such that every marginal distribution is uniform on the interval [0, 1]. The interest of copulas comes from the fact that (Sklar's theorem), if H is an n-dimensional distribution function with margins $F_1, ..., F_n$, then there exists an n-copula C such that for all $x \in \mathbb{R}^n$,

$$H(x_1, ..., x_n) = C(F_1(x_1), ..., F_n(x_n)). \tag{21.10}$$

The copula captures the dependence structure of the distribution. An important family of copulas is the family of Archimedean copula, given by the following expression :

$$C(u_1, ..., u_n) = \psi\left[\sum_{i=1}^{n} \phi(u_i)\right]. \tag{21.11}$$

where ϕ, called the generator, has the same properties that a *QAMML* generator.
This second limiting case shows that *QAMML* shares the properties and limitations of archimedeans copulas in the modeling of an *frv* $\underline{X}$ (see the GQAMML section).

21.2 Gateaux density

A *fcdf* is an incomplete tool without an associate density, but as the *QAMML* distributions deal directly with infinite nature of functional data, we cannot use the classical multivariate density function:

$$h(x_1, ..., x_n) = \frac{\partial^n}{\partial x_1 \ldots \partial x_n} H(x_1, \ldots, x_n). \tag{21.12}$$

To solve this problem we propose to use a concept of the functional analysis : the *Gâteaux differential* which is a generalization of directional derivative. Let $\underline{X}$ be a *frv*, $F_{\underline{X},\mathcal{D}}$ its *fcdf* and u a function of $L^2(\mathcal{D})$. Then for $h \in L^2(\mathcal{D})$ we define the *Gâteaux density of* $F_{\underline{X},\mathcal{D}}$ at u and in the direction of h by:

$$f_{\underline{X},\mathcal{D},h}(u) = \lim_{\varepsilon \to 0} \frac{F_{\underline{X},\mathcal{D}}\left(u + h \cdot \varepsilon\right) - F_{\underline{X},\mathcal{D}}\left(u\right)}{\varepsilon} = DF_{\underline{X},\mathcal{D}}(u; h) \tag{21.13}$$

where $DF_{\underline{X},\mathcal{D}}(u;h)$ is the *Gâteaux differential* of $F_{\underline{X},\mathcal{D}}$ at u in the direction $h \in V$.
It is easy to show that, if $F_{\underline{X},\mathcal{D}}$ is a *QAMML fcdf*, u and h are two functions of $L^2(\mathcal{D})$, then the corresponding *Gâteaux density of* $F_{\underline{X},\mathcal{D}}$ computed in u, in direction of h is given by:

$$f_{\underline{X},\mathcal{D},h}(u) = \frac{1}{|\mathcal{D}|} \cdot \psi' \left[\frac{1}{|\mathcal{D}|} \int_{\mathcal{D}} \phi \left(G\left[t;u\right] \right) dt \right] \cdot \left\{ \int_{\mathcal{D}} \phi' \left(G\left[t;u\right] \right) \cdot g\left[t;u\right] h(t) dt \right\}. \tag{21.14}$$

We can show that, if we use the statistical dispersion $\sigma(t)$ of the functional data, then $f_{\underline{X},\mathcal{D},\sigma}(u) = P[\underline{X} = u]$.

21.3 GQAMML distributions

QAMML shares the limitations of archimedeans copulas (see section 1), but the archimedean copulas of dimension $n > 2$, can capture dependence structures from independence until the complete positive dependence between variables. Thus, if for $s, t \in \mathcal{D}$, there is a negative dependence between $X(s)$ and $X(t)$, the *QAMML* will not be able to model the situation. But the bidimensional archimedean copulas can deal with this kind of dependence, using the same generator, but with larger domain for the parameter. Then we define the *Generalized Quasi-Arithmetic Mean of Margins Limit (GQAMML)* $\mathbb{F}_{\underline{X},\mathcal{D}}(u)$ as follows. Let $\underline{X}$ be a *frv* defined on $\mathcal{D}$, $u \in L^2(\mathcal{D})$, $\{\mathcal{D}_p, \mathcal{D}_n\}$ a partition of $\mathcal{D}$ such :

- $\forall s, t \in \mathcal{D}_p$, there is a positive dependence between $X(s)$ and $X(t)$,
- $\forall s, t \in \mathcal{D}_n$, there is a positive dependence between $X(s)$ and $X(t)$,
- $\forall s \in \mathcal{D}_p$ and $\forall t \in \mathcal{D}_n$, there is a negative dependence between $X(s)$ and $X(t)$.

Then

$$\mathbb{F}_{\underline{X},\mathcal{D}}(u) = \psi \left(\frac{|\mathcal{D}_p|}{|\mathcal{D}|} \phi \left[F_{\underline{X},\mathcal{D}_p}(u) \right] + \frac{|\mathcal{D}_n|}{|\mathcal{D}|} \phi \left[F_{\underline{X},\mathcal{D}_n}(u) \right] \right) \tag{21.15}$$

where ϕ is the generator of an bidimensional archimedean copulas.
Of course, using the chain rule, the *Gâteaux density of* $\mathbb{F}_{\underline{X},\mathcal{D}}$ is given by

$$\mathbf{f}_{\underline{X},\mathcal{D},\sigma}(u) = \psi' \left(\frac{|\mathcal{D}_p|}{|\mathcal{D}|} \phi \left[F_{\underline{X},\mathcal{D}_p}(u) \right] + \frac{|\mathcal{D}_n|}{|\mathcal{D}|} \phi \left[F_{\underline{X},\mathcal{D}_n}(u) \right] \right)$$
$$\left\{ \frac{|\mathcal{D}_p|}{|\mathcal{D}|} \phi' \left[F_{\underline{X},\mathcal{D}_p}(u) \right] f_{\underline{X},\mathcal{D}_p,\sigma}(u) + \frac{|\mathcal{D}_n|}{|\mathcal{D}|} \phi' \left[F_{\underline{X},\mathcal{D}_n}(u) \right] f_{\underline{X},\mathcal{D}_n,\sigma}(u) \right\} \tag{21.16}$$

21.4 CQAMML distributions

In functional data analysis, we know that, some times, when we treat smooth data, there is a lot of information in the derivatives of the data. Of course we can apply the *GQAMML* distributions to the concerned derivative, but we can also consider jointly the distribution of the different derivatives. Then we define the *Complete Quasi-Arithmetic Mean of Margins Limit (CQAMML)* $\mathbb{F}_{i\,\underline{X},\mathcal{D}}^{j}(u)$ (with $i < j$) as follows. Let $\underline{X}$ be a *frv* defined on $\mathcal{D}$ with j successive derivatives, $u \in L^2(\mathcal{D})$ with j successive derivatives:

$$\mathbb{F}_{i\,\underline{X},\mathcal{D}}^{j}(u) = C\left(\mathbb{F}_{\underline{X}^{[i]},\mathcal{D}}\left(u^{[i]}\right), \ldots, \mathbb{F}_{\underline{X}^{[j]},\mathcal{D}}\left(u^{[j]}\right)\right) \tag{21.17}$$

where :

- $\underline{X}^{[i]}$ and $u^{[i]}$ are the ith derivatives for $\underline{X}$ and u,
- C is a n-dimensional copula.

Note that the copula C is not necessarily an archimedean copula. The density of the *CQAMML* distribution is a classical joint density used with the *Gâteaux densities* of the different *GQAMML* distributions.

21.5 Supervised classification

To illustrate the interest of the *QAMML* families of distribution we propose to use it in a supervised classification application. To perform the classification we use the *Gâteaux density of a QAMML distribution* to build a bayesian classifier:

$$P(\omega_i|u) = \frac{\mathbf{f}_{\omega_i,\mathcal{D},h}(u) \cdot P(\omega_i)}{P(u)} \tag{21.18}$$

where $P(\omega_i|u)$ is the probability that u belong to the ith group, $\mathbf{f}_{\omega_i,\mathcal{D},h}(u)$ the adequate *Gâteaux density*, and $P(u)$ the probability of u (but this latter is constant for all cluster, so it is not necessary to compute it).
We compute the parameters of each cluster using the classical maximum likelihood, and the cluster of u is the cluster with the highest probability $P(\omega|u)$.
The chosen dataset is the well known spectrometric data from Tecator. The data consist in 100 channels of spectrum absorbance (wavelength from 850 nm to 1050 nm). The goal is to distinguish the data with more than 20% of fat content, from the data with less than 20% of fat content. We have performed a 10-fold cross validation on the data, the first derivative, the second derivative using the GQAMML distributions, and jointly on the different derivatives using the CQAMLL distributions, and this with the following parametrization :

- Surface of distributions G : Normal distribution,
- QAMML and GQAMML generators : Clayton generator,
- CQAMML copula : Normal copula.

Table 21.1 Results of the 10-fold cross validations

Distributions	misclassifications
$F_{\underline{X},\mathcal{D}}$	31.4%
$F_{\underline{X}',\mathcal{D}}$	9.4%
$F_{\underline{X}'',\mathcal{D}}$	5.5%
$\mathbb{F}^1_{0\underline{X},\mathcal{D}}$	16.5%
$\mathbb{F}^2_{1\underline{X},\mathcal{D}}$	4%
$\mathbb{F}^2_{0\underline{X},\mathcal{D}}$	9.4%

The table 21.1 shows the results, and we can see that the best results are given using the distribution of the second derivative, and also considering jointly the distribution of the first and second derivative, but it is well known that the second derivative of these data contains the more interesting information to distinguish the clusters. We can also remark that when we use directly the functionnal data jointly with the derivatives, the quality of the classification decrease, but we know that original functions contain only slight differences between the two groups.

21.6 Conclusions

The good results of the supervised classification example show that our new families of parametric distributions for functional data can be used in classifications task in FDA. These distributions can be used also in unsupervised classification with existing algorithms. And a lot of parametrization can be chosen using existing copulas in the different level of the QAMML families, and other choices for the distributions of the surface of distributions can be done. So a great field of experimentation is open with the QAMML families of distributions for functional data.

References

[1] Aczel J.: Lectures on Functional Equations and Their Applications, Academic Press, Mathematics in Science and Engineering, New York and London.(1966).

[2] Cuvelier E. and Noirhomme-Fraiture M.: Classification de fonctions continues à l'aide d'une distribution et d'une définies dans un espace de dimension infinie, Conférence EGC07- Namur, Belgique. (2007).

[3] Cuvelier. E. and Noirhomme-Fraiture M.: A probability distribution of functional random variable with a functional data analysis application, ICDM 06 Conference - MCD 06 Workshop, Hong-Kong. (2006).
[4] Cuvelier E. and Noirhomme-Fraiture M.: An approach to stochastic process using quasi-arithmetic means, International Symposium on Applied Stochastic Models and Data Analysis (ASMDA), Crete - Chania. (2007).
[5] Joe, H.: Multivariate models and dependence concepts, Chapman and Hall, London. (1997).
[6] Kolmogorov A.: Sur la notion de moyenne, Rendiconti Accademia dei Lincei, vol. **12**, pages 388-391, number 6 (1930).
[7] Lusternik L. A. and Sobolev V. J.: Elements of Functional Analysis, Hindustan Publishing Corpn., Delhi. (1974).
[8] Nelsen R.B.: An introduction to copulas, Springer, London. (1999).

Chapter 22
Point-wise Kriging for Spatial Prediction of Functional Data

Pedro Delicado, Ramón Giraldo and Jorge Mateu

Abstract We propose a methodology to carry out spatial prediction when measured data are curves. Our approach is based on both the kriging predictor and the functional linear point-wise model theory. The spatial prediction of an unobserved curve is obtained as a linear combination of observed functions. We employ a solution based on basis function to estimate the functional parameters. A real data set is used to illustrate the proposals.

22.1 Introduction

Spatial prediction models for many types of data (univariate, multivariable and space-time) have been proposed. It is possible to consider other geostatistical settings in which instead of univariate, space-time or multivariate data set, the observations consist of a sample of random functions collected in different sites of a region. Many dynamic processes in environmental sciences obey smooth functional forms. An example is meteorology, where curves of climatological variables are obtained in weather stations of a country (Ramsay and Silverman, 2005). Statistical methods to model data sets based on curves are included in the term functional data analysis (FDA). Functional

Pedro Delicado
Departament d'Estadística I Investigació Operativa, Universitat Politècnica de Catalunya, Barcelona. Campus Nord Edifici C5, Despatx 214, C/ Jordi Girona 1-3, 08034, Barcelona, Spain, e-mail: pedro.delicado@upc.es

Ramón Giraldo
Universidad Nacional de Colombia, Departamento de Estadística, Ciudad Universitaria, Bogotá, Colombia, e-mail: rgiraldoh@unal.edu.co

Jorge Mateu
Departamento de Matemáticas, Universitat Jaume I, Campus Riu Sec, E-12071, Castellón, Spain, e-mail: mateu@mat.uji.es

versions for many branches of statistics have been given. Examples of such methods include exploratory analysis (Ramsay and Silverman, 2005), linear models (Cardot et al., 1999), non parametric models (Ferraty and Vieu, 2006) or multivariate techniques (Silverman, 1995; Ferraty and Vieu, 2003). Our goal in this work is to propose a geostatistical methodology useful to analyze functional data. The paper is focused on spatial prediction of functional data. We take into account the geographical coordinates of sampling points in order to estimate spatial association between observed curves and to carry out parameters estimation. The kriging predictor proposed is based on the functional linear model concurrent philosophy (Ramsay and Silverman, 2005), that is, the influence of the functional covariates on a functional response is simultaneous or point-wise.

22.2 Point-wise kriging for functional Data

Let $\left\{\chi_s(t), t \in T,\ s \in D \subset \mathbf{R}^d\right\}$ be a random field where observations are functions defined on some compact set T of $\mathbf{R}$. Assume we observe a sample of curves $\chi_{s_1}(t), \cdots, \chi_{s_n}(t)$ defined for $t \in T,\ s_i \in D, i = 1, \cdots, n$. It is usually assumed that these curves belong to a separable Hilbert space $\boldsymbol{H}$ of square integrable functions defined on T. We assume for each $t \in T$ that we have a second-order stationary and isotropic random process, that is, the mean and variance functions are constant and the covariance depends only on the distance between sampling points. Formally we assume that:

- $E(\chi_s(t)) = m(t)$, for all $t \in T, s \in D$.
- $\mathrm{Cov}(\chi_s(t), \chi_s(u)) = \sigma(t,u)$, $t, u \in T, s \in D$. If $t = u$, $V(\chi_s(t)) = \sigma^2(t)$.
- $\mathrm{Cov}(\chi_{s_i}(t), \chi_{s_j}(u)) = C(h; t, u)$, where $h = \|s_i - s_j\|$.
  If $t = u$, $\mathrm{Cov}(\chi_{s_i}(t), \chi_{s_j}(t)) = C(h; t)$.
- $\frac{1}{2}\mathrm{V}(\chi_{s_i}(t) - \chi_{s_j}(u)) = \gamma(h; t, u)$, where $h = \|s_i - s_j\|$.
  If $t = u$, $\frac{1}{2}\mathrm{V}(\chi_{s_i}(t) - \chi_{s_j}(t)) = \gamma(h; t)$.

The function $\gamma(h;t)$, as a function of h, is called semivariogram of $\chi(t)$. For a non-sample site s_0, we propose a family of linear predictors for $\chi_{s_0}(t)$, $t \in T$, given by

$$\hat{\chi}_{s_0}(t) = \sum_{i=1}^{n} \lambda_i(t)\chi_{s_i}(t),\ \ \lambda_1(t), \ldots, \lambda_n(t) : T \to \mathbf{R}. \tag{22.1}$$

For each $t \in T$, the predictor (22.1) has the same expression as an ordinary kriging predictor. This modeling approach is coherent with the functional linear concurrent model (FLCM) (Ramsay and Silverman, 2005) which the influence of each covariate on the response is *simultaneous* or *point-wise.* FLCM is defined as $Y(t) = \alpha(t) + \beta_1(t)X_1(t) + \cdots + \beta_q(t)X_q(t) + \varepsilon(t)$. In this model the response $Y(t)$ and each covariate $X_j(t), j = 1, \cdots, q$, are functions of the same argument and $X_j(t)$ only influences $Y(t)$ through its value at time t. Estimation of functional parameters $\alpha(t), \beta_j(t), j = 1, \cdots, q$, is carried out by solving

$$\underset{\alpha(\cdot),\ldots,\beta_q(\cdot)}{Min} \; E\|\hat{\boldsymbol{Y}}(t) - \boldsymbol{Y}(t)\|^2.$$

In our context the covariates are the observed curves in n sites of a region and the functional response is an unobserved function on an unsampled location. Consequently our optimization problem is

$$\underset{\lambda_1(\cdot),\ldots,\lambda_n(\cdot)}{Min} \; E\|\hat{\chi}_{s_0}(t) - \chi_{s_0}(t)\|^2$$

or equivalently by using Fubinit's Theorem

$$\underset{\lambda_1(\cdot),\ldots,\lambda_n(\cdot)}{Min} \int_T E\left(\hat{\chi}_{s_0}(t) - \chi_{s_0}(t)\right)^2 dt. \tag{22.2}$$

In a classical univariate geostatistical setting we assume that the observations are a realization of a random field $\left\{Z(s) : s \in D, D \in \mathbf{R}^d\right\}$. The kriging predictor is defined as $\sum_{i=1}^n \lambda_i Z(s_i)$ and the best linear unbiased predictor (BLUP) is obtained by minimizing $\sigma_{s_0}^2 = V(\hat{Z}(s_i) - Z(s_i))$ subject to $\sum_{i=1}^n \lambda_i = 1$. On the other hand in multivariable geostatistics (Myers, 1982; Ver Hoef and Cressie, 1993; Wackernagel, 1995) the data consist of $\{\mathbf{Z}(s_1), \cdots, \mathbf{Z}(s_n)\}$, that is, we have observations of a spatial vector-valued process $\{\mathbf{Z}(s) : s \in D\}$, where $\mathbf{Z}(s) \in \mathbf{R}^m$ and $D \subseteq \mathbf{R}^d$. In this context $V(\hat{\boldsymbol{Z}}(s_0) - \boldsymbol{Z}(s_0))$ is a matrix and the BLUP of m variables on an unsampled location s_0 is obtained by minimizing $\sigma_{s_0}^2 = \sum_{i=1}^m V\left(\hat{Z}_i(s_0) - Z_i(s_0)\right)$ subject to constraints that guarantees unbiasedness conditions, that is, minimizing the trace of the mean-squared prediction error matrix subject to some restrictions given by the unbiasedness condition (Myers,1982). The optimization problem given in (22.2) is an extension of the minimization criterium given by Myers (1982) to the functional context, by replacing the summation by an integral and the random vectors $[Z_1(s_0), \cdots, Z_m(s_0)]$ and $[\hat{Z}_1(s_0), \cdots, \hat{Z}_m(s_0)]$ by the functional variables $\chi(t)$ and $\hat{\chi}(t)$ with $t \in T$. The predictor (22.1) is unbiased if $E(\hat{\chi}_{s_0}(t)) = \mu(t)$, for all $t \in T$, that is, if $\sum_{i=1}^n \lambda_i(t) = 1$. Consequently, in order to find the BLUP, the n functional parameters in the predictor proposed are given by the solution of the following optimization problem:

$$\underset{\lambda_1(\cdot),\ldots,\lambda_n(\cdot)}{Min} \int_T V\left(\hat{\chi}_{s_0}(t) - \chi_{s_0}(t)\right) dt, \; s.t. \sum_{i=1}^n \lambda_i(t) = 1, \text{ for all } t \in T. \tag{22.3}$$

In order to solve the problem (22.3) we give a solution based on basis functions. We assume that each observed function can be expressed in terms of K basis functions by

$$\chi_{s_i}(t) = \sum_{l=1}^{K} a_{il} B_l(t) = \boldsymbol{a}_i^T \boldsymbol{B}(t),\ i = 1, \cdots, n, \tag{22.4}$$

and the functional parameters in (22.1) can be expressed by means of

$$\lambda_i(t) = \sum_{l=1}^{K} b_{il} B_l(t) = \boldsymbol{b}_i^T \boldsymbol{B}(t). \tag{22.5}$$

Then using (22.4) and (22.5) the expression (22.1) is given by

$$\begin{aligned} \hat{\chi}_{s_0}(t) &= \sum_{i=1}^{n} \boldsymbol{b}_i^T \boldsymbol{B}(t) \boldsymbol{a}_i^T \boldsymbol{B}(t) \\ &= \sum_{i=1}^{n} \boldsymbol{b}_i^T \boldsymbol{B}(t) \boldsymbol{B}^T(t) \boldsymbol{a}_i. \end{aligned} \tag{22.6}$$

Parameters estimation is carried out by solving the optimization problem (22.3) replacing (22.6) in that expression and using a coregionalization linear model (LMC) (Wackernagel, 1995) to estimate covariances between coefficients of fitted basis functions.

22.3 Example

A well-known application of FDA in an environmental context is the functional modeling of temperature and precipitation curves observed at 35 weather stations of Canada (Ramsay and Silverman, 2005). We use temperature curves of this data set to provide an applied context for our proposal. We use 45 Fourier basis functions to smooth each observed curve. Point-wise kriging using the expression (22.1) was used to predict a temperature curve on a site do not considered in the original data. This site is located in Slave Lake, Alberta, near to Edmonton station in Figure 22.1. As a first stage of the analysis a LMC was fitted to the multivariable random field composed by the coefficients of the Fourier basis used to smooth each sampled curve. Based on the LMC obtained the functional parameters $\lambda_i(t), i = 1, \cdots, 35$ were estimated (Figure 22.2). An estimated functional parameter has considerably greater magnitude than others (curve with values around 0.6). This functional parameter correspond to Edmonton (Figure 22.1), the nearest station to Slave Lake in the considered set of curves. Other stations near to

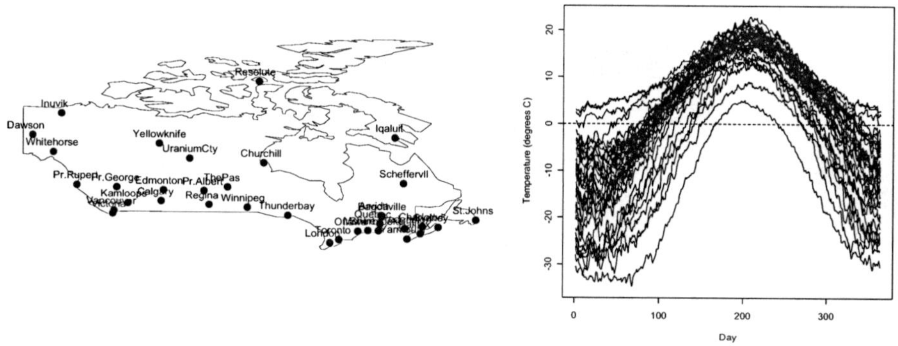

Fig. 22.1 Averages (over 30 years) of daily temperature curves (*right panel*) observed at 35 Canadian weather stations (*left panel*).

Slave Lake and consequently with great influence on the temperature prediction of this place are Yellowknife (weights around 0.2, Figure 22.2), and Uranium City and Pr George (values around 0.1, Figure 22.2). This result is coherent with the kriging philosophy, that is, sites closer to the prediction location have greater influence than others more far apart. Sum of estimated functional parameters is equal to 1 for all t (Figure 22.2). With this result we verify graphically unbiasedness constraint. A plot of the temperature prediction on Slave Lake appears in Figure 22.2 (*right panel*). From this figure it is remarkable that the predicted curve shows a seasonal behavior similar to the smoothed curves. In addition predicted values can be considered consistent with real values reported for this weather station (http://www.climate.weatheroffice.ec.gc.ca/climateData/).

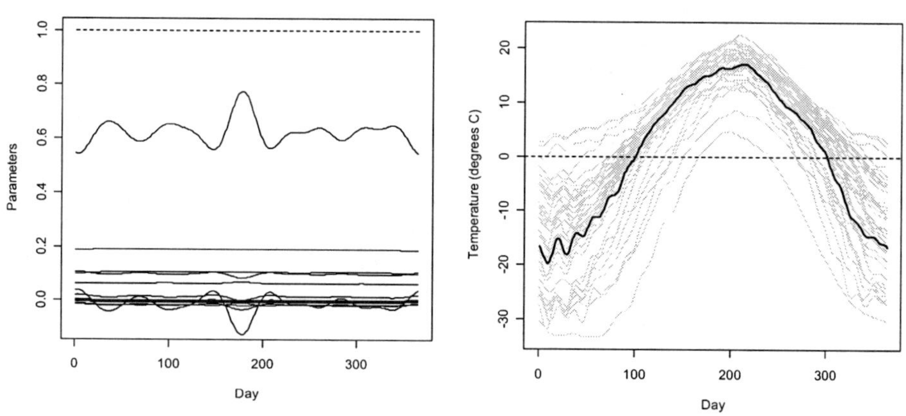

Fig. 22.2 Estimated functional parameters (*left, clear lines*), sum of functional estimated parameters (*left, dark line*), smoothed temperature curves (*right, clear lines*) and temperature prediction function on unsampled site (*right, dark line*).

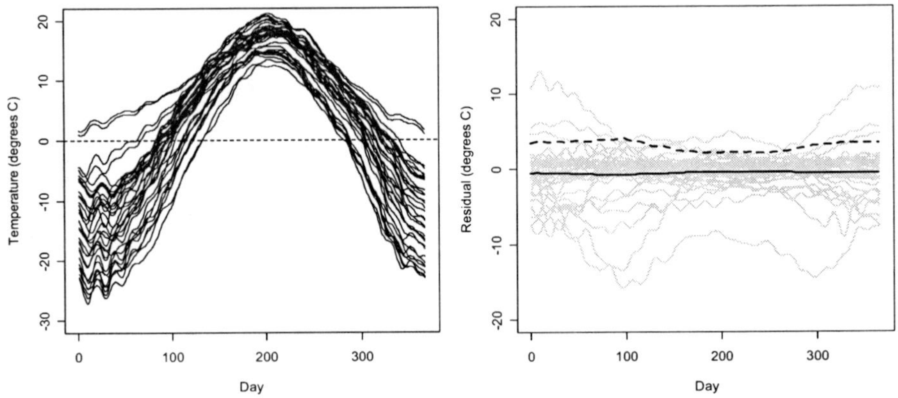

Fig. 22.3 *Right panel*: Point-wise kriging predictions based on cross-validation. *Left panel*: Cross-validation residuals (*clear lines*), residual mean (*dark line*) and residual standard deviation (*dashed line*) of Canadian temperature data set.

To verify the goodness-of-fit of the point-wise kriging prediction model, we use cross-validation methods. Each smoothed curve $\chi_{s_i}(t)$, $i = 1, \cdots, 35$ was removed, and further predicted from remaining data. A graphical comparison among observed (after smoothing) and predicted curves (Figures 22.2 and 22.3) shows that predicted curves are more smoothed than observed ones, as well as that the predicted data set has less variance specially towards both at the beginning and the end of the year, that is, in wintertime which Canadian weather is most variable (Figure 22.3). This was not surprising on the one hand because kriging is a smoothing method and on the other hand due to the considered temperature data set includes weather stations with very different temperature magnitudes (Figure 22.1). Resolute and Iqaluit stations in the Arctic (Figure 22.1) as well as Inuvik in the northwest (200 kilometers from Arctic circle), with very cold winters and short summers, have different magnitudes that other weather stations considered in our data set as some marine stations as Victoria, Vancouver or Prince Rupert in the southwest of the country.

Figure 22.3 (right panel) shows cross-validation residuals. The plot indicates reasonable or good prediction for a high proportion of places (residuals around cero). However there are some stations with large positive or negative residual curves. This is due to the fact that the temperature functions at Resolute, Inuvik, Iqaluit, Dawson, Churchill, Prince Rupert and St Johns are not well predicted by the model because of both these have extreme temperature values and are spatially very separated of remaining ones (Figure 22.1). We can also observe in Figure 22.3 that although there are some outliers, the residual mean indicates that there was not evidence of biased predictions. The residual variance is non-homogeneous through the year (Figure 22.3) as consequence of reasons above mentioned.

References

[1] Cardot, H., Ferraty, F. and Sarda, P.: Functional linear model. Statistics and Probability Letters. **45**, 11-22 (1999).

[2] Ferraty, F. and Vieu, P.: Curves discrimination. A non parametric functional approaches. Computational Statistics & Data Analysis. **44**, 161-173 (2003).

[3] Ferraty, F. and Vieu, P.: Non parametric functional data analysis. Theory and practice. New York, Springer. (2006).

[4] Myers, D.: Matrix formulation of cokriging. Mathematical Geology. **14**(3), 249-257 (1982).

[5] Ramsay, J. and Silverman, B.: Functional Data Analysis, second edition. New York, Springer. (2005).

[6] ilverman, B.: Incorporating parametric effects into functional principal components. Journal Royal Statistical Society, Series B. **57**, 673-689 (1995).

[6] Ver Hoef, J. and Cressie, N.: Multivariable spatial prediction. Mathematical Geology. **25**(2), 219-240 (1993).

[7] Wackernagel, H.: Multivariable geostatistics. An introduction with applications. Berlin, Springer-Verlag. (1995).

Chapter 23
Nonparametric Regression on Functional Variable and Structural Tests

Laurent Delsol

Abstract The aim of this talk is to highlight the usefulness of kernel methods in regression on functional variables. After reminding some asymptotic properties of the kernel estimator of the regression operator, we introduce a general framework to construct various innovative structural tests (no-effect, linearity, single-index, ...). Various bootstrap procedures are implemented on datasets in order to emphasize the pertinence of such structural testing methods.

23.1 Introduction

For many years statisticians have worked on models designed for multivariate random variables. However, the improvement of measuring apparatus provides data discretized on a thinner and thinner grid. Consequently, these data become intrinsically functional. Spectrometric curves, satellite images, annual electricity consumptions or sounds records are few examples, among others, of such intrinsically functional variables. This has led to a new statistical way of thinking in which we are interested in models where variables may belong to a functional space. To get more references on the state of art in functional statistics the reader may refer to the synthetic books, Ramsay and Silverman (2002-2005) that gather a large scope of statistical methods adapted to functional data study, while Bosq (2000) focuses on dependent functional random variables. More recently, nonparametric kernel methods have been adapted to the functional case with the ideas introduced by (18) in the context of regression on functional variables. The monography of Ferraty and Vieu (2006) gives an overview of some recent advances with nonpara-

Laurent Delsol
Institut de Mathématiques, Université de Toulouse et CNRS (U.M.R. 5219), 118 route de Narbonne, 31062 cedex 9 Toulouse, France, e-mail: delsol@cict.fr

metric kernel methods.

In this talk one focuses more precisely on a functional regression model where the response variable Y is real-valued while the explanatory variable X belongs to a functional space $\mathcal{E}$. In other words one considers the following model

$$Y = r(X) + \varepsilon \tag{23.1}$$

where the regression operator r is unknown. Such models have already been widely studied in the linear case (i.e. when r is linear) and still are topical issues (see for instance Ramsay and Dalzell (1991), Cardot *et al.* (2003) and Crambes *et al.* (2007)). In the nonparametric case (i.e. only under regularity assumptions on r) the first results come from Ferraty and Vieu (2002) in which a generalization of the well-known Nadaraya-Watson estimator is introduced. Many papers have been devoted to prove asymptotic properties of this estimator. The first part of this talk makes a short review on some of these results and presents specificities of the use of kernel methods with functional data.

In the multivariate case, many structural testing procedures have been proposed based on nonparametric methods, empirical likelihood ratio, orthogonal projection ... (see for instance Azzalini an Bowman (1993), Härdle and Mammen (1993), Härdle and Kneip (1999), Eubank (2000), Horowitz and Spokoiny (2001), Fan and Yao (2003), Chen *et al.* (2003), Chen *et al.* (2006), González-Manteiga *et al.* (2002) or Lavergne and Patilea (2007) and the references therein). In the second part of this talk we are interested in the potential use of nonparametric tools created for functional regression to extend structural testing procedures from the multivariate case to the functional case. Despite the abundant literature devoted to functional regression models and structural testing procedures in multivariate regression, there are very few papers on structural testing procedures in functional regression. As far as we know the existing literature is reducted to papers dealing with tests for no-effect (see for instance Gadiaga and Ignaccolo (2005)), tests of $\mathcal{H}_0 : \{r = r_0\}$ (where r_0 is a known operator) in the functional linear model (see for instance Cardot *et al.* (2003)) and an heuristic goodness-of-fit test (see Chiou *et al.* (2007)). There is no test to check if the regression model is linear or not, and more generally to test if the true regression operator belongs to a given family of operators. In this talk we present a general structural testing procedure adapted to functional regression that allows to check if r belongs to a given family $\mathcal{R}$. The proposed approach consists in a comparison between a general nonparametric estimator and a particular one that converges quicker under the null hypothesis. This idea is similar to the one used in Härdle and Mammen (1993) and González-Manteiga *et al.* (2002) in the multivariate case where a nonparametric and a parametric estimator are compared to check for a parametric model. This work extends the previous abundant existing literature on structural testing procedures to the case of a functional explanatory variable. Indeed, the general assumptions used through this paper allow to cover a large scope of testing procedures such as, for instance, tests for no effect, tests for linearity, tests for functional single index model, tests for dimension reduction.

23.2 Nonparametric estimation

We aim to present some asymptotic properties of the following generalisation of the Nadaraya-Watson estimator to functional data introduced in (18):

$$\hat{r}(x) = \frac{\sum_{i=1}^{n} Y_i K\left(\frac{d(X_i,x)}{h_n}\right)}{\sum_{i=1}^{n} K\left(\frac{d(X_i,x)}{h_n}\right)}, \tag{23.2}$$

where K is a kernel function, h_n the smoothing parameter, x a fixed element of $\mathcal{E}$ and d a semimetric on $\mathcal{E}$.
Before giving any asymptotic result we have to introduce the notion of small ball probability $F_x(s) := \mathbb{P}(d(X,x) \leq s)$ that has a key importance in asymptotic properties of kernel methods adapted to functional data. The quantity $F_x(h_n)$ is the equivalent, for the functional case, of the quantity $h_n^d f(x)$ (standard in the multivariate case when the density f is continuous) and do not need the existence of a positive density with regard to a specific measure. The nature of the functional variable X, of the center x and the choice of the semimetric used have a direct effect on the shape of these probabilities and hence on asymptotic properties of the kernel estimator (23.2). The use of a projection semimetric may be seen as an alternative to the curse of dimensionality.
The following results give almost complete convergence, asymptotic normality and $\mathbb{L}^q$ errors of the kernel estimator. Our contribution concerns Theorems 2 and 3 in which we explicit the function ψ_m and give explicit expressions of the constants V and B.

Theorem 1 *Under some assumptions one gets (see Ferraty and Vieu (2006) for more references):*

$$\hat{r}(x) - r(x) = O\left(h_n^{\beta}\right) + O\left(\sqrt{\frac{\log(n)}{nF_x(h_n)}}\right) \quad a.co.$$

Theorem 2 *Under some assumptions one gets (see Masry (2005), Ferraty et al. (2007) and Delsol (2008)):*

$$\frac{\sqrt{nF_x(h_n)}}{V}\left(\hat{r}(x) - r(x) - Bh_n\right) \to \mathcal{N}(0,1).$$

Theorem 3 *Under some assumptions one gets (see Dabo-Niang and Rhomari (2003), and Delsol (2007)):*

$$\mathbb{E}\left[|\hat{r}(x) - r(x)|^{2m}\right] = \sum_{k=0}^{m} \frac{V^{2k} B^{2(m-k)} (2m)!}{(2(m-k))!k!2^k} \frac{h_n^{2(m-k)}}{(nF_x(h_n))^k} + o\left(\frac{1}{(nF_x(h_n))^m}\right),$$

$$\mathbb{E}\left[|\hat{r}(x) - r(x)|^{2m+1}\right] = \frac{V^{2m+1}}{(nF_x(h_n))^{m+\frac{1}{2}}} \psi_m\left(\frac{Bh_n\sqrt{nF_x(h_n)}}{V}\right) + o\left(\frac{1}{(nF_x(h_n))^{m+\frac{1}{2}}}\right),$$

23.3 Structural tests

We are now interested in constructing a structural testing procedure. In other words, we want to check if r belongs to a given family of operators $\mathcal{R}$. We propose to check the null hypothesis $\mathcal{H}_0 : \{\exists r_0 \in \mathcal{R}, \mathbb{P}(r(X) = r_0(X)) = 1\}$ against the local alternative $\mathcal{H}_{1,n} : \left\{\inf_{r_0 \in \mathcal{R}} \|r - r_0\|_{\mathbb{L}^2(wdP_X)} \geq \eta_n\right\}$. We consider the following test statistic constructed from the ideas of Härdle and Mammen (1993) and González-Manteiga *et al.* (2002):

$$T_n^\star = \int \left(\sum_{i=1}^{n} (Y_i - r_0^\star(X_i)) K\left(\frac{d(x, X_i)}{h_n}\right)\right)^2 w(x) dP_X(x),$$

where w is a weight function with bounded support W and $r_0^\star$ is a particular estimator, depending on the family $\mathcal{R}$, constructed from a dataset $D^* := (X_i, Y_i)_{n+1 \leq i \leq N}$ independent of $D := (X_i, Y_i)_{1 \leq i \leq n}$. It is a generalisation of the statistic proposed by González-Manteiga *et al.* (2002) to the functional case. We note $m_n = N - n$ and introduce two variables that do not depend on the fact that $\mathcal{H}_0$ holds or not and provide asymptotic bias and variance terms:

$$T_{1,n} = \int \sum_{i=1}^{n} K^2\left(\frac{d(X_i, x)}{h_n}\right) \varepsilon_i^2 w(x) dP_X(x),$$

$$T_{2,n} = \int \sum_{1 \leq i \neq j \leq n} K\left(\frac{d(X_i, x)}{h_n}\right) K\left(\frac{d(X_j, x)}{h_n}\right) \varepsilon_i \varepsilon_j w(x) dP_X(x).$$

It is now possible to state the next theorem concerning the asymptotic distribution of $T_n^\star$.

Theorem 4 *Under general assumptions one gets:*

- *Under* $(\mathcal{H}_0)$, $\frac{1}{\sqrt{Var(T_{2,n})}}\left(T_n^\star - \mathbb{E}\left[T_{1,n}\right]\right) \xrightarrow{\mathcal{L}} \mathcal{N}(0,1)$,
- *Under* $(\mathcal{H}_{1,n})$, $\frac{1}{\sqrt{Var(T_{2,n})}}\left(T_n^\star - \mathbb{E}\left[T_{1,n}\right]\right) \xrightarrow{\mathcal{L}} +\infty$.

Because of the general assumptions used, our approach allows to construct various innovative structural tests:

- If $\mathcal{R} = \{r_0\}$, Theorem 4 may be applied with $r_0^* = r_0$ and $n = N$ when r_0 is Hölderian on a neighborhood of W. This complete former works Cardot *et al.* (2003) (linear functional model) and Gadiaga and Ignaccolo (2005) (with projection arguments).
- If $\mathcal{R} = \{r_0, \exists C, r_0 \equiv C\}$, our test statistic may be used with $r_0^* \equiv m_n^{-1} \sum_{i=n+1}^{N} Y_i$ if $n\Phi^{\frac{1-l}{2}}(h_n) = o(m_n)$ (where $l \in \left[0, \frac{1}{2}\right]$ is involved in one of our conditions). We get a no-effect test of the variable X on Y that completes the results established by Gadiaga and Ignaccolo (2005) in the case of a known constant C and Cardot *et al.* (2003) in the linear functional model.
- If $\mathcal{R} = \{r_0 : \mathcal{E} \to \mathbb{R},\ \text{linear}\}$, our results may be used, under some assumptions, taking for r_0^* the estimator studied in Crambes *et al.* (2007). We get the first linearity test proposed for a regression model with functional covariate.
- Let $V : \mathcal{E} \to \mathbb{R}^q$ known. If $\mathcal{R} = \{r_0, \exists \psi : \mathbb{R}^q \to \mathbb{R},\ r_0 = \psi \circ V\}$, our approach may be used, under some assumptions, taking for r_0^* the kernel estimator constructed from $(V(X_i), Y_i)_{n+1 \le i \le N}$. We get an innovative test that allows to check if the effect of an explanatory functional variable is indeed the effect of the vector $V(X)$ constituted from some features of this curve (for instance minima, maxima, inflection points,...).
- If $\mathcal{E}$ is an Hilbert space and if $\mathcal{R} = \{r_0, \exists \theta \in \mathcal{E}, \exists \psi : \mathbb{R} \to \mathbb{R},\ r_0 = \psi(<.,\theta>)\}$, the results given by Ait-Saïdi *et al.* (2008) show that in certain cases θ_{CV} may be chosen by cross-validation and we can take for r_0^* the kernel estimator constructed from $(<X_i, \theta_{CV}>, Y_i)_{n+1 \le i \le N}$.

23.4 Bootstrap procedures and simulations

Instead of computing quantiles from the asymptotic law, we propose various residual-based bootstrap procedures. We call $\hat{r}^*$ the kernel estimator constructed from the sample D^*. We propose to repeat the following procedure for b in $\{1, \ldots, N_{boot}\}$ to construct N_{boot} bootstrap values of T_n^*. We compute successively the values of:

1. Estimated residuals: $\hat{\varepsilon_i} = Y_i - \hat{r}(X_i), 1 \leq i \leq n$ and $\hat{\varepsilon_i} = Y_i - \hat{r}^*(X_i)$, $n+1 \leq i \leq N$.
2. Estimated centered residuals: $\hat{\hat{\varepsilon_i}} = \hat{\varepsilon_i} - \overline{\hat{\varepsilon}}_n, 1 \leq i \leq n$, $\hat{\hat{\varepsilon_i}} = \hat{\varepsilon_i} - \overline{\hat{\varepsilon}}_{N-n}$, $n+1 \leq i \leq N$.
3. Bootstrap residuals:
   a) Resampling: $(\tilde{\varepsilon}_i^b)_{1\leq i\leq n}$, respectively $(\tilde{\varepsilon}_i^b)_{n+1\leq i\leq N}$, are drawn with replacement from $\left\{\hat{\hat{\varepsilon_i}}, 1 \leq i \leq n\right\}$, respectively from $\left\{\hat{\hat{\varepsilon_i}}, n+1 \leq i \leq N\right\}$.
   b) Naive bootstrap: $(\tilde{\varepsilon}_i^b)_{1\leq i\leq n}$, respectively $(\tilde{\varepsilon}_i^b)_{n+1\leq i\leq N}$, are drawn from the empirical distribution of $\left(\hat{\hat{\varepsilon_i}}\right)_{1\leq i\leq n}$, respectively $\left(\hat{\hat{\varepsilon_i}}\right)_{n+1\leq i\leq N}$.
   c) Wild bootstrap: $\tilde{\varepsilon}_i^b = \hat{\hat{\varepsilon_i}} U_i, 1 \leq i \leq N$, where $(U_i)_{1\leq i\leq n}$ are i.i.d.$\sim P_W$, independent of $(X_i, Y_i)_{1\leq i\leq N}$ and fulfill $\mathbb{E}[U_1] = 0$, $\mathbb{E}\left[U_1^j\right] = 1, j = 2, 3$.
4. Bootstrap responses $\tilde{Y}_i^b = r_0^*(X_i) + \tilde{\varepsilon}_i^b, 1 \leq i \leq N$.
5. Bootstrap test statistic $\tilde{T}_n^{b*}$ computed from the sample $\left(X_i, \tilde{Y}_i^b\right)_{1\leq i\leq N}$.

Finally, if α is the level of the test, we reject assumption $\mathcal{H}_0$ if our test statistic T_n^* is greater than the value of the empirical $(1-\alpha)$-quantile of the family $\left(\tilde{T}_n^{b*}\right)_{1\leq b\leq N_{boot}}$.

We compare level and power of these bootstrap procedures on simulation studies. For instance, in the case of a no-effect test, we simulate 300 curves

$$X_i(t) = a_i cos(2\pi t) + b_i sin(3\pi t) + c_i (t - 0.45)(t - 0.75) e^{d_i t},$$

with $a_i \sim \mathcal{U}([-1;1]), b_i \sim \mathcal{N}(1;1), c_i \sim \mathcal{U}([1;5])$ and $d_i \sim \mathcal{U}([-1.5;1.5])$ and consider the following model where $\varepsilon_i \sim \mathcal{N}(0;1)$

$$Y_i = \gamma(a_i + b_i + c_i + d_i) + 2 + \varepsilon_i.$$

We split this dataset in three independent datasets of size 100. The first one corresponds to D, the second one to D^* while the third one is used to approximate the integral. In the following table we give the probabilities of rejecting the no-effect assumption for various values of γ computed on 10000 tests with $N_{boot} = 100$. R represents the empirical signal-to-noise ratio. We propose three wild bootstrap procedures constructed from three distributions P_W.

γ	Resampling	Naive Boot.	Wild Boot. 1	Wild Boot. 2	Wild Boot. 3	R
0	0.0618	0.0463	0.0614	0.0437	0.0534	1
0.2	0.1609	0.1275	0.1601	0.1133	0.1373	1.14
0.4	0.5180	0.4603	0.5437	0.4447	0.4999	1.56
0.59	0.8293	0.7806	0.8517	0.7836	0.8231	2.21
0.8	0.9557	0.9420	0.9659	0.9445	0.9591	3.23
1	0.9862	0.9775	0.9921	0.9827	0.9886	4.47

References

[1] Ait-Saïdi, A., Ferraty, F., Kassa, R. and Vieu, P.: Cross-validated estimations in the single functional index model (to appear) (2008).

[2] Azzalini, A., Bowman, A.: On the use of Nonparametric Regression for Checking Linear Relationships. J.R. Statist. Soc. B. **55,2** 549-557 (1993).

[3] Bosq, D.: Linear Processes in Function Spaces: Theory and Applications. Lecture Notes in Statistics **149** Springer-Verlag, New York. (2000).

[4] Cardot, H., Ferraty, F.,Mas, A. and Sarda, P.: Testing Hypothesys in the Functional Linear Model. Scandinavian Journal of Statistics. **30** 241-255 (2003).

[5] Cardot, H., Ferraty, F. and Sarda, P.: Spline Estimators for the Functional Linear Model. Statistica Sinica. **13** (3) 571-591 (2003).

[6] Chao, C.J.: Testing for no effect in nonparametric regression via spline smoothing technics. Ann. Inst. Statist. Math. bf 46 (2) 251-265 (1994).

[7] Chen, S.X., Hardle, W. and Li, M.: An empirical likelihood goodness-of-fit test for time series. J. R. Statist. Soc.- Series B. **65** 663-678 (2003).

[8] Chen, S.X. and Van Keilegom I.: A goodness-of-fit test for parametric and semiparametric models in mulireponse regression. Inst de Statistique, U.C.L., Discussion paper. **0616** (2006).

[9] Chiou, J.M. and Müller H.-G.: Diagnostics for functional regression via residual processes. Computational Statistics & Data Analysis. **51**, (10) 4849-4863 (2007).

[10] Crambes, C., Kneip, A. and Sarda, P.: Smoothing splines estimators for functional linear regression. submitted. (2007).

[11] Dabo-Niang, S. and Rhomari, N.: Estimation non paramétrique de la régression avec variable explicative dans un espace métrique. (French. English, French summary) [Kernel regression estimation when the regressor takes values in metric space]. C. R. Math. Acad. Sci. Paris. **336** (1) 75–80 (2003).

[12] Dauxois, J., Nkiet, G.M. and Romain, Y.: (2001) Projecteurs orthogonaux, opérateurs associés et Statistique multidimensionnelle (in french). Annales de l'ISUP. **45** (1) 31-54.

[13] Delsol, L.: Advances on asymptotic normality in nonparametric functional Time Series Analysis (submitted). (2008).

[14] Delsol, L.: Régression non-paramétrique fonctionnelle: Expressions asymptotiques des moments. Annales de l'I.S.U.P. **LI** (3) 43-67 (2007).

[15] Delsol, L., Ferraty, F., Vieu, P.: Structural test in regression on functional variables (submitted). (2008).

[16] Eubank, R.L.: Testing for No Effect by Cosine Series Methods. Scandinavian Journal of Statistics. **27**, (4) 747-763 (2000).

[17] Fan, J. and Yao, Q.: Nonlinear Time Series: Nonparametric and Parametric methods. Springer, New York (2003).

[18] Ferraty, F. and Vieu, P.: Dimension fractale et estimation de la régression dans des espaces vectoriels semi-normés. Compte Rendus de l'Académie des Sciences Paris. **330**, 403-406 (2000).

[19] Ferraty, F., Mas, A. and Vieu, P.: Advances on nonparametric regression for fonctionnal data. ANZ Journal of Statistics In print. (2007).

[20] Ferraty F. and Vieu P.: Nonparametric modelling for functional data. Springer-Verlag, New York. (2006).

[21] Gadiaga, D. and Ignaccolo, R.: Test of no-effect hypothesis by nonparametric regression. Afr. Stat. **1**, (1) 67-76 (2005).

[22] González-Manteiga, W., Quintela-del-Río, A. and Vieu, P.: A note on variable selection in nonparametric regression with dependent data. Statistics and Probability Letters. **57** 259-268 (2002).

[23] Gozalo, P.L.: A consistent model specification test for nonparametric estimation of regression function models. Econometric Theory. **9** (3) 451-477 (1993)

[24] Hall, P.: Central Limit Theorem for Integrated Square Error of Multivariate Non-parametric Density Estimators. Academic Press Inc. **0047-259X/84** (1984) .

[25] Härdle, W. and Kneip, A.: Testing a Regression Model When We Have Smooth Alternatives in Mind. Scandinavian Journal of Statistics. **26** 221-238 (1999)

[26] Härdle, W. and Mammen, E.: Comparing Nonparametric Versus Parametric Regression Fits. The Annals of Statistics. **21**, (4) 1926-1947 (1993).

[27] Hart, J.: Nonparametric Smoothing and Lack-of-fit Tests. Springer, New York. (1997).

[28] Horowitz, J.L. and Spokoiny, V.G.: An adaptative, rate-optimal test of a parametric mean-regression model against a nonparametric alternative. Econometrica. **69** 599-632 (2001).

[29] Lavergne, P. and Patilea, V.: Un test sur la significativité des variables explicatives en régression non-paramétrique. JDS Angers 2007. (2007).

[30] Masry, E.: Nonparametric regression estimation for dependent functional data: asymptotic normality Stochastic Process. Appl. **115** (1) 155-177 (2005).

[31] Ramsay, J. and Dalzell, C.: Some tools for functional data analysis. J. R. Statist. Soc. B. **53** 539-572 (1991).

[32] Ramsay, J. and Silverman, B.: Applied functional data analysis: Methods and case studies. Spinger-Verlag, New York. (2002).

[33] Ramsay, J. and Silverman, B.: Functional Data Analysis (Second Edition). Spinger-Verlag, New York. (2005).

Chapter 24
Vector Integration and Stochastic Integration in Banach Spaces

Nicolae Dinculeanu

24.1 Introduction

The Stochastic Integral $H \cdot X$ with respect to a real-valued process X, as constructed by Dellacherie and Meyer (1975-1980), is obtained by extending a linear functional from the space if simple processes to the space of all bounded predictable processes H.

This Stochastic Integral is not a genuine integral, in the sense that it is not an integral with respect to a measure. It is desirable, as in the classical Measure Theory, to have a space of integrable processes, with a norm on it for which it is a Banach space, and to have an integral for the integrable processes, which would be the Stochastic Integral. Also, desirable would be to have Vitali and Lebesgue-type convergence theorems. Such a goal is legitimate and many attempts have been made to fulfill this goal.

We present a measure-theoretic approach of the Stochastic Integral $H \cdot X$ by using a vector measure I_X associated with this process X, where both processes H and X have their values in Banach spaces. A particular case was previously considered by Kussmaul (1977) in case both processes H and X are real-valued.

In order to be able to adopt a measure-theoretic approach for the Stochastic Integral, we have to construct, first, an integration theory for vector-valued functions with respect to vector-valued measures with finite semivariation.

Nicolae Dinculeanu
University of Florida Gainesville, Florida 32611, USA, e-mail: `dinculeanunicola@bellsouth.net`

24.2 Vector integration

There are four stages in the development of the integration theory. Each stage is based on the preceding one.

24.2.1 The classical integral

The framework for this stage is a measure space $(X, \sum, \mu)$ with μ a positive measure. As it is well known, the following notions are defined: $\mu-$measurability and $\mu-$integrability of real-valued functions, the space $L^1(\mu)$ of $\mu-$integrable functions, an integral $\int |f|\mu$ for the $\mu-$integrable functions, a seminormal $||f||_1 = \int |f| d\mu$ on the space $L^1(\mu)$ for which it is complete and the space of $\sum-$step functions is dense in $L^1(\mu)$. Moreover, the Vitali and the Lebesgue convergence theorems are valid. All these features will be found in the following stages.

24.2.2 The Bochner integral

We have the same framework $(X, \sum, \mu)$ with $\mu \geq 0$, but the functions have values in a Banach space F. A function $f : X \to F$ is $\mu-$measurable if it is the pointwise limit of a sequence of $\sum-$step functions.

The function f is Bochner-integrable with respect to μ if it is $\mu-$measurable and if the function $|f|$ is $\mu-$integramble in the sense of stage 1. The space of Bouchner-integrable functions $f : X \to F$ is denoted by $L^1_F(\mu)$ and we define the seminorm $||f||_1 = \int |f| d\mu$ for $f \in L^1_F(\mu)$. The space $L^1_F(\mu)$ is complete for this seminorm and the $\sum-$step functions are deuse in $L^1_F(\mu)$. Moreover, the Vitali and the Lebesgue convergence theorems remain valid. It remains to define the Bochner integral. For a $\sum-$step function $f = \sum \varphi_{A_i} x_i$ with $A_i \in \sum$ disjoint and $x_i \in F$, we define the integral $\int f d\mu$ by

$$\int f d\mu = \sum \mu(A_i) x_i \in F.$$

For such a function we have

$$|\int f d\mu| \leq \int |f| d\mu = ||f||_1,$$

hence the linear mapping $f \mapsto \int f d\mu$, with values in F, is continuous on the subspace of step functions, for the seminorm $||f||_1$. We extend this functional by continuity to the whole space $L^1_F(\mu)$. The value of this extension for a functions $f \in L^1_F(\mu)$ is denoted by $\int f d\mu$ and is called the Bochner integral of f with respect to μ. We have $\int f d\mu \in F$.

24.2.3 Integration with respect to a vector-measure with finite variations

The framework for this stage is a measurable space $(X, \sum)$, three Banach spaces E, F, G such that $E \subset L(F, G)$ continuously and a $\sigma-$additive measure $m : \sum \to E$. We assume that m has finite variation, which is the same to assume that there is a finite, positive, $\sigma-$additive measure γ dominating m, i.e., such that $|m(A)| \leq \gamma(A)$ for $A \in \sum$. There is a smallest finite, positive, $\sigma-$additive measure dominating m, denoted by $|m|$ and called the variation of m.

Measurability, and integrability of a function $f : X \to F$ is, by definition, measurability and integrability with respect to the variation $|m|$, in the sense of stage 2. We denote by $L^1_F(m) := L^1_F(|m|)$, the space of $m-$integrable functions, and we define the seminorm $||f||_1 = \int |f| d|m|$, for $f \in L^1_F(\mu)$. The space $L^1_F(m)$ is complete, the $\sum-$step functions are dense and the Vitali and Lebesgue convergence theorems are valid.

The integral $\int f dm$ for $f \in L^1_F(m)$ is an element of G and is defined first, as in stage 2, for $\sum-$step functions and then extended by continuity to the whole space $L^1_F(m)$. We notice that $\int f dm \in G$ for $f \in L^1_F(\mu)$.

24.2.4 Integration with respect to a vector-measure with finite semivariation

This stage is more complicated, but seems to be custom-made for application to the Stochastic Integral. We have the same framework as in stage 3: a measurable space $(X, \sum)$, three Banach spaces $E \subset L(F, G)$ and a $\sigma-$additive measure $m : \sum \to E$. This measure is not necessarily with finite variation, but we associate to it a family $(m_z)_{z \in G^*}$, of vector measures $m_z : \sum \to F^*$ with finite or infinite variation $|m_z|$, in the following way: For $z \in G^*$, the measure $m_z : \sum \to F^*$ is defined for each set $A \in \sum$ by the equality

$$< x, m_z(A) > = < m(A)x, z >, \text{ for } x \in F.$$

The semivariation $\tilde{m}$ (or $\tilde{m}_{F,G}$) of m with respect to the embedding $E \subset L(F, G)$ is defined by

$$\tilde{m}(A) = \sup_{|z|\leq 1} |m_z|(A)|, \text{ for } A \in \sum.$$

We assume the semivariation $\tilde{m}$ is finite. Then all measures m_z have finite variation $|m_z|$. For each measure $m_z : \sum \to F^* = L(F, \mathbb{R})$ we apply the theory of stage 3, and we obtain a space $L^1_F(m_z)$. For each function $f \in \bigcap_{z \in G^*} L^1_F(m_z)$ we define the seminorm

$$\tilde{m}(f) = \sup_{|z|\leq 1} \int |f|d|m_z| \leq +\infty.$$

The functions f with $\tilde{m}(f) < \infty$ are called $m-$integrable functions and the space of $m-$integrable functions $f : X \to F$ is denoted by $\mathcal{F}_F(\tilde{m})$. We have, evidently $\mathcal{F}_F(\tilde{m}) \subset \bigcap_{z \in G^*} L^1_F(m_z)$ and $\tilde{m}(f)$ is a seminorm on $\mathcal{F}_F(\tilde{m})$, for which it is complete and Vitali and Lebesgue-type convergence theorems are valid.

We define now the integral for functions $f \in \mathcal{F}_F(\tilde{m})$. If $f \in \mathcal{F}_F(\tilde{m})$, then $f \in L^1_F(m_z)$ for each $z \in G^*$ and

$$|\int f dm_z| \leq \int f|d|m_z| \leq |z|\tilde{m}(f),$$

hence this mapping belongs to G^{**}. We denote this mapping by $\int f dm$ and we call it the integral of f with respect to m. We have

$$\int f dm \in G^{**},$$

$$< \int f dm.z >= \int f dm_z, \text{ for } z \in G^*,$$

and

$$|\int f dm| \leq \tilde{m}(f).$$

24.3 The stochastic integral

The framework for this section is a probability space $(\Omega, \mathcal{F}, P)$, a filtration $(\mathcal{F}_t)_{t\in R_+}$ satisfying the usual conditions and three Banach spaces $E \subset L(F, G)$. If $1 \leq p < \infty$ we denote $L^p_F = L^p_F(P)$.

We define the ring $\mathcal{R}$ of subsets of $\mathbb{R}_+ \times \Omega$ consisting of predictable rectangles of the form $\{0\} \times A$ with $A \in \mathcal{F}_0$ and $(s,t] \times A$ with $s < t$ and $A \in F_s$. The $\sigma-$algebra $\mathcal{P}$ generated by $\mathcal{R}$ is called the predictable $\sigma-$algebra. We consider a vector-valued process $X : R_+ \times \Omega \to E$. We assume that X is cadlag (i.e. continue à droite, limits à gauche), adapted (i.e. X_t is $\mathcal{F}_t-$measurable for $t \in \mathbb{R}_+$), and that $X_t \in L^p_F$ for each $t \in \mathcal{R}_+$. From the embedding $E \subset L(F,G)$ we deduce the embedding $L^p_E \subset L(F, L^p_G)$.

We associate to the process X a stochastic measure $I_X : R \to L^p_E \subset L(F, L^p_G)$, by $I_X(\{0\} \times A) = X_0 1_A \in L^p_E$, for $A \in F_0$ and $I_X((s,t] \times A) = (X_t - X_s)1_A \in L^p_E$, for $s \leq t$ and $A \in \mathcal{F}_s$. The measure I_X is additive (but not necessarily $\sigma-$additive) and does not necessarily have finite semivariation.ă

We say that the process X is $p-$summable (with respect to (F, L^p_G), if the measure I_X can be extended to a $\sigma-$additive measure $I_X : P \to L^p_E \subset L(F, L^p_G)$, with finite semivariation $(\tilde{I}_X)_{F,L^p_G}$ (or with respect to (F, L^p_G)). We assume that X is $p-$summable with respect to (F, L^p_G). We shall define the stochastic integral $H \cdot X$ with respect to X, of certain predictable processes $H : R_+ \times \Omega \to F$. The stochastic integral $H \cdot X$ is itself a process with valued in G. For this purpose we shall apply stage 4 of the preceding section, by replacing $(X, \sum, \mu)$ with $(R_+ \times \Omega, P, I_X)$ and $E \subset L(F,G)$ with $L^p_E \subset L(F, L^p_G)$.

Instead of $z \in G^*$ from stage 4, we consider here $z \in (L^p_G)^* = L^q_{G^*}$ with $\frac{1}{p} + \frac{1}{q} = 1$. For $z \in L^q_{G^*}$ we consider the measure $(I_X)_z : P \to F^* = L(F, \mathbb{R})$ with finite variation $|(I_X)_z|$, the space $L^1_F((I_X)_z)$ and the integral $\int H d(I_X)_z \in R$ for $H \in L^1((I_X)_z)$.

The measure $(I_X)_z$ satisfies the equality:

$< y, (I_X)_z(A) >= \int < I_X(A)(\omega)y, z(\omega) > dP(\omega)$, for $A \in P$ and $y \in F$, where the bracket in the integral represents the duality between G and G^*, and the bracket outside the integral represents the duality between F and F^*.

According to stage 4 of the preceding section, we define the space

$F_F(I_X) \subset \bigcap_{z \in L^q{}_G} L^1_F\Big((I_X)_z\Big)$, and the seminorm

$$\tilde{I}_X(H) = \sup \int |H| d|(I_X)_z|, \text{ for } H \in F_F(\tilde{I}_X).$$

The space $F_F(\tilde{I}_X)$ is complete for this seminorm. Moreover, we can define for each $H \in F_F(\tilde{I}_X)$ the integral $\int H dI_X \in (L^P_G)^{**}$. We notice that for each $H \in F_F(\tilde{I}_X)$ and each $t \in R_+$ we have $1_{[0,t]}H \in F_F(\tilde{I}_X)$.

Then we denote

$$\int_{[0,t]} H dI_X = \int 1_{[0,t]} H dI_X \in (L^P_G)^{**}.$$

We obtain a family

$$\left(\int_{[0,t]} H dI_X\right)_{t \in R_+}$$

of elements of $(L^p_G)^{**}$.

We are interested in processes H for which $\int_{[0,t]} HdI_X \in L^p_G$ (rather than $(L^p_G)^{**}$). In this case we denote by the same symbol the equivalence class $\int_{[0,t]} HdI_X$ in L^p_G, as well as any random variable belonging to this equivalence class. We obtain in this way a process $(\int_{[0,t]} HdI_X)_{t\in R_+}$ with values in G. This process is always adapted to the filtration $(F_t)_{t\in R_+}$ but it is not necessarily cadlag. Then, we denote by $L^1_{F,G}(X)$ the set of processes $H \in F_{F,G}(\tilde{I}_X)$ satisfying the following two conditions:

a) $\int_{[0,t]} HdI_X \in L^P_G$, for every $t \in R_+$.

b) The process $(\int_{[0,t]} HdI_X)_{t\in\mathbb{R}_+}$ has a cadlag modification.

The processes $H \in L^1_{F,G}(X)$ are said to be integrable with respect to X.

If $H \in L^1_{F,G}(X)$, then any cadlag modification of the process $(\int_{[0,t]} HdI_X)_{t\in\mathcal{R}_+}$ is called the stochastic integration of H with respect to X and is denoted by $H \cdot X$ or $\int HdX$:

$(H \cdot X)_t(\omega) = (\int HdX)_t(\omega) = (\int_{[0,t]} HdI_X)(\omega)$, *a.s.*

It follows that the stochastic integral is defined up to an evanescent process.

Examples of p−summable processes:

1.) If E and G are Hilbert spaces and $X : R_+ \times \Omega \to E \subset L(F,G)$ is a square-integrable martingale, then X is 2−summable.

2.) If $X : R_+ \times \Omega \to E$ is a cadlag, adapted process with integrable variations, then X is 1−summable for any embedding $E \subset L(F,G)$ and the stochastic integral can be computed pathwise as a Stieltjes integral: $(H \cdot X)_t(\omega) = \int_{[0,t]} H_s(\omega)dX_s(\omega)$, a.s. for $t \in R_+$.

3.) If $X : R_+ \times \Omega \to E \subset L(F,G)$ is a cadlag, adapted process with p−integrable semivariation relative to (F,G) and if c_0 does not belong to E and G, then X is p−summable and the stochastic integral can be computed pathwise as a Stieltjes integral:

$$\left(H \cdot X\right)_t(\omega) = \int_{[0,t]} H_s(\omega)dX_s(\omega), \text{ a. s. for } t \in R_+.$$

References

[1] Dellacerie, C.: and Meyer, P.A. Probabilités et Potentiel. Herman Paris (1975-1980).

[2] Dinculeanu, N.: Vector Integration and Stochastic Integration in Banach Spaces. Wiley (2000).

[3] Kussmaul, A.U.: Stochastic Integration and Generalized Martingales. Pittman, London(1977).

Chapter 25
Multivariate Functional Data Discrimination Using ICA: Analysis of Hippocampal Differences in Alzheimer's Disease

Irene Epifanio and Noelia Ventura

Abstract Recently, independent component analysis (ICA) has been successfully used for classification of univariate curves, Epifanio (2008). Extending this methodology to the multivariate functional case, an analysis of hippocampal differences in Alzheimer's disease is carried out.

25.1 Introduction

Early diagnosis of Alzheimer's disease (AD) is a topic of great importance. As more effective pharmacological therapies become available, the administration of these agents to individuals who are subtly impaired may render the treatments more effective. Mild cognitive impairment (MCI) has been proposed and commonly accepted as a diagnostic entity within the continuum of cognitive decline towards AD in old age [Grundman *et al.* 2004, Petersen, 2004]. Longitudinal studies suggest that hippocampal volume loss predicts cognitive decline [Jack, *et al.* 1999, Mungas *et al.* 2001]. Volumetric measurements are simple features, but structural changes at specific locations cannot reflected in them. If morphological changes could be established, then this should enable researchers to gain an increased understanding about condition. This explains why shape analysis has thus become of increasing interest to the neuroimaging community, Styner *et al.* (2003).

We analyse the information extracted from magnetic resonance (MR) scans in 28 subjects for three groups: controls, patients with MCI, and patients with early AD. The main objective is to understand the way in which their hip-

Irene Epifanio
Dpt. Matemàtiques, Universitat Jaume I, Castelló, SPAIN, e-mail: `epifanio@uji.es`

Noelia Ventura
Dpt. Psicologia Bàsica, Clínica i Psicobiologia, Universitat Jaume I, Castelló, SPAIN, e-mail: `venturan@uji.es`

pocampi differ. The available information is translated in a (multivariate) functional form, as explained in Section 2, and used in a functional discriminant analysis. This methodology uses ICA, and is explained in Section 3. Finally, results are presented in Section 4, together with some conclusions and future developments.

25.2 Brain MR scans processing

Twenty-eight subjects participated in this study: 12 healthy elders (five males and seven females, mean age 70.17±3.43), 6 patients with MCI (two males and four females, mean age 75.50±3.33), and 10 patients with early AD (one male and nine females, mean age 71.50±4.35). All subjects were recruited from the Neurology Service at La Magdalena Hospital and the Neuropsychology Service at the Universitat Jaume I. All experimental procedures complied with the guidelines of the ethical research committee at the Universitat Jaume I. Written informed consent was obtained from every subject or their appropriate proxy prior to participation. Selection for the participant group was made after careful neurological and neuropsychological assessment. The neuropsychological test battery involved Digit Span, Similarities, Vocabulary, and Block Design of the WAIS-III; Luria's Watches test, and Poppelreutert's Overlapping Figure test. MRI studies were performed on a 1.5T General Electric system. A whole brain high resolution 3D-Gradient Echo (FSPGR) T1-weighted anatomical reference scan was acquired (TE 4.2 ms, TR 11.3 ms, FOV 24 cm; matrix = 256×256×124, 1.4 mm-thick coronal images).

Hippocampi are traced on contiguous coronal slices (or sections) following the guidelines of Watson *et al.* (1992), and Hasboun et al. [Hasboun *et al.*, 1996]. Each hippocampus is described by around 30 slices. The hippocampus segmentation was done by a double tracer, blinded to the clinical data of the study subjects. The first tracing was done manually by an expert rater with the VOXAR program (v4.2) and the second tracing was done manually with the MRIcro software by other expert tracer, giving nearly equal segmentations. So, we consider only one of the segmentations, the second one. Total time for the segmentation of one hippocampus was approximately 40 minutes. Fig. 25.1 (a) shows an example of one coronal slice, with the right and left hippocampus drawn in white, whereas Fig. 25.1 (b) displays a sagital view of one of the hippocampus.

As aforementioned, volumen is a discriminatory feature for this problem. Therefore, we think that area could be a good descriptor for each slice. Area of right and left hippocampus in each slice is computed (it can be estimated as the number of pixels of each hippocampal segmented slice). Therefore, for each subject we have two functional data, where the argument is not time, as usual, but the space, the coronal axis. We observe the right and

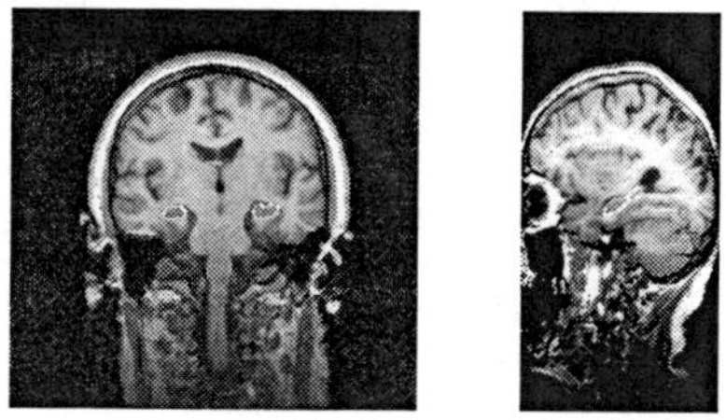

(a) (b)

Fig. 25.1 Hippocampal outlines in a coronal (a) and sagital (b) slice.

left hippocampal area only in each coronal slice (with 1.4mm of separation between them), although they are really continuous functions.

25.3 Methodology: ICA and linear discriminant analysis

Epifanio has studied recently several shape descriptors for classification of univariate functional data Epifanio (2008), where existing literature on functional discrimination is also discussed. In this paper, shape descriptors have been exhaustively compared with classical and the most recent advances in functional data classification. Coefficients of independent component analysis (ICA) components were one of them. They can be computed easily, and provide better than or similar results to those from existing techniques. Furthermore, they can be extended easily to the multivariate functional case. Although, Epifanio (2008) can be seen for details, here a brief summary is given.

Assume that we observe n linear mixtures $x_1(t)$, ..., $x_n(t)$ of n independent components $s_j(t)$,

$$x_i(t) = \sum_{j=1}^{n} a_{ij} s_j(t), \text{ for all } i. \tag{25.1}$$

In practice, we have discretized curves $(\{x_i(t_k); k = 1, ..., m\})$, therefore we can consider the $m \times n$ data matrix $X = \{x_i(t_k)\}$ to be a linear combination of independent components, i.e. $X = SA$, where columns of S contain the independent components and A is a linear mixing matrix. ICA attempts to "un-mix" the data by estimating an un-mixing matrix W where $XW = S$. Under this generative model the measured "signals" in X will tend to be "more Gaussian" than the source components (in S) due to the Central Limit Theorem. Thus, in order to extract the independent components/sources we search for an un-mixing matrix W that maximizes the nongaussianity of the sources.

We compute ICA for functions in the training set. The coefficients in this base (S) can be easily obtained by least squares fitting, Ramsay and Silverman (2005). If $y = \{y(t_k)\}_{k=1}^{m}$ is a discretized function, its coefficients are: $(S^T S)^{-1} S^T y$, where T indicates the transposed matrix. These coefficients constitute the feature vector used in the classification stage. We assume that all functions are observed at the same points. In any case, this is not a restrictive issue, since we can always fit a basis and estimate the functions at the desired points.

Before the application of the ICA algorithm, data preprocessing is useful, Hyvärinen *et al.* (2001). Smoothing of the data is useful for reducing noise. Another very useful thing to do is to reduce previously the dimension of the data by principal component analysis (PCA), thus reducing noise and preventing overlearning. Therefore, we compute the PCA first, retaining a certain number of components, and then estimate the same number of independent components as the PCA reduced dimension. FastICA algorithm, with the default parameters, is used for obtaining ICA (`http://www.cis.hut.fi/projects/ica/fastica/`).

When having multivariate functional data, we can concatenate observations of the functions into a single long vector, as done for computing bivariate functional PCA, Ramsay and Silverman (2002).

Coefficients in ICA base are used in a classical linear discriminant analysis. The number of independent components used is selected by leave-one-out cross-validation. We can also compute a linear discriminant function $\alpha(t)$ based on ICA as made in [Ramsay and Silverman, 2002, Ch. 8] with PCA. The linear discriminant values can be expressed in terms of the ICA coefficients and discriminant scores a (a vector of the same length as the number of functions in the ICA basis): $a(S^T S)^{-1} S^T X$. At the same time, we can approximate $\int \alpha(t) x_i(t) dt$ by $\sum_{k=1}^{m} \alpha(t_k) x_i(t_k)$ if we consider the separation between points as one. Therefore, we estimate $\alpha(t)$ at points t_k as $a(S^T S)^{-1} S^T$.

This methodology is applied to a known bivariate functional data: the bone shape data of [Ramsay and Silverman,2002, Ch. 8], where the best results using PCA yielded 26 errors (19 false positives and 7 false negatives). In our case, the number of errors was reduced to 20 (19 false positives and 1 false negative) using only the coefficients for one independent component.

25.4 Results of the hippocampus study

Firstly, the classical features, right and left hippocampal volumes, are computed. The misclassifications by a linear discriminant analysis and leave-one-out cross-validation are 5 if volumes of right and left hippocampi are considered. This number is increased if they are considered separately: 6 and 7 for the left and right hippocampi, respectively.

Secondly, bivariate functional data compiling areas of slices for the left and right hippocampi are considered. As the coronal length of each hippocampus is variable, for having a common axis (33 slices), we complete the raw data by adding zeros when hippocampal surface is finished. Moreover, for all subjects, the first and last slice are zero. Data are smoothed by 31 Fourier basis functions, and different smoothing parameter λ (we consider $\lambda = 0, 0.01, 1$), where the roughness penalty is the integrated squared second derivative. In order to take into account the phase variation (some hippocampi only appear in 24 slices), we carry out a registration process, applying the function *registerfd* of the package *fda*, using the minimum eigenvalue of a cross-product matrix as the continuous registration criterion and the mean function as the target function, Ramsay and Silverman (2005).

Using the methodology presented in Section 3, with λ and number of independent components chosen by leave-one-out cross-validation, the number of misclassifications is 5, with one component and $\lambda = 0.1$. This result does not improve that of the volume. We think that this because the same argument (axis) is used in the registration of the right and left hippocampi, and maybe their behaviour is not the same. Therefore, we consider areas of slices for the left and right hippocampi separately, as univariate functions, and repeat the procedure. The number of misclassifications for the left hippocampi is 3, with five components and $\lambda = 0.1$, whereas it is 6 for the right hippocampi, with two components and $\lambda = 0$. Figures 25.2 (a) and (b) display the mode of variability corresponding to the resulting $\alpha(t)$s, for the left and right hippocampi, respectively, with the vertical lines. The solid curve is the mean. The dashed, dashdotted and dotted curves represent the mean of the controls, patients with MCI, and patients with early AD, respectively. The first linear discriminant explains 95.7% and 91.9% of the variance between groups, for the left and right hippocampi, respectively.

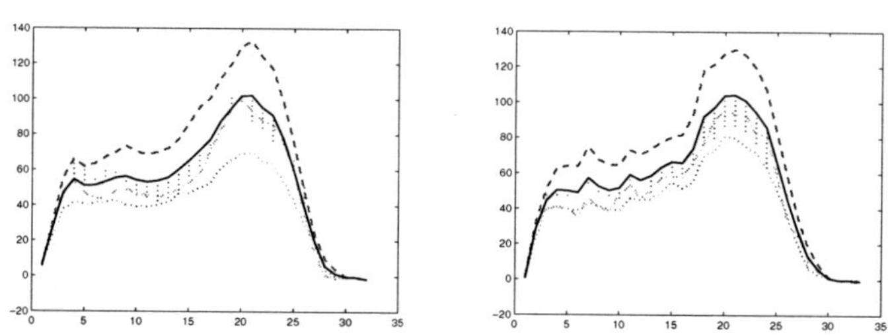

(a) (b)

Fig. 25.2 The mode of variability corresponding to $\alpha(t)$s, for the left (a) and right (b) hippocampi. See the text for details.

Both for the right and left hippocampus, the zone where the linear discriminant functions are bigger in absolute value corresponds to the head of the hippocampi (hippocampi can be divided in three parts: head, body and tail) Hasboun *et al.* (1992). This agrees with the conclusions obtained in other studies with other methodologies, Wang *et al.* (2003). We consider the point where the corresponding $\alpha(t)$s takes its maximum absolute value. Inverting the corresponding warping function, we find the left and right slices of each subject (rounding to the nearest integer) corresponding to those maximum values. In order not to base the following analysis only in one slice, we also consider the previous and subsequent slice to the determined slice. Therefore, three slices for the right and three slices for the left hippocampi are considered for each subject, corresponding to the zone of the hippocampal head indicated by the inversion of the warping functions. The mean of the areas of these slices are shown in Fig. 25.3 (b), together with the volumes for the right and left hippocampi (Fig. 25.3 (a)). We can see how it is possible to discriminate better between groups (the number of misclassifications with the hippocampal head areas is 3 by leave-one-out).

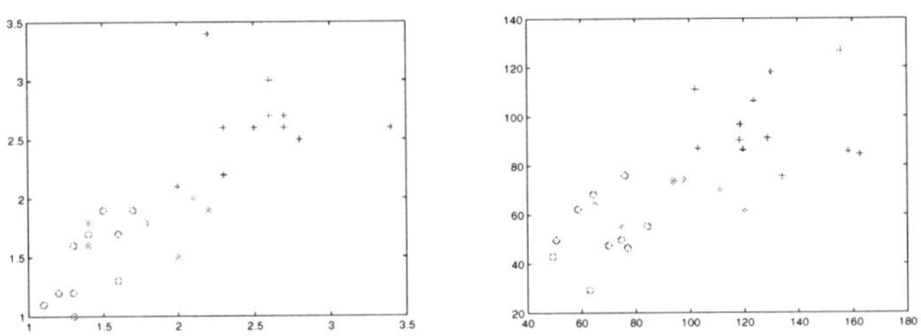

(a) (b)

Fig. 25.3 Left vs. right hippocampal volumes (a), and mean of the areas of the determined hippocampal head slices (b). Crosses, stars and circles indicate the controls, patients with MCI, and patients with early AD, respectively.

Finally, we parameterize by arc length the outlines of each of the three determined slices with 45 points. The different slices are translated to the origin in such a way its centroid coincides with the origin. The tracing begins counterclockwise in the most eastern outline point in the same row as the centroid, using *bwtraceboundary* of the image toolbox of MatLab. Twenty-five Fourier basis are used to represent these functions. Averages of the three considered slices per individual are calculated for right and left hippocampi. Therefore, we have two pairs of functions $\{X(t), Y(t)\}$ for each individual, one pair for the right and another pair for the left hippocampus, i.e., a total of four functional data per individual. Using these four functions jointly with ICA, only 2 misclassifications are achieved with 3 independent components, which are very promising results.

This point is very interesting, since if segmentation was reduced only to the hippocampal head, the segmentation time would be shorter. Furthermore, it is easier to implement an automatic segmentation only for the hippocampal head, which will decrease even more that time, and will eliminate the variability due to the subjectivity of the manual tracer.

The study should be repeated with a larger database in order to achieve valid medical conclusions, although the methodology could be used without modifications. Other point to study could be the use of ICA in other situations, such as functional logistic regression.

Acknowledgements This work is supported by CICYT MTM2005-08689-C02-02, TIN2006-10134, and Bancaixa P11B2004-15. The authors thanks V. Belloch and C. Ávila for their support.

References

[1] Epifanio, I. : Shape Descriptors for Classification of Functional Data. Technometrics, to appear. Available: http://www3.uji.es/~epifanio/RESEARCH/epifanio08.pdf. (2008).

[2] Grundman, M. et al.: Mild Cognitive Impairment can be distinguished from Alzheimer disease and normal aging for clinical trials. Arch. Neurol. **61**, 59–66 (2004).

[3] Petersen, R.C.: Mild cognitive impairment as a diagnostic entity. J. Intern. Med. **256**, 183–194 (2004).

[4] Jack, C.R. Jr. et al.: Prediction of AD with MRI-based hippocampal volume in mild cognitive impairment. Neurology. **52**, 1397–1403 (1999).

[5] Mungas, D. et al.: MRI predictors of cognition in subcortical ischemic vascular disease and Alzheimer's disease. Neurology. **57**, 2229–2235 (2001).

[6] Styner, M. et al.: Boundary and Medial Shape Analysis of the Hippocampus in Schizophrenia. Medical Image Analysis Journal.**8**(3), 197–203 (2003).

[12] Watson, C. et al.: Anatomic basis of amygdaloid and hippocampal volume measurement by magnetic resonance imaging. Neurology. **42**(9), 1743–1750 (1992).

[8] Hasboun, D. et al.: MR Determination of Hippocampal Volume: Comparison of Three Methods. AJNR Am. J. Neuroradiol. **17** 1091–1098 (1996).

[9] Ramsay, J.O. and Silverman, B.W.: Functional Data Analysis, Springer. (2005).

[10] Hyvärinen, A. et al.: Independent component analysis, Wiley. (2001).

[11] Ramsay, J.O. and Silverman, B.W.: Applied Functional Data Analysis, Springer. (2002).

[12] Wang, L. et al.: Changes in hippocampal volume and shape across time distinguish dementia of the Alzheimer type from healthy aging. NeuroImage. **20** 667–682 (2003).

Chapter 26
Influence in the Functional Linear Model with Scalar Response

Manuel Febrero, Pedro Galeano and Wenceslao González-Manteiga

Abstract This paper studies how to identify influential curves in the functional linear model in which the response is scalar and the predictor is functional and how to measure their effects on the estimation of the model and on the forecasts, when the model is estimated by the principal components method. For that, we introduce and analyze two statistics that measure the influence of each curve on the functional slope estimate of the model, which are generalizations of the measures proposed for the standard regression model by Cook (1977) and Peña (2005), respectively.

26.1 Introduction

The collection of data which consists of repeated measurements of the same subject densely taken over an ordered grid of points belonging to a finite length interval is becoming progressively frequent. Data of these characteristics are usually called functional data, because even though the recording points are really discrete, we may assume that the entire function has been completely observed. It is well known that multivariate statistical methods are not well suited for functional data for several reasons. For instance, multivariate statistical methods ignore the time correlation structure of functional

Manuel Febrero
Departamento de Estadística e Investigación Operativa Universidad de Santiago de Compostela, Spain, e-mail: mfebrero@usc.es

Pedro Galeano
Departamento de Estadística e Investigación Operativa Universidad de Santiago de Compostela, Spain, e-mail: pgaleano@usc.es

Wenceslao González-Manteiga
Departamento de Estadística e Investigación Operativa Universidad de Santiago de Compostela, Spain, e-mail: wenceslao@usc.es

data. Thus, there exists a demand for suitable procedures to analyze such data. The books of Ramsay and Silverman (2004, 2005) and Ferraty and Vieu (2006) are texts of reference and summarize several methods and case studies for handling functional data from different approaches.

In the recent literature, functional linear models in which the predictors and/or the response are of a functional nature have received considerable attention. This paper deals with the functional linear model with scalar response in which the predictor is functional. Several approaches have been proposed for estimating the functional parameter of the functional linear model with scalar response. For instance, Hastie and Mallows (1993), Marx and Eilers (1999), Cardot, Ferraty and Sarda (2003) and Ramsay and Silverman (2005) have analyzed the use of restricted basis functions and penalization methods. Ferraty and Vieu (2006) have proposed the use of nonparametric estimates based on kernels. Cardot, Ferraty and Sarda (1999) proposed a least-squares estimate based on functional principal components, which has been further analyzed in Cardot, Ferraty and Sarda (2003), Hall and Hosseini-Nasab (2006), Cai and Hall (2006) and Hall and Horowitz (2007), among others.

As any other statistical data, influential observations may be sometimes found in functional datasets. The aim of this paper is to analyze influence in the functional linear model with scalar response. In particular, we study how to identify curves with larger influence on the estimation of the functional parameter of the model and how to measure their effects on the estimation. For that, we propose two statistics that seems to be useful in detecting which curves have strong influence on the estimated slope. These statistics are the generalization to functional data of the measures proposed by Cook (1977) and Peña (2005) for the standard linear regression model. We use bootstrap methods to calibrate the distribution of these statistics, which allow us to detect the presence of influential observations. No much is known about influence in functional models. Only Shen and Xu (2007) and Chiou and Müller (2007) have introduced functional versions of the Cook distance in the case in which the predictors are real and the responses are functional, and in the case in which both the predictors and the responses are functional, respectively. Both models are different that the one considered here.

As mention previously, there are several ways to estimate the functional linear model with scalar response. The approach taken in this paper is based on the functional principal components technique, which has become very popular. Although it is well known that this estimator may be rough even for large sample sizes and alternative more smoother estimates have been proposed, we show that the estimator based on functional principal components provides a natural framework to analyze influence. Nevertheless, the results derived in this paper can be generalized to alternative smoothing estimators in a simple way.

The rest of this abstract is as follows. Section 2 presents the functional linear model with scalar response and reviews estimation based on the functional principal components. Section 3 proposes to analyze influence from a functional point of view by proposing two measures of influence which are the generalization of the measures proposed by Cook (1977) and Peña (2005) for the standard linear regression model.

26.2 The functional linear model with scalar response

The functional linear model with scalar response faces the problem of estimating the relationship between a real variable y and a square integrable random function X, defined in the same probability space. For that, let us assume that a set of pairs of the form $(X_1, y_1), \ldots, (X_n, y_n)$ is observed, where the exploratory variables, $X_1, \ldots, X_n$, and the response variables, $y_1, \ldots, y_n$, are independent and identically distributed realizations of the stochastic process X and the real variable y, respectively. Additionally, we assume that both variables are centered and that X is valued in $L^2(T)$, the separable Hilbertian space of square integrable functions defined on the closed interval $T = [a, b] \subset \Re$.

The functional linear model with scalar response assumes that the relationship between X_i and y_i is given by:

$$y_i = \langle X_i, \beta \rangle + \varepsilon_i = \int_T X_i(t)\, \beta(t)\, dt + \varepsilon_i, \tag{26.1}$$

where β is a square integrable function defined on T, $\langle \cdot, \cdot \rangle$ denotes the usual inner product on $L^2(T)$, and the errors ε_i, $i = 1, \ldots, n$, have $E[\varepsilon_i] = 0$ and constant variance $E[\varepsilon^2] = \sigma^2$, and are independent of the functions X_i.

The functional slope β is the unknown parameter of the functional linear model with scalar response (26.1) and has to be estimated from the set of pairs $(X_1, y_1), \ldots, (X_n, y_n)$. For that, we use the principal components approach of Cardot, Ferraty and Sarda (1999). The functional principal components of $X_1, \ldots, X_n$ are the orthonormal eigenfunctions of the sample covariance operator Γ_n, which maps any function x in $L^2(T)$ into another function in $L^2(T)$, as follows:

$$\Gamma_n x = \frac{1}{n} \sum_{i=1}^{n} \langle X_i, x \rangle X_i = \frac{1}{n} \sum_{i=1}^{n} \left(\int_T X_i(s)\, x(s)\, ds \right) X_i. \tag{26.2}$$

The orthonormal eigenfunctions are denoted by v_k, $k = 1, 2, \ldots$, and have associated eigenvalues $\lambda_1 \geq \lambda_2 \geq \ldots \geq \lambda_{n-1} \geq 0 = \lambda_n = \cdots$, such that $\Gamma_n v_k = \lambda_k v_k$, for $k \geq 1$, and $\langle v_k, v_l \rangle = \delta_{kl}$, the Dirac delta function such that $\delta_{kl} = 1$, for $k = l$ and $\delta_{kl} = 0$, for $k \neq l$. The eigenfunctions v_k, $k = 1, 2, \ldots$ form an orthonormal basis of the functional space $L^2(T)$. Consequently, the functional variables X_i and the slope β can be written in terms of the eigenfunctions v_k, $k = 1, 2, \ldots$, using the Karhunen-Loève expansion as follows:

$$X_i = \sum_{k=1}^{\infty} \gamma_{ik} v_k,$$
$$\beta = \sum_{k=1}^{\infty} \beta_k v_k,$$

where $\beta_k = \langle \beta, v_k \rangle$ and $\gamma_{ik} = \langle X_i, v_k \rangle$, respectively, for $k = 1, 2, \ldots$ and $i = 1, \ldots, n$. Note that, in fact, $\gamma_{ik} = 0$, for $k \geq n$, so that, model (26.1) can be rewritten as follows:

$$y_i = \langle X_i, \beta \rangle + \varepsilon_i = \left\langle \sum_{k=1}^{n} \gamma_{ik} v_k, \sum_{k=1}^{\infty} \beta_k v_k \right\rangle + \varepsilon_i = \sum_{k=1}^{n} \gamma_{ik} \beta_k + \varepsilon_i.$$

Cardot, Ferraty and Sarda (1999) proposed to estimate β by taking $\beta_k = 0$, for $k \geq k_n + 1$, where k_n is some positive integer such that $k_n < n$ and $\lambda_{k_n} > 0$, and estimating the coefficients β_k for $k = 1, \ldots, k_n$ by minimizing the residual sum of squares given by:

$$RSS\left(\beta_{(k_n)}\right) = \sum_{i=1}^{n} \left(y_i - \sum_{k=1}^{k_n} \gamma_{ik} \beta_k \right)^2 = \left\| y - \gamma_{(k_n)} \beta_{(k_n)} \right\|^2,$$

where $y = (y_1, \ldots, y_n)'$, $\beta_{(k_n)}$ is the $k_n \times 1$ vector $\beta_{(k_n)} = (\beta_1, \ldots, \beta_{k_n})'$ and $\gamma_{(k_n)}$ is the $n \times k_n$ matrix whose k-th column is the vector $\gamma_{\cdot k} = (\gamma_{1k}, \ldots, \gamma_{nk})'$, which is usually called the k-th principal component score and verifies $\widehat{var}\left(\gamma_{\cdot k}\right) = (1/n) \sum_{i=1}^{n} \gamma_{ik}^2 = \lambda_k$ and $\widehat{cov}\left(\gamma_{\cdot k}, y\right) = (1/n) \sum_{i=1}^{n} \gamma_{ik} \gamma_{il} = 0$, for $k \neq l$. Using standard arguments, the least-squares estimate of $\beta_{(k_n)}$ is given by:

$$\widehat{\beta}_{(k_n)} = \left(\gamma'_{(k_n)} \gamma_{(k_n)} \right)^{-1} \gamma'_{(k_n)} y,$$

where $\gamma_{(k_n)}{}' \gamma_{(k_n)}$ is a $k_n \times k_n$ diagonal matrix whose (k,k)-th element is $n\lambda_k$, and $\gamma_{(k_n)}{}' y$ is a $k_n \times 1$ vector whose k-th element is $n \times \widehat{cov}\left(\gamma_{\cdot k}, y\right)$. Thus, $\widehat{\beta}_{(k_n)}$ can be written as follows:

$$\widehat{\beta}_{(k_n)} = \left(\frac{\widehat{cov}\left(\gamma_{\cdot 1}, y\right)}{\lambda_1}, \ldots, \frac{\widehat{cov}\left(\gamma_{\cdot k_n}, y\right)}{\lambda_{k_n}} \right)',$$

which allows us to define the least-squares estimate of the slope β, denoted by $\widehat{\beta}$, and based on the functional principal components of $X_1, \ldots, X_n$, as follows:

$$\widehat{\beta} = \sum_{k=1}^{k_n} \widehat{\beta}_k v_k = \sum_{k=1}^{k_n} \frac{\widehat{cov}\left(\gamma_{\cdot k}, y\right)}{\lambda_k} v_k. \tag{26.3}$$

Cardot, Ferraty and Sarda (1999) showed that under several conditions the estimator $\widehat{\beta}$ converges in probability and almost surely to β. Further analysis on the asymptotic and finite sample properties of $\widehat{\beta}$ can be found in Cai and Hall (2006), Hall and Hosseini-Nasab (2006) and Hall and Horowitz (2007).

26.3 Influence measures for the functional linear model

Once that the model (26.1) has been estimated, it is necessary to asses the appropriateness of the model using the residuals of the fit. For that, note that the vector of fitted values can be written as $\widehat{y} = H_{(k_n)} y$, where:

$$H_{(k_n)} = \gamma_{(k_n)} \left(\gamma'_{(k_n)} \gamma_{(k_n)} \right)^{-1} \gamma'_{(k_n)}, \tag{26.4}$$

is called the hat matrix. Note that no matrix inversion is necessary in order to obtain $H_{(k_n)}$ because it can be written as follows:

$$H_{(k_n)} = \Gamma_{(k_n)} \Gamma'_{(k_n)}, \tag{26.5}$$

where $\Gamma_{(k_n)}$ is the $n \times k_n$ matrix whose k-th column is the vector:

$$\Gamma_{\cdot k} = \left(\gamma_{1k}/\sqrt{n\lambda_k}, \ldots, \gamma_{nk}/\sqrt{n\lambda_k} \right)' = \gamma_{\cdot k}/\sqrt{n\lambda_k}.$$

Therefore, the residuals of the fit are given by $e = y - \widehat{y} = \left(I - H_{(k_n)} \right) y$. The relationship between ε and e can be established by substituting y by its true value $\gamma_{(n)}\beta_{(n)} + \varepsilon$. This leads to:

$$e = \left(I - H_{(k_n)} \right) \left(\gamma_{(n)}\beta_{(n)} + \varepsilon \right) = \gamma_{(k_n+1:n)}\beta_{(k_n+1:n)} + \left(I - H_{(k_n)} \right) \varepsilon,$$

where $\gamma_{(k_n+1:n)}$ is the $n \times (n - k_n)$ matrix whose columns are the vectors $\gamma_{\cdot k}$, for $k = k_n + 1, \ldots, n$ and $\beta_{(k_n+1:n)} = (\beta_{k_n+1}, \ldots, \beta_n)'$. The last relationship shows that the residuals are biased. Nevertheless, as shown by Cardot, Ferraty and Sarda (2003) and Hall and Hosseini-Nasab (2006), the bias can be neglected if n is large enough and k_n has been chosen suitably. Thus, the relationship between ε and e strongly depends on the matrix $I - H_{(k_n)}$. In fact, as the vector ε has zero mean and covariance $\sigma^2 I$, then the vector e has mean $\gamma_{(k_n+1:n)}\beta_{(k_n+1:n)}$ and covariance $\sigma^2 \left(I - H_{(k_n)} \right)$. In order to overcome the heteroscedasticity of e, it is preferable to work with the internally Studentized residuals, which are defined as follows:

$$r_i = \frac{e_i}{s_R \sqrt{1 - H_{(k_n),ii}}}, \qquad i = 1, \ldots, n$$

where $H_{(k_n),ii}$ is the (i,i)-th element of the diagonal of the matrix $H_{(k_n)}$ and s_R^2 is the functional residual variance given by:

$$s_R^2 = \frac{e'e}{Trace\left(I - H_{(k_n)}\right)} = \frac{e'e}{n-k_n},$$

which attempts to estimate the variance of the error term, σ^2.

Using the the internally Studentized residuals we can carry out diagnostics on the functional linear model (26.1) such as the ones usually considered for the standard linear regression model. In this paper, we focus on the identification of influence curves. For that we consider two measures. The first one is the Cook distance introduced by Cook (1977) for determining the influence of a data point in linear regression. The functional Cook distance for the model (26.1) can be defined as follows:

$$D_i = \frac{\left(\widehat{y} - \widehat{y}_{(-i,k_n)}\right)'\left(\widehat{y} - \widehat{y}_{(-i,k_n)}\right)}{k_n s_R^2}, \tag{26.6}$$

where $\widehat{y}_{(-i,k_n)}$ is the prediction of the response vector y using the cutoff k_n and excluding the i-th observation (X_i, y_i) in the estimation. The second one is the Peña distance introduced by Peña (2005) for determining the influence of a data point in the standard linear regression model. This distance is based on determining how each point is influenced by the rest of points. Here, we adapt the measure proposed by Peña (2005) to the functional principal components estimation. For that, for each curve, we define the vector:

$$s_i = \left(\widehat{y}_i - \widehat{y}_{(-1,k_n),i}, \ldots, \widehat{y}_i - \widehat{y}_{(-n,k_n),i}\right)', \qquad i = 1, \ldots, n$$

where $\widehat{y}_i$ is the i-th component of the vector $\widehat{y}$ and $\widehat{y}_{(-i,k_n),h}$ is the h-th component of $\widehat{y}_{(-i,k_n)}$, for $h = 1, \ldots, n$. Then, the measure of the influence of the i-th curve is measured as follows:

$$S_i = \frac{s_i' s_i}{k_n s_R^2 H_{(k_n),ii}}, \qquad i = 1, \ldots, n$$

which is the squared norm of the vector s_i standardized.

The distribution of both statistics is of a complicated form. Thus, in order to determinate the presence of influential observations, we propose a smoothed bootstrap method to approximate percentiles of the distribution of both statistics. The performance of both measures and the proposed bootstrap method will be analyzed by means of several Monte Carlo experiments and will be illustrated by means of a real data example.

References

[1] Cai, T. T. and Hall, P.: Prediction in functional linear regression. Annals of Statistics. **34**, 2159-2179 (2006).

[2] Cardot, H., Ferraty, F. and Sarda, P.: Functional linear model. Statistics and Probability Letters. **45**, 11-22 (1999).

[3] Cardot, H., Ferraty, F. and Sarda, P.: Spline estimators for the functional linear model. Statistica Sinica. **13**, 571-591 (2003).

[4] Chiou, J. M. and Müller, H. G.: Diagnostics for functional regression via residual processes. Computational Statistics and Data Analysis. **51**, 4849-4863 (2007).

[5] Cook, D. R.: Detection of influential observations in linear regression. Technometrics. **19**, 15-18. (1977).

[6] Ferraty, F. and Vieu, P.: Nonparametric functional data analysis. Springer-Verlag, New York (2006)

[7] Hall, P. and Hosseini-Nasab, M.: On properties of functional principal components analysis. Journal of the Royal Statistical Society, Series B. **68**, 109-126 (2006).

[8] Hall, P. and Horowitz, J. L.: Methodology and convergence rates for functional linear regression. Annals of Statistics. **35**, 70-91 (2007).

[9] Hastie, T. and Mallows, C.: A discussion of "A statistical view of some chemometrics regression tools" by I. E. Frank and J. H. Friedman. Technometrics. **35**, 140-143 (1993).

[10] Marx, B. D. and Eilers, P. H.: Generalized linear regression on sampled signals and curves: a p-spline approach. Technometrics. **41**, 1-13 (1999).

[11] Peña, D.: A new statistic for influence in linear regression. Technometrics. **47**, 1-12 (2005).

[12] Ramsay, J. O. and Silverman, B. W.: Applied Functional Data Analysis. Springer-Verlag, New York (2004).

[13] Ramsay, J. O. and Silverman, B. W.: Functional data analysis, 2nd edition. Springer-Verlag, New York (2005).

[14] Shen, Q. and Xu, H.: Diagnostics for linear models with functional responses. Technometrics. **49**, 26-33 (2007).

Chapter 27

Is it Always Optimal to Impose Constraints on Nonparametric Functional Estimators? Some Evidence on the Smoothing Parameter Choice

Jean-Pierre Florens and Anne Vanhems

Abstract The objective of this work is to analyze the impact of imposing linear constraints on a nonparametric estimator. Such a framework is basic and has been studied many times, with various applications in Economics. Our purpose is to analyze whether or not it is worth imposing some economic linear constraint in a nonparametric setting for well-specified problems. In particular, we investigate some optimal choice for smoothing parameters.

27.1 Introduction

Taking into account structural constraints to estimate an interest parameter lies at the heart of many problems in economics. Various examples of economic restrictions can be found in the literature, like concavity, monotonicity of functions, equilibrium conditions. A general overview on how to include economic restrictions can be found in Matzkin 1994 or 2003. Nonparametric estimation and testing methods for econometric models have also been investigated widely. Imposing linear constraints like monotonicity or concavity has in particular been studied by Delecroix, Simioni and Thomas-Agnan 1996, Mammen and Thomas-Agnan 1999, Delecroix and Thomas-Agnan 2000. And more recently, Hall and Huang 2001 or Blundell and Horowitz 2004 have suggested general kernel-types estimators to impose shape constraints on a regression function.

Generally speaking, consider the estimation of a function φ satisfying some economic constraint: $A(\varphi) = 0$ or $A(\varphi) > 0$. The functional φ is usually

Jean-Pierre Florens
University of Toulouse 1 and Toulouse Business School, France, e-mail: florens@cict.fr

Anne Vanhems
University of Toulouse 1 and Toulouse Business School, France, e-mail: a.vanhems@esc-toulouse.fr

given by some economic background, like demand function, or production function and it can be defined as the best approximation of Y satisfying a structural constraint, that can be defined for example by:

- $A(\varphi)(z) = \partial_i \varphi(z) = 0$, dimension reduction
- $A(\varphi)(z) = \varphi'(z) \leq 0$, monotonicity constraint
- $A(\varphi)(z) = \varphi''(z) \leq 0$, convexity constraint
- Slutsky constraint in demand function theory
- Linear compact constraints:
  $A(\varphi) = \int \varphi(z)k(z)dz, \quad A(\varphi)(t) = \int \varphi(z)k(z\,|t)dz$

These constraints, usually given by the economic theory or identification issues (like for example in nonseparable models), have different properties: linear or nonlinear, defined through an integral or a differential operator.

However, whatever are the properties of the constraint you impose on your functional of interest, it raises several issues that may prevent the econometrician to impose it. First of all, a constrained function defined on infinite dimension space is not easily tractable and is usually discretized for computational purposes. For example, monotonicity or concavity constraints are often imposed on a fixed grid of points in order to make to constraint more tractable. Second, the operator A may depend on unknown parameters, like the unknown distribution of the dataset, and then needs to be estimated and approximated too. A third issue may be identification and overidentification problem of the estimated solution. This issue is classical in ill-posed inverse problems and the usual way to deal with it is to regularize the solution using regularization methods like Tikhonov or Landweber-Friedman regularization.

At last, the cost to pay for estimating the constraint may prevent the econometrician to impose it, if it worsens too severely the rate of convergence. The objective of this work is to analyze the impact of these three different issues in the estimation and asymptotic properties of a constrained estimator in order to answer the question of the usefulness of imposing a constraint in an nonparametric context. The kind of constraint we have in mind can be very general like Slutsly constraint in demand theory, which nonlinear, differential. Nevertheless, in this paper, we will focus on a very particular class of constraints, which are linear integral and compact. The next step will be of course to extend this restrictive class.

Estimating nonparametrically a functional under shape constraint is closely linked to the general framework of inverse problems. Therefore, we intend to investigate the usefulness of imposing a linear constraint to estimate nonparametrically a regression function in well-specified inverse problems. An inverse problem is said well-specified when the dataset is driven by the true underlying model (satisfying the model). We implicitly assume that the theoretical underlying economic model is true. Then, intuitively, imposing a constraint may not be always necessary, in particular when the initial estimator converges sufficiently quickly. In this work, we investigate some rule to choose optimally the smoothing parameters of the estimated functionals of interest. Due to the particular shape of the constraint we impose, our work is also closely linked to the estimation of projection operator, studied for example in Johannes 2005.

Our paper is organized as follows. The next section is devoted to the definition of our model, and identification and overidentification of our interest parameter. We mainly deal with linear integral constraints (in the case of equality constraint) and provide some analytic expression of the constrained solution. We then discuss the choice of the regularization method and provide some asymptotic results.

27.2 General constrained solutions

Consider a random vector $(Z,Y) \in \mathbb{R}^k \times \mathbb{R}^q$, P the probability distribution on (Z,Y), and the following model:

$$\begin{cases} Y &= m(Z) + U \\ \mathbb{E}(U \mid Z) = & 0 \end{cases} \tag{27.1}$$

Solving this problem leads to the classical solution $m(z) = \mathbb{E}(Y \mid Z = z)$. The function m is well defined on $L^2(Z)$, the Hilbert space of square integrable functions $\varphi(Z)$ with respect to P. It is frequent to impose some additional conditions that are usually given by economic context, like belonging to a subspace C of L^2_Z. C can characterize monotonic functions, convex functions, or more complex constraint like Slutsky condition. The object of interest becomes the best-approximate solution φ of (27.1) on the set C. If the subset C is convex and closed, there exists a unique solution: $\varphi = P_C(m)$ where P_C is defined as the orthogonal projector onto C.

In what follows, we focus on a particular form of constraint. The set C is characterized through some linear operator A:

$$C = \{\varphi \in L^2(Z); A\varphi = 0\}$$

Remark that C defines the Null Space of A, denoted by $N(A)$. Therefore, the interest parameter to be studied is solution of the following equation: $\varphi = P_{N(A)}(m)$. Depending on the regularity properties of A, and on whether or not A is known, we expect to derive different properties for the solution φ and its estimator. In principle, the operator A can either be linear or nonlinear, compact or noncompact, defined as an integral or a differential operator. In what follows, we will assume that:

Assumption $[A1]$: (i) A is a linear integral compact operator, from $L^2(Z)$ into L^2.

(ii) A is not injective, but its adjoint A^* is injective.

An example of constraint satisfying the previous assumption is: $A(\varphi)(t) = \int \varphi(z)k(z \mid t)dz = \mathbb{E}(\varphi(Z) \mid T = t)$ where k is the density function of $Z|T$.

We recall that the dual operator A^* is defined by the following property: $\forall (\varphi, \psi) \in L^2(Z) \times L^2, \langle A\varphi, \psi \rangle_{L^2} = \langle \varphi, A^*\psi \rangle_{L^2(Z)}$. Note that the definition of A^* is closely linked to the topology considered (this is an important issue in practice for computation).

Existence and uniqueness: Due to the properties of A, $N(A)$ is closed and convex, and there exists a unique solution to the projection problem. This solution is defined by:

$$\varphi = P_{N(A)}(m) = \left(I - (A^*A)^{\dagger} A^*A\right) m \tag{27.2}$$

where $A^{\dagger}$ is the generalized Moore-Penrose inverse of A (see Engl, Hanke and Neubauer 2000 for more details). This operator is the unique linear extension of the inverse of A restricted on $N(A)^{\perp}$.

27.3 Regularized estimated solutions

The joint distribution of (Y, Z) is unknown and needs to be estimated using dataset. Consider a *iid* sample of this random vector: $(y_i, z_i)_{i=1,\ldots,n}$. A nonparametric estimator of the conditional expectation is denoted by $\widehat{m}$. Thanks to assumption $[A1]$, any projection on the Null Space of A is uniquely defined and we consider the following estimated solution:

$$\widehat{\varphi} = P_{N(A)}(\widehat{m}) = \left(I - (A^*A)^{\dagger} A^*A\right) \widehat{m} \tag{27.3}$$

This estimator $\widehat{\varphi}$ can be very interesting to analyze as the true estimator associated to the true solution φ (defined by equation (27.2)). Both functions are the real interest parameters of the constrained model, they are perfectly well-defined, but unfortunately not easily tractable for two reasons. First, the operator (A^*A) to inverse lies in a infinite dimension space and the generalized inverse operator $(A^*A)^{\dagger}$ has to be numerically approximated. Second, the operator A itself may need to be estimated. Let consider both cases.

Inversion of $\mathbf{A^*A}$. A strategy to deal with this ill-posedness is to define a regularized operator $r_\alpha(A^*A)$ converging to $(A^*A)^{\dagger}$ as α decreases to zero. More precisely, r_α is assumed to be piecewise continuous real function defined on $[0; c]$ for $c > 0$ such that there exists a constant c with $\left|\sigma^2 r_\alpha(\sigma^2)\right| \leq c$ and $\lim_{\alpha -> 0} r_\alpha(\sigma^2) = \frac{1}{\sigma^2}$ for all $\sigma^2 \in (0; c]$. We will assume classical conditions on the regularization scheme r_α in order to prove all the results.

Various examples of regularization methods can be applied. We can think in particular of the Landweber Friedman recursive scheme:

$$\varphi_{\alpha,k} = (I - bA^*A)\varphi_{\alpha,k-1}, k = 1, ..., \frac{1}{\alpha} - 1$$

with b fixed constant, $b < \frac{1}{\|A^2\|}$. The smoothing parameter α determines the number of iterations on the algorithm. Another classical method is Tikhonov regularization: $r_\alpha(A^*A) = (\alpha I + A^*A)^{-1}$

Therefore, we define a regularized solution φ_α by:

$$\varphi_\alpha = (I - r_\alpha(A^*A)A^*A)\, m = P^\alpha_{N(A)}(m)$$

Estimation of A. Up to now, we haven't discuss the case where A is unknown, depending on the law of distribution of the dataset, or even simply numerically untractable. In both cases, we need to replace the true operator A by some estimation or approximation $\widehat{A}$. As we mentioned in introduction, the exists different kinds of constraints. Consider for example $A(\varphi)(t) = I\!E(\varphi(Z)\,|T = t)$. In this case, the operator A depends on the law of distribution of (Z, T) and a natural estimation of the constraint is to replace the conditional expectation by a kernel estimate, or series estimator. For the monotonous or convexity constraint case, the operator A is known but not easily computable. An approximation is then given by discretizing the constraint on a fixed grid of points.

Depending on whether the operator A is estimated or not, we will consider either

$$\widehat{\varphi}_\alpha = (I - r_\alpha(A^*A)A^*A)\,\widehat{m} = P^\alpha_{N(A)}(\widehat{m})$$

or

$$\widehat{\varphi}_\alpha = \left(I - r_\alpha\left(\widehat{A}^*\widehat{A}\right)\widehat{A}^*\widehat{A}\right)\widehat{m} = P^\alpha_{N(\widehat{A})}(\widehat{m})$$

These last expressions define explicit regularized estimators for our solution φ. The asymptotic properties will depend in particular on the choice of the smoothing parameter α.

27.4 Asymptotic behavior

The objective of this part is to derive some consistency and asymptotic rate results using the estimated expressions of φ derived above.

One specific feature of the model is the case of well-specified problem. It means that the underlying economic model is true and the conditional expectation automatically satisfies the constraint, that is $m \in N(A)$ and $\varphi = m$. The difficulty of the problem comes from the estimation part since the nonparametric regression may not satisfy the constraint. We will pay a particular attention to that case. Intuitively, if the model is well-specified, we should at least expect the same rate of convergence as the conditional expectation estimator or even better.

By definition,

$$\begin{aligned}\|\widehat{\varphi}-\varphi\| &= \left\|P_{N(A)}(\widehat{m}-m)\right\| \\ &\leq \|\widehat{m}-m\|\end{aligned}$$

Therefore, the "true estimator" $\widehat{\varphi}$ will always converge to φ, with a rate of convergence that is quicker or equal to the initial nonparametric rate. This rate is in principal unknown, apart from a few exceptions. We can think in particular of a projection on a finite dimension space, that will lead to a parametric rate of convergence. Other examples will lead to no gain in rates of convergence, and the true estimator will converge at the initial nonparametric rate. Generally speaking, in what follows, we will assume that: $\|\widehat{\varphi}-\varphi\|^2 = O\left(\frac{1}{n^{2a}}\right), a>0$.

By construction, $\widehat{\varphi}_\alpha$ is the regularized computable version of the "true" estimator $\widehat{\varphi}$. An important issue to deal with is to check whether or not the regularized estimation error $(\widehat{\varphi}_\alpha-\varphi)$ achieves the same rate of convergence as $(\widehat{\varphi}-\varphi)$. This implies in particular an optimal choice of the smoothing parameters α. Such an investigation could lead to practical adaptative methods for choosing α, which are crucial issues for practitioners.

We have:

$$\begin{aligned}\widehat{\varphi}_\alpha-\varphi &= \left[I-r_\alpha\left(\widehat{A}^*\widehat{A}\right)\widehat{A}^*\widehat{A}\right]\widehat{m}-\varphi \\ &= P^\alpha_{N(\widehat{A})}\widehat{m}-\varphi \\ &= \underbrace{\left[P^\alpha_{N(\widehat{A})}-P^\alpha_{N(A)}\right]\widehat{m}}+\underbrace{\left[P^\alpha_{N(A)}-P_{N(A)}\right]\widehat{m}}+\underbrace{\widehat{\varphi}-\varphi}\end{aligned}$$

However, there exists a tradeoff between these three terms, and depending on the regularization bias (second term) and the properties of $\widehat{A}$ (first term), we get different rates of convergence. In particular, we show that it may not always be useful to impose the constraint on $\widehat{m}$, it depends on the slow rate of convergence of $\widehat{A}$ to A.

At last, we give some data driven rule to select the smoothing parameter α.

References

[1] Blundell, R. and Horowitz, J.L.: Shape restrictions and endogeneity in the analysis of consumer behavior, *Preprint.* (2004).

[2] Blundell, R. and Horowitz, J.L.: A nonparametric test of exogeneity, *Review of Economic Studies, Vol 74, pp. 1035-1058*(2007).

[3] Carrasco, M., Florens J.P. and Renault, E.: *Linear inverse problems in structural econometrics estimation based on spectral decomposition and regularisation*, forthcoming Handbook of Econometrics, vol 6 (2003).

[4] Delecroix, M., Simioni, M., Thomas-Agnan, C.: Functional estimation under shape constraints, *Nonparametric Statistics, Vol 6, pp. 69-89*(1996).

[5] Delecroix, M., Thomas-Agnan, C.: *Spline and kernel regression under shape restriction*, in Smoothing and regression: approaches, computation and application, John Wiley&Sons, Inc. (2000).

[6] Engl, H.W., Hanke, M., Neubauer, A.: *Regularization of Inverse Problems*, Kluwer Academic Publishers. (2000).

[7] Florens, J-P.: *Inverse problems in structural econometrics: the example of instrumental variables*, in Advances in Economics and Econometrics: Theory and Application - Eight World Congress, ed. by M. Dewatripont, L. P. Hansen, and S. J. Turnovsky, vol. 36 of Econometric Society Monographs. Cambridge University Press. (2001).

[8] Hall, P., Huang, L-S: Nonparametric kernel regression subject to monotonicity constraints, *The Annals of Statistics, Vol29, n° 3,pp.624-647*(2001).

[9] Johannes, J.: Estimation of the null space of an operator: Application to an hedonic price model, *Preprint*(2005).

[10] Mammen, E.: Estimating a smooth monotone regression function, *Annals of Statistics, Vol19, n° 2, pp. 724-740*(1991a).

[11] Mammen, E.: Nonparametric regression under qualitative smoothness assumptions, *Annals of Statistics, Vol19, n° 2, pp. 741-759*(1991b).

[12] Mammen, E., Thomas-Agnan, C.: Smoothing splines and shape restrictions, *Scandinavian Journal of Statistics, Vol26, pp. 239-252*(1999).

[13] Matzkin, R. L.: *Restrictions of economic theory in nonparametric methods*, in Handbook of Econometrics, Vol 4, edited by R.F. Engel and D.L. McFadden. (1994).

[14] Matzkin, R. L.: Nonparametric estimation of nonadditive random functions, *Econometrica, Vol. 71, n° 5, pp. 1339-1375* (2003).

Chapter 28
Dynamic Semiparametric Factor Models in Pricing Kernels Estimation

Enzo Giacomini and Wolfgang Härdle

Abstract Dynamic semiparametric factor models (DSFMs) smooth in time and space simultaneously, approximating complex dynamic structures by basis functions and a time series of loading coefficients. In this paper DSFMs are used to estimate in a time varying approach the term structure from state price densities and pricing kernels obtained from German option data.

28.1 Introduction

Option prices are a valuable source of information concerning risk assessments from investors about future financial payoffs. The information is summarized in the state price densities (SPD), the continuous counterpart from Arrow-Debreu security prices. Under no arbitrage assumptions the state price densities q - corresponding to a risk neutral measure Q - are derived from option prices as in Breeden and Litzenberger (1978). In contrast to the state price density, the historical density p describes the random variations of the underlying price.

According to standard economic theory, risk averse investors facing financial risk have preference-indifference relations represented by a concave utility function u. Equilibrium and non-arbitrage arguments, as in Merton (1973), show that u is related to the state price and historical densities, allowing to

---

Enzo Giacomini
Humboldt-Universität zu Berlin Spandauer Str. 1 10178 Berlin, Germany, e-mail: `giacomini@wiwi.hu-berlin.de`

Wolfgang Härdle
Enzo Giacomini
Humboldt-Universität zu Berlin Spandauer Str. 1 10178 Berlin, Germany, e-mail: `haerdle@wiwi.hu-berlin.de`

conclude its functional form from q and p. Part of this relation is given by the pricing kernel (28.2).

In this paper we investigate, in a *time varying* approach, pricing kernels from DAX and ODAX data and their *term structure*. The complex dynamic structure from pricing kernels across different maturities is approximated and analysed by *dynamic semiparametric factor models* (DSFMs).

28.2 Pricing kernels

A flexible approach is to assume a complete market where the diffusion process

$$\frac{dS_t}{S_t} = \mu(S_t, t)dt + \sigma(S_t, t)dB_t$$

describes the price of a security, $t \in [0, T]$ and B_t is a standard Brownian motion defined on a probability space $(\Omega, \mathcal{F}, P)$. The arbitrage-free price at time $t \leq s \leq T$ from a payoff $\Psi(S_s)$ is given by

$$E^Q \left[e^{-r\tau}\Psi(S_s) \middle| \mathcal{F}_t \right] = E^P \left[e^{-r\tau}\Psi(S_s)\frac{\zeta_s}{\zeta_t} \middle| \mathcal{F}_t \right]$$

where r is interest rate, $\tau = s - t$ time to maturity, $\mathcal{F}_t = \sigma(S_n, 0 \leq n \leq t)$ represents the information available at t and $\zeta_t = \left.\frac{dQ}{dP}\right|_{\mathcal{F}_t}$. The pricing kernel (PK) is defined as:

$$M_{t,\tau} = e^{-r\tau}\frac{\zeta_s}{\zeta_t}. \tag{28.1}$$

We assume the existence of a representative investor with utility function u that solves the Merton optimization problem. In equilibirum the PK is path independent and equal to the marginal rate of substitution:

$$\frac{u'(S_T)}{u'(S_t)} = M_{t,\tau} = e^{-r\tau}\frac{q_t(S_T)}{p_t(S_T)}. \tag{28.2}$$

Here q_t, p_t denote the risk neutral and historical density at time t.

Breeden and Litzenberger (1978) showed how $q_t(S_T)$ may be obtained from option prices. Ait-Sahalia and Lo (1998) used the estimate:

$$\widehat{q}_t(S_T) = e^{r\tau} \left. \frac{\partial^2 C_{BS}\{S_t, K, \tau, r_t, \widehat{\sigma}_t(\kappa, \tau)\}}{\partial K^2} \right|_{K=S_T} \tag{28.3}$$

where $C_{BS}(S, K, \tau, r, \sigma) = S\Phi(d_1) - Ke^{-r\tau}\Phi(d_2)$ is the Black-Scholes price of a call option with strike K and maturity τ. Here $\Phi(x)$ is the standard normal cdf, $d_1 = \left\{\log\left(\frac{S}{K}\right) + (r + \frac{1}{2}\sigma^2)\tau\right\} / (\sigma\sqrt{\tau})$, $d_2 = d_1 - \sigma\sqrt{\tau}$ and $\widehat{\sigma}_t(\kappa, \tau)$ is a nonparametric estimator for the implied volatility at moneyness $\kappa_t = \frac{K}{S_t}e^{-r_t\tau}$ and maturity τ.

Implied volatilities may be estimated from observed option prices. On each day $t = 1, \ldots, T$ there are J_t options traded. Each intra-day trade $j = 1, \ldots, J_t$ corresponds to an implied volatility $\sigma_{t,j}$ and a pair of moneyness and maturity $X_{t,j} = (\kappa_{t,j}, \tau_{t,j})^\top$. Figure 1 depicts the implied volatilities corresponding to trades on ODAX in day 20000502 (dates are written as year, month, day).

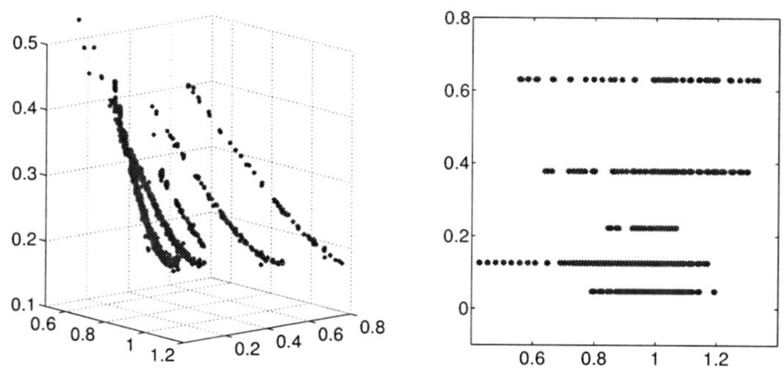

Fig. 28.1 Observed implied volatilities (left), data design (right), ODAX on 20000502

28.3 Pricing kernels estimation with DSFM

Dynamic semi-parametric factor models (DSFM), Fengler et al. (2007), employ the time series structure of implied volatilities regressing log implied volatilities $Y_{t,j} = \log \sigma_{t,j}$ on $X_{t,j}$ using smooth basis functions m_l, $l = 0, \ldots, L$ weighted with factor loadings $z_{t,l}$:

$$Y_{t,j} = \sum_{l=0}^{L} z_{t,l} m_l(X_{t,j}) + \varepsilon_{t,j} \tag{28.4}$$

where $\varepsilon_{t,j}$ is noise and $z_{t,0} \equiv 1$.

For each t the observations $\mathcal{Y}_t = \{Y_{t,j} : 1 \le j \le J_t\}$ may be viewed as discretized values from a smooth surface $\mathcal{S}_t$, therefore interest might be placed on $\mathcal{S}_t$, which has a linkage to functional data analysis. The usual approach as in Cont and da Fonseca (2002) is to obtain an estimate of $\mathcal{S}_t$ from $\mathcal{Y}_t$ and build up a factor model. DSFM avoids an initial fit of $\mathcal{S}_t$ that may suffer from design-sparseness by transfering the discrete representation directly to the functions m_l.

Following Borak et al. (2007), the basis functions are expanded using a series estimator for functions $\psi_k : \mathbb{R}^2_+ \to \mathbb{R}$, $k = 1\ldots, K$ and coefficients $\gamma_{l,k} \in \mathbb{R}$

$$m_l(X_{t,j}) = \sum_{k=1}^{K} \gamma_{l,k}\psi_k(X_{t,j}).$$

Defining the matrices $Z = (z_{t,l})$, $\Gamma = (\gamma_{l,k})$ we obtain the least square estimators as

$$(\widehat{\Gamma}, \widehat{Z}) = \arg \min_{\Gamma \in \mathcal{G}, Z \in \mathcal{Z}} \sum_{t=1}^{T} \sum_{j=1}^{J} \left\{Y_{t,j} - z_t^\top \Gamma \psi(X_{t,j})\right\}^2$$

where $z_t = (z_{t,0}, \ldots, z_{t,L})^\top$, $\psi(x) = \{\psi_1(x), \ldots, \psi_K(x)\}^\top$, $\mathcal{G} = \mathcal{M}(L+1, K)$, $\mathcal{Z} = \{Z \in \mathcal{M}(T, L+1) : z_{t,0} \equiv 1\}$ and $\mathcal{M}(a, b)$ is the set of all $(a \times b)$ matrices. The implied volatility (IV) at time t is estimated as

$$\widehat{\sigma}_t(\kappa, \tau) = exp\left\{\widehat{z}_t^\top \widehat{m}(\kappa, \tau)\right\} \tag{28.5}$$

where $\widehat{m} = (\widehat{m}_0, \ldots, \widehat{m}_L)^\top$ are the estimators for the basis functions in (28.4) with $\widehat{m}_l(x) = \widehat{\gamma}_l^\top \psi(x)$ and $\gamma_l = (\gamma_{l,1}, \ldots, \gamma_{l,K})^\top$. Using (28.3), the state price density may be approximated by

$$\widehat{q}_t(\kappa, \tau, \widehat{z}_t, \widehat{m}) = \tag{28.6}$$

$$\varphi(d_2) \left\{ \frac{1}{K\widehat{\sigma}_t\sqrt{\tau}} + \frac{2d_1}{\widehat{\sigma}_t}\frac{\partial \widehat{\sigma}_t}{\partial K} + \frac{K\sqrt{\tau}d_1 d_2}{\widehat{\sigma}_t}\left(\frac{\partial \widehat{\sigma}_t}{\partial K}\right)^2 + K\sqrt{\tau}\frac{\partial^2 \widehat{\sigma}_t}{\partial K^2} \right\}\Bigg|_{K=S_T}$$

where $\varphi(x)$ is the standard normal pdf. As in Ait-Sahalia and Lo (2000) we define an estimate $\widehat{M_t}(\kappa, \tau)$ of the PK as the ratio between the estimated SPD and the estimated p:

$$\widehat{M_t}(\kappa, \tau, \widehat{z}_t, \widehat{m}) = e^{-r_t\tau}\frac{\widehat{q}_t(\kappa, \tau, \widehat{z}_t, \widehat{m})}{\widehat{p}_t(\kappa, \tau)}. \tag{28.7}$$

It is our interest to examine the dynamic structure of (28.6) and (28.7).

28.4 Empirical results

Here IVs, SPDs and PKs are estimated from intraday DAX and ODAX data from 20010101 to 20020101 corresponding to 253 trading days. The implied volatilities are estimated with DSFM as in (28.5) with $L = 3$. The number of dynamic functions is chosen based on

$$RV(L) = \frac{\sum_{t=1}^{T} \sum_{j=1}^{J_t} \left\{ Y_{t,j} - \sum_{l=0}^{L} \widehat{z}_{t,l} \widehat{m}_l(X_{t,j}) \right\}^2}{\sum_{t=1}^{T} \sum_{j=1}^{J_t} (Y_{t,j} - \bar{Y})^2}$$

where $\bar{Y} = \frac{\sum_{t=1}^{T} \sum_{j=1}^{J_t} Y_{t,j}}{\sum_{t=1}^{T} J_t}$. The value $1 - RV(L)$ may be interpreted as the ratio of variation explained by the model to total variation. Table 28.1 shows that the addition of the fourth or fifth dynamic function results in small model fit improvement.

L	$1 - RV(L)$
1	0.772
2	0.966
3	0.978
4	0.979
5	0.978

Table 28.1 Number of dynamic basis functions and explained variation

Tensor B-splines, quadratic in τ and cubic in κ directions placed on 8×6 knots, are used for the series estimators of $\widehat{m}_l$. We note that, as in Borak et al. (2007), the order of the splines and number of knots have negligible influence on $RV(L)$. The loading factors series $\{\widehat{z}_{t,l}\}$ are depicted in Figure 28.2.

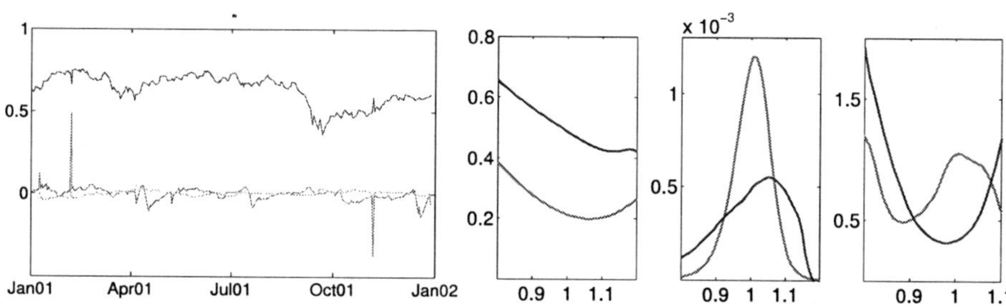

Fig. 28.2 Left: Loading factors $\widehat{z}_{t,l}$, $l = 1, 2, 3$ (top to bottom). Right: $\widehat{\sigma}_t$ (left), $\widehat{q}_t$ (middle) and $\widehat{M}_t$ (right), $\tau = 20$ days for $t = 20010824$ where $\widehat{z}_{t,1} = 0.68$ (red) and $t = 20010921$, $\widehat{z}_{t,1} = 0.36$ (blue)

The historical density $\widehat{p}_t$ is estimated with GARCH(1,1) from the last 240 observations. From (28.6) and (28.7) we obtain sequences of 253 SPDs and

PKs over a grid of moneyness and maturities. Figure 28.3 shows one shot of these sequences at day 20010710.

Risk averse utilities u are concave. Hence, (28.2) implies that under risk aversion pricing kernels are monotone decreasing in moneyness. DAX PKs are *not decreasing*, i.e. present *risk proclivity* for some levels of moneyness, hence we verify the *empirical pricing kernel paradox*.

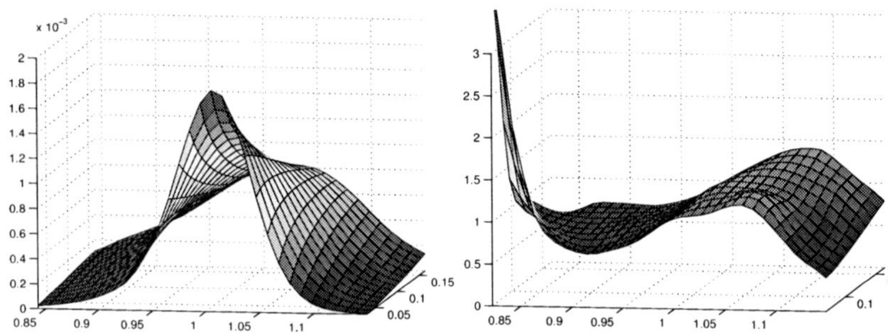

Fig. 28.3 Estimated SPD (left) and PK (right) across κ and τ at $t = 20010710$

Figure 28.2 displays the effects of large variations in $\widehat{z}_{t,1}$ around 20010911 on IV, SPD and PK. Figure 28.4 shows that skewness and excess kurtosis of $\widehat{q}_t(S_T)$ are correlated with factor loadings $\widehat{z}_{t,1}$ and $\widehat{z}_{t,3}$ for different maturities.

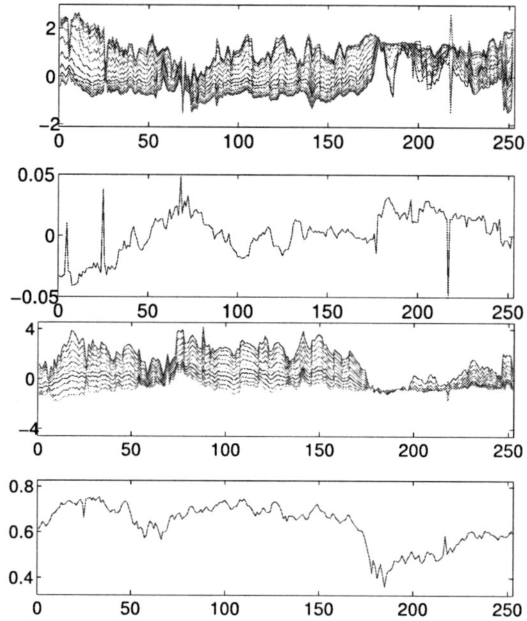

Fig. 28.4 Left: SPD skewness for $\tau = 18, (2), 50$ days (top), $\widehat{z}_{t3}$ (bottom). Right: SPD excess kurtosis for $\tau = 18, (2), 40$ days (top), $\widehat{z}_{t1}$ (bottom)

References

[1] Ait-Sahalia, Y. and Lo, A.: Nonparametric estimation of state-price densities implicit in financial asset prices. Journal of Finance. **53**, 499–547(1998).

[2] Ait-Sahalia, Y. and Lo, A.: Nonparametric risk management and implied risk aversion. Journal of Econometrics. **94**, 9–51 (2000).

[3] Borak, S., Härdle, W., Mammen, E., and Park, B.: Time series modelling with semiparametric factor dynamics. Discussion paper, SFB **649** - Humboldt-Universität zu Berlin, 2007–23 (2007).

[4] Breeden, D. and Litzenberger, R.: Prices of state-contingent claims implicit in options prices. Journal of Business. **51**, 621–651 (1978).

[5] Cont, R. and da Fonseca, J.: The Dynamics of Implied Volatility Surfaces. Quantitative Finance. **2**(1), 45–60 (2002).

[6] Fengler, M., Härdle, W., and Mammen, E.: A semiparametric factor model for implied volatility surface dynamics. Journal of Financial Econometrics. **5**, 189–218 (2007).

[7] Merton, R.: Theory of rational option pricing. The Bell Journal of Economics and Management Science. **4**, 141–183 (1973).

Chapter 29
The Operator Trigonometry in Statistics

Karl Gustafson

Abstract An operator trigonometry developed chiefly by this author during the past 40 years has interesting applications to statistics, and provides new geometrical understandings of statistical efficiency, canonical correlations, and other statistical bounds and inequalities.

29.1 The origins of the operator trigonometry

The operator trigonometry was created in 1967–1969 by this author. The original motivating application was a question about multiplicative perturbation of contraction semigroup generators. Specifically, if A is a generator, when is BA a generator? For example, B may represent a time change operator when A generates a Markov process.

In retrospect, it was good fortune that this question naturally brought out two entities which became cornerstones of the general operator trigonometry:

$$\inf_{0\neq x\in D(A)} \mathrm{Re}\frac{\langle Ax, x\rangle}{||Ax||||x||} \equiv \mu_1(A), \tag{1}$$

and

$$\inf_{-\infty<\varepsilon<\infty} ||\varepsilon B - I|| \equiv \nu_1(B). \tag{2}$$

In (1) A is an arbitrary strongly accretive densely defined closed operator in a Banach X : $\mathrm{Re}\langle Ax, x\rangle \geqq m||x||^2$, $m > 0$. Here the notation $\langle y, x\rangle$ denotes a semi-inner product on the space X. In (2) B is an arbitrary strongly accretive bounded operator on X. Given A the infinitesimal generator of a strongly continuous contraction semigroup, one knows by the Hille-Yosida-Phillips-

Karl Gustafson
Colorado University, USA, e-mail: gustafs@euclid.colorado.edu

Lumer theory that A is dissipative, i.e., $-A$ is accretive, $\text{Re}\langle -Ax, x\rangle \geqq 0$, and $\mu_1(-A) \geqq 0$. For all accretive multiplicative perturbing bounded operators B, one knows from convexity properties of operator norms on $B(X)$ that $\nu_1(B) \leqq 1$. When A is strongly dissipative, i.e., $-A$ is strongly accretive, and B is strongly accretive, then the result I obtained was that the multiplicative perturbation BA still generates a contraction semigroup, if and only if,

$$\nu_1(B) \leqq \mu_1(-A). \tag{3}$$

29.2 The essentials of the operator trigonometry

In the following, let us for convenience, in discussing the operator trigonometry, assume that A and B are both strongly accretive bounded operators in $B(X)$. Then both entities in (1), (2), (3) are positive real numbers $\nu_1(B)$ and $\mu_1(A)$ strictly between 0 and 1. Notice that we have changed the sign on A so that for the general operator trigonometry as we describe it here, we will always be in the context of both A and B strongly accretive bounded operators. Also, although the original question was in a context of Banach spaces and semi-inner products, we will also here specialize to the case of Hilbert spaces X and the usual inner product $\langle y, x\rangle$.

Intuition led me in 1967 to interpret (1) as geometrically characterizing the largest angle that A could turn a vector x to Ax. I called this angle $\phi(A)$, the operator angle of A. This intuition is readily seen as that coming from the Schwarz inequality. Motivated by the Rayleigh Ritz variational theory of eigenvalues, in 1969 I decided to call μ_1 by a similar name, specifically:

$$\mu_1(A) \equiv \cos\phi(A) \equiv \text{the first antieigenvalue of } A. \tag{4}$$

To amplify why I coined the term antieigenvalue, assume for the moment that the infimum in (1) is attained by some vector x. I called such x a corresponding first antieigenvector of A. Those are the vectors most turned by A, in contradistinction to eigenvectors, which are not turned at all.

It seemed to me that it would be nice if (3) could be made completely trigonometric. To achieve that caused me in 1968 to obtain an important result, which I called the Min-Max Theorem, reflecting how I proved it. The result of this Theorem is that (2) became trigonometric:

$$\nu_1(B) \equiv \sin\phi(B) = (1 - \cos^2\phi(B))^{1/2}. \tag{5}$$

Then the requirement (3) becomes fully trigonometric:

$$\sin\phi(B) \leqq \cos\phi(A). \tag{6}$$

In the case that A and B are selfadjoint positive definite bounded operators on a Hilbert space, I found that

$$\cos\phi(A) = \frac{2\sqrt{mM}}{m+M}, \quad \sin\phi(A) = \frac{M-m}{M+m}, \tag{7}$$

where m and M are the minima and maxima of the spectrum $\sigma(A)$. For (A) a finite dimensional symmetric positive definite $n \times n$ matrix with eigenvalues $0 < \lambda_1 \leqq \lambda_2 \leqq \cdots \leqq \lambda_n$, the expressions in (7) become:

$$\cos\phi(A) = \frac{2\sqrt{\lambda_1\lambda_n}}{\lambda_n+\lambda_1}, \quad \sin\phi(A) = \frac{\lambda_n-\lambda_1}{\lambda_n+\lambda_1}. \tag{8}$$

In that case there are exactly two antieigenvectors:

$$x_\pm = \pm\left(\frac{\lambda_n}{\lambda_1+\lambda_n}\right)^{1/2} x_1 + \left(\frac{\lambda_1}{\lambda_1+\lambda_n}\right)^{1/2} x_n, \tag{9}$$

where x_1 and x_n are norm-one eigenvectors from the eigenspaces corresponding to λ_1 and λ_n, respectively.

29.3 The operator trigonometry in statistics

Skipping details of the intervening history of the operator trigonometry from 1969 to 1999, I had early surmised that there would be applications to statistics. I even mentioned this in my antieigenvalues lecture in 1969 at the Third Symposium on Inequalities at UCLA, which became the 1972 paper Gustafson (1972). In that paper I allude to the fact that μ_1 is a ratio of moments. However, in the early 1970's my interests turned principally to quantum scattering theory, and in the 1980's, to computational fluid dynamics. Only in the 1990's did I return to further systematic development of the operator trigonometry, and applications, for example, to numerical linear algebra and quantum mechanics. In 1999 I wrote the paper Gustafson (1999), which connects the operator trigonometry to statistics, and sent it to a journal not well known to me. There one of the two referees peremptorially rejected the paper. Shortly thereafter my ideas and connections in Gustafson (1999) surfaced as of considerable interest within the matrix statistics community. I announced the main results of Gustafson (1999) in my survey of the operator trigonometry given at the Second International Conference on Unconventional Models of Computation in Brussels in December 2000, which became the paper Gustafson (2001). I also mentioned the results for statistics in the paper Gustafson (1999), submitted in 1999. Those fundamental connections between my operator trigonometry and statistics established in

Gustafson (1999) finally appeared in the paper Gustafson (2004). See also my later papers Gustafson (2005-2007a) for further applications to statistics.

29.4 Operator trigonometry in general

For more background on the operator trigonometry and its applications in other domains, I suggest the two books Gustafson (1997) and Gustafson *et al.*(1997) and my recent review article Gustafson (2006). In particular, the latter contains more than 60 citations to work on the operator trigonometry by me, by my two former Ph.D. students D.K.M. Rao and M. Seddighin, and 40 citations to related work by others. One might also see the recent paper Gustafson (2007b).

29.5 Conclusions

In this lecture, I will first present the essentials of the operator trigonometry, giving additional insights. Then I will describe and come up-to-date on the ideas, connections, and applications to statistics in Gustafson (2005-2007a). These include the new geometrical understandings of statistical efficiency, canonical correlations, Hotelling correlations, and other statistical bounds and inequalities due to Durbin, Watson, Bloomfield, Knott, Khatri, Rao, Ando, Styan, Puntanen, Drury, Liu, Lu, Bartmann, Eaton, and further back, Hotelling and Von Neumann, among others.

References

[1] Gustafson, K.: Antieigenvalue Inequalities in Operator Theory, Inequalities III (O. Shisha, ed.), Academic, 115–119 (1972).

[2] Gustafson, K.: On Geometry of Statistical Efficiency. (1999, preprint).

[3] Gustafson, K.: An Unconventional Linear Algebra: Operator Trigonometry, in Unconventional Models of Computation (I. Antoniou, C. Calude, M. Dinneen, eds.), Springer, 48–67 (2001).

[4] Gustafson, K.: An Extended Operator Trigonometry. LAA **319**, 117–135 (2000).

[5] Gustafson, K.: Operator Trigonometry of Statistics and Econometrics. LAA. **354**, 151–158 (2004).

[6] Gustafson, K.: The Geometry of Statistical Efficiency, Research Letters Inf. Math. Sci. **8**, 105–121 (2005).

[7] Gustafson, K.: The Trigonometry of Matrix Statistics, International Statistical Review. **74**, 187–202 (2006).

[8] Gustafson, K.: The Geometry of Statistical Efficiency and Matrix Statistics, J. of Applied Math. and Decision Sciences, to appear. (2007a).
[9] Gustafson, K.: Lectures on Computational Fluid Dynamics, Mathematical Physics, and Linear Algebra, World Scientific, Singapore. (1997).
[10] Gustafson, K. and Rao, D.K.M.: Numerical Range, Springer. (1997).
[11] Gustafson, K: Noncommutative Trigonometry, Operator Theory: Advances and Applications. **1967**, 127–155 (2006).
[12] Gustafson, K: Noncommutative Trigonometry and Quantum Mechanics, in Advances in Deterministic and Stochastic Analysis (N. Chuong, P. Ciarlet, P. Lax, D. Mumford, D. Phong, eds.), World Scientific, 341–360 (2007b).

Chapter 30
Selecting and Ordering Components in Functional-Data Linear Prediction

Peter Hall

Abstract For some fifty years the problem of basis choice, in linear prediction problems based on high-dimensional data, has been under discussion. From some viewpoints the debate is no closer to resolution today than in the past. We shall discuss the issues involved, describe theoretical results which shed light on the debate, and introduce methodology that is appropriate in cases where non-standard techniques can be effective.

30.1 Introduction

The problem of component choice in regression-based prediction has a long history. The main cases where important choices have to be made are functional data analysis, or FDA, and problems in which the explanatory variables are relatively high-dimensional vectors, for example when sample size is smaller than dimension. The setting of FDA is arguably the most prominent; principal component analysis has become a common method for prediction in functional linear regression.

In the functional-data context the number of components can also be interpreted as a smoothing parameter, and so the viewpoint is a little different from that for conventional linear regression. However, arguments for and against different component-choice methods are relevant to both settings, and have received significant recent attention. We shall discuss a theoretical result, applicable in a variety of settings, which to some extent justifies the standard approach. Although the result is of minimax type, it is not asymptotic in nature; it holds for each sample size.

Peter Hall
Department of Mathematics and Statistics, The University of Melbourne Melbourne, VIC 3010, AUSTRALIA, e-mail: `p.hall@ms.unimelb.edu.au`

Nevertheless, there are clearly instances where prediction bases that are alternative to the conventional one, are beneficial. For example, this can occur when the explanatory random function, x say, is atypical of most of the functions that are actually observed. We shall discuss these issues, and suggest methodology for tackling them.

30.2 Linear prediction in a general setting

Suppose we observe independent and identically distributed data pairs $(X_1, Y_1), \ldots, (X_n, Y_n)$, generated as (X, Y) by the model

$$Y = \alpha + \langle \beta, X \rangle + \varepsilon \,. \tag{1}$$

Here Y, α and ε are scalars, X and β are vectors or functions in a space $\mathcal{F}$, $\langle \cdot, \cdot \rangle$ denotes an inner product on $\mathcal{F}$, and the experimental error ε has zero mean and finite variance and is independent of the explanatory variable X. For a given value x of X, we wish to estimate the conditional mean of Y, given that $X = x$:

$$\mu(x) = E(Y \mid X = x) = \alpha + \langle \beta, x \rangle \,.$$

Of course, the case of FDA is of greatest interest to us. There, β and X are both functions defined on a region $\mathcal{R}$, say, and the inner product represents an integral:

$$\langle \beta, X \rangle = \int_{\mathcal{R}} \beta(t)\, X(t)\, dt \,. \tag{2}$$

However, the problem we are addressing arose in relatively conventional linear regression long before FDA came on the scene. In k-variate linear regression, $\beta = (\beta^{(1)}, \ldots, \beta^{(k)})\mathrm{T}$ and $X = (X^{(1)}, \ldots, X^{(k)})\mathrm{T}$ are both vectors, and the inner product in (1) is given by vector multiplication: $\langle \beta, X \rangle = \beta \mathrm{T} X$.

A conventional approach to estimating $\mu(x)$ is based on estimators $\hat{\phi}_1, \hat{\phi}_2, \ldots$ of the respective orthonormal eigenvectors $\phi_1, \phi_2, \ldots$ that arise in the canonical decomposition of the covariance function $K(s,t) = \mathrm{cov}\{X(s), X(t)\}$, of X. (In the latter formula, and below, the value of t in $X(t)$ refers to the component index of X if X is a vector, and to the argument of X if X is a function.) The quantities $\hat{\phi}_1, \hat{\phi}_2, \ldots$ are generally ordered in terms of an empirical measure of their importance, by asking that the respective eigenvalue estimators $\hat{\theta}_j$ form a decreasing sequence:

$$\hat{\theta}_1 \geq \hat{\theta}_2 \geq \ldots \,. \tag{3}$$

In very high-dimensional settings only the first few of the eigenvectors $\hat{\phi}_j$ are employed for prediction. Therefore, ordering according to (3), and using a technique such as cross-validation to choose a frequency cut-off, leads to a method for dimension reduction.

30.3 Arguments for and against the principal component basis

The arguments summarised in the previous section underpin the principal-component approach to selecting, and ordering, regressor variables. They go back at least to work of Kendall (1957, p. 75). Ordering in the way prescribed by (3) is a reasonable first choice, but it is open to question because it is based solely on the data X_i. In particular, it takes no account of x in the function $\mu(x)$. Surely the methodology should at least consider x; for example, it should take x into account when ranking the eigenvectors $\hat{\phi}_j$ for the purpose of estimating $\mu(x)$. It might also be appropriate to estimate $\mu(x)$ using a choice of orthonormal eigenvectors different from $\hat{\phi}_1, \hat{\phi}_2, \ldots$, attuned to both x and the sequence of X_i's.

The queries raised in the previous paragraph have, of course, a history. Cox (1968, p. 272) argued that x should be taken into account, noting that:

> A difficulty [with the conventional approach] seems to be that there is no logical reason why the dependent variable should not be closely tied to the least important principal component.

Cox then went on to consider an alternative approach, in which "simple combinations, not the principal components, can be used as regressor variables."

Mosteller and Tukey (1977, p. 397) argued that this type of approach might be unnecessarily pessimistic, since it took the unreasonable view that nature connived to render principal components contrary:

> A malicious person who knew our x's and our plan for them could always invent a y to make our choices look horrible. But we don't believe nature works that way — more nearly that nature is, as Einstein put it (in German), "tricky, but not downright mean." And so we offer a technique that frequently helps...

Mosteller and Tukey went on to discuss principal component methods, and endorsed the component ranking determined by (3) to the extent that "separating big components from little components can be effective." However, they also tacitly acknowledged Cox's (1968) alternative approach (without mentioning the paper), by suggesting, among other things, that a statistician could "choose new linear combinations of the few largest principal components so as to make them more interpretable."

Kendall's (1957) relatively rudimentary approach remains very popular today, particularly in functional data analysis. The approach has long had adherents, including Spurrell (1963) and Hocking (1976).

A number of these issues, and others, are addressed at greater length by Cook (2007), in the paper on which his 2005 Fisher Lecture was based; and by the paper's discussants, Christensen (2007), B. Li (2007) and L. Li and Nachtsheim (2007). In particular, Cook (2007), like us, juxtaposes the quotations above from Cox (1968) and Mosteller and Tukey (1977). He argues that Hotelling (1957) and Hawkins and Fatti (1984) expressed a viewpoint in sympathy with that of Cox, but that remarks of Fisher (1924) can be interpreted as favouring the position of Mosteller and Tukey. (However, Cook (2007) does not mention Mosteller and Tukey's (1977) subsequent comments, which lend support to Cox's (1968) position.)

30.4 Theoretical argument in support of the ordering (3) of the principal component basis

Let $\mathcal{X} = \{X_1, \ldots, X_n\}$ denote the set of explanatory variables in the dataset introduced in section 2. Take $1 \leq p \leq n$, and take $\psi_1, \ldots, \psi_p$ to be any subsequence of $\hat{\phi}_1, \ldots, \hat{\phi}_n$. (The quantities $\psi_{p+1}, \psi_{p+2}, \ldots$ can be defined arbitrarily, subject to $\psi_1, \psi_2, \ldots$ forming a complete orthonormal basis.) In particular, $\psi_j = \hat{\phi}_{r_j}$, say, for $1 \leq j \leq p$, where $r_1, \ldots, r_p$ are distinct members of the set $\{1, \ldots, n\}$. Define (conditional) mean squared error by:

$$\mathrm{MSE} = E\big[\{\hat{\mu}(x) - \mu(x)\}^2 \mid \mathcal{X}\big] .$$

Then the following minimax property can be established.

Theorem 1. *If the only information we have about β is that its norm, $\|\beta\| = \langle \beta, \beta \rangle^{1/2}$, equals a given constant C (or alternatively, that $\|\beta\| \leq C$), then the choice of $r_1, \ldots, r_p$ that produces, for each value of C and p, the least mean squared error when this quantity assumes the largest value that it is permitted for the class of slopes β satisfying $\|\beta\| = C$ (or, respectively, $\|\beta\| \leq C$), is $r_j = j$ for each j. This is identical to the choice that orders the eigenvectors $\hat{\phi}_j$ canonically by insisting on a decreasing ranking of eigenvalues, as at* (3).

If we have additional information about β, over and above that permitted in Theorem 1, then we may be able to reduce mean squared error by re-ordering the basis. For example, suppose the problem is one of linear regression for functional data; that the region $\mathcal{R}$, in the inner product at (2), is the interval $\mathcal{I} = [-1, 1]$; and that we know that β is an even function on $\mathcal{I}$. Then we can ignore any basis functions $\hat{\phi}_j$ that are odd functions, since their contribution to accurate prediction will be zero. Equivalently, when we rank the functions $\hat{\phi}_j$, we should rank last all $\hat{\phi}_j$'s that are odd functions.

Of course, in practice we shall not know that β is exactly even, and moreover, none of the $\hat{\phi}_j$'s will be exactly odd. Nevertheless, the insight provided by the idealised example above can be helpful in practice. For instance, suppose $x(t)$ equals the number of miles travelled by a fleet of trucks in week t of the year, and Y is the total amount of fuel used by the fleet during the year. We expect β to be significantly influenced by seasonal effects, arising for example because of the impact of highway conditions on fuel consumption, and we anticipate that those effects will be approximately symmetrically distributed about the middle of the year. Therefore, if the period of one year is re-centred and rescaled so that it starts at -1 and ends at $+1$, then β is likely to be close to an even function on $\mathcal{I}$. In this setting, prediction could potentially benefit from ordering the components $\hat{\phi}_1, \hat{\phi}_2, \ldots$ so that those that were close to being odd functions were indexed relatively late.

30.5 Other approaches to basis choice

In addition to the context discussed above, where we make use of information from outside the dataset, it is sometimes possible to choose a more effective basis by using only information only from the dataset itself. For example, this is the case if the regressand, x, in the prediction problem is atypical of the explanatory variables $X_1, \ldots, X_n$. In instances like this it can be useful, and effective, to construct the basis using only those variables X_i that are "close to" x in some sense, and to use cross-validation to make the selection.

To illustrate this point we give a simple, finite-dimensional example, where "close to" is interpreted in the sense of simple Euclidean distance. In practice, alternative distances measures can be more effective.

Assume that $X = U\,\psi_U + V\,\psi_V$, where the functions ψ_U and ψ_V are orthonormal, the random variables U and V are independent, U has a zero-one distribution with $P(U=0) = P(U=1) = \frac{1}{2}$, and V has a continuous distribution, for instance normal $\mathrm{N}(0,1)$. A realisation of X has, with probability $\frac{1}{2}$, the form $x = v\,\psi_V$, where v is a scalar. The corresponding value of $\mu(x)$ is $b_V v$, where $b_V = \int_{\mathcal{R}} \beta\,\psi_V$.

For this version of x, $\|X - x\|^2 = U^2 + (V - v)^2$. Hence, if $0 < \delta < 1$ then $\|X - x\| \le \delta$ entails $U = 0$ and $X = V\,\psi_V$. Therefore, if we restrict attention to data X_i for which $\|X_i - x\| \le \delta < 1$ then the reduced dataset will include only functions of the form $X_i = V_i\,\psi_V$ for nonzero V_i's. In consequence, an adaptive basis-choice method which constructs the basis only from data X_i that are within δ of x will, with probability converging to 1 exponentially fast, correctly produce the singleton $\{\psi_V\}$ as the basis for the class of random functions X satisfying $\|X - x\| \le \delta$.

Once ψ_V is concisely identified, identification of ψ_U will quickly follow, and it is then clear that the Karhunen-Loève expansion of X involves no other components. Moreover, the principal components $X_{iU} = \int_{\mathcal{R}} X_i\,\psi_U$ and $X_{iV} = \int_{\mathcal{R}} X_i\,\psi_V$ are now explicitly known, and we see that the functional linear regression problem has degenerated, as it ideally should, to the simple linear regression problem where we observe triples (X_{iU}, X_{iV}, Y_i) generated as $Y_i = \alpha + b_U X_{iU} + b_V X_{iV} + \varepsilon_i$ for $1 \le i \le n$, where $b_W = \int_{\mathcal{R}} \beta \psi_W$ for $W = U, V$. The identification of X_{iU} and X_{iV}, and of the regression model, occurs with probability exponentially close to 1, and does not occur if we work with the standard principal component basis.

While this example is naive in its simplicity, it correctly focuses on atypicality as a major issue in alternative basis choice. If the regressand, x, is sufficiently atypical of a substantial number of the actual explanatory data X_i, then empirical methods can be used to choose a basis that is more appropriate for prediction starting with x.

References

[1] Christensen, R.: Comment: Fisher Lecture: Dimension reduction in regression. Statistical Science. **22**, 27–31 (2007).
[2] Cook, R.D.: Fisher Lecture: Dimension reduction in regression. Statistical Science. **22**, 1–26 (2007).
[3] Cox, D.R.: Notes on some aspects of regression analysis. J. Roy. Statist. Soc. Ser. A. **131**, 265–279 (1968).
[4] Fisher, R. A.: The influence of rainfall on the yield of wheat at Rothamsted. Philos. Trans. Roy. Soc. London. Ser. B. **213** 8–142 (1924).
[5] Hawkins, D. M. and Fatti, L. P.: Exploring multivariate data using the minor principal components. The Statistician. **33**, 325–338 (1984).
[6] Hocking, R. R.: The analysis and selection of variables in linear regression. Biometrics. **32**, 1–49 (1976).
[7] Hotelling, H.: The relationship of the newer multivariate statistical methods to factor analysis. British J. Statist. Psychology. **10**, 69–79 (1957).
[8] Kendall, M.G.: A Course in Multivariate Analysis. Griffin, London. (1957).
[9] Li, B.: Comment: Fisher Lecture: Dimension reduction in regression. Statistical Science. **22**, 32–35 (2007).
[10] Li, L. and NachtsheimA, C.J.: Comment: Fisher Lecture: Dimension reduction in regression. Statistical Science. **22**, 36–39 (2007).
[11] Mosteller, F. and Tukey, J.W.: Data Analysis and Regression: A Second Course in Statistics. Addison-Wesley, Reading, MA. (1977).
[12] Spurrell, D.J.: Some metallurgical applications of principal components. Applied Statist. **12**, 180–188 (1963).

Chapter 31
Bagplots, Boxplots and Outlier Detection for Functional Data

Rob Hyndman and Han Lin Shang

Abstract We propose some new tools for visualizing functional data and for identifying functional outliers. The proposed tools make use of robust principal component analysis, data depth and highest density regions. We compare the proposed outlier detection methods with the existing "functional depth" method, and show that our methods have better performance on identifying outliers in French male age-specific mortality data.

31.1 Introduction

Although the presence of outliers has a serious effect on the modeling and forecasting of functional data, the problem has so far received little attention. In this paper, we propose the functional bagplot and a functional boxplot in order to visualize functional data and to detect any outliers present.

Recently, two papers have considered the problem of outlier detection in functional data. Hyndman & Ullah (2007) used a method based on robust principal components analysis and the integrated squared error from a linear model while Febrero et al. (2007) considered functional outlier detection using functional depth, a likelihood ratio test and smoothed bootstrapping. The method of Hyndman & Ullah involves several parameters to be specified and so is perhaps too subjective for regular use, while the method of Febrero et al. involves fewer decisions by users but is time consuming to compute and is not able to detect some types of outliers. We propose a new method that

Rob Hyndman
Department of Econometrics and Business Statistics, Monash University Clayton, VIC 3800, Australia, e-mail: Rob.Hyndman@buseco.monash.edu.au

Han Lin Shang
Department of Econometrics & Business Statistics, Monash University Clayton, VIC 3800, Australia, e-mail: Han.Shang@buseco.monash.edu.au

uses robust principal components analysis, but is simpler to apply than that of Hyndman & Ullah (2007).

Suppose we have a set of curves $\{y_i(x)\}$, $i = 1, \dots, n$, which are realizations on the functional space $\mathcal{I}$. We are interested in visualizing these curves for large n using functional equivalents of boxplots and bagplots, and we are interested in identifying outliers in the observed curves.

To illustrate the ideas, we will consider annual French male age-specific mortality rates (1899–2003) shown in Figure 31.1. These data were used by Hyndman & Ullah (2007) who obtained them from the Human Mortality Database (2007). The mortality rates are the ratio of death counts to population exposure in the relevant period of age and time. The data were first scaled using natural logarithms. The colours reflect the years of observation in "rainbow" order, with the oldest curves in red and the most recent curves in purple. There are some apparent outliers (in yellow and green) which show an unusual increase in mortality rates between ages 20 and 40. These are mainly due to the First and Second World Wars, as well as the Spanish influenza which occurred in 1918.

Before proceeding further, we need to define the notion of ordering a set of curves. López-Pintado & Romo (2007) proposed the use of "generalized band depth" to order a set of curves. The generalized band depth of a curve is the proportion (computed using Lebesgue measure) of times that the curve is entirely contained in the band defined by J curves from the sample. They suggest using $J = 2$ and propose that the "median" should be defined as the curve with the highest depth. See also Ferraty & Vieu (2006, p.129) for some related discussion.

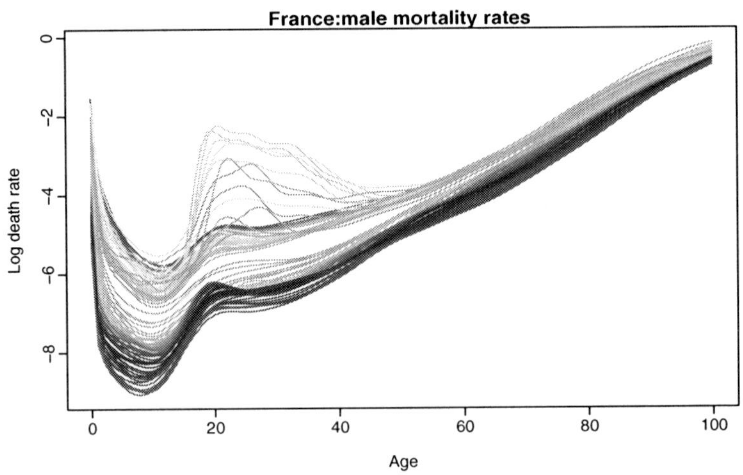

Fig. 31.1 Functional time series plot for the French male mortality data (1899–2003).

While ordering by depth is useful in some contexts, we prefer an alternative approach to ordering obtained using a principal component decomposition of the set of observed curves. If we let

$$y_i(x) = \mu(x) + \sum_{k=1}^{n-1} z_{i,k}\phi_k(x),$$

where $\{\phi_k(x)\}$ represents the eigenfunctions, then we can use an ordering method from multivariate analysis based on the principal components scores $\{z_{i,k}\}$.

The simplest procedure is to consider only the first two scores, $\boldsymbol{z}_i = (z_{i,1}, z_{i,2})$. Then an ordering of the curves is defined using the ordering of $\{\boldsymbol{z}_i; i = 1, \ldots, n\}$. For example, bivariate depth can be used (Rousseeuw et al., 1999). Alternatively, the value of the kernel bivariate density estimate at $\boldsymbol{z}_i$ can be used to define an ordering.

There are two major advantages in ordering via the principal component scores: (1) it leads to a natural method for defining visualization methods such as functional bagplots and functional boxplots; and (2) it seems to be better able to identify outliers in real data (as we will see in the application).

Outliers will usually be more visible in the principal component space than the original (functional) space (Filzmoser et al., 2008). Thus finding outliers in the principal component scores does no worse than searching for them in the original space. Often, it is the case that the first two principal component scores suffice to convey the main modes of variation (Hall et al., 2007). We have found empirically that the first two principal component scores are adequate for outlier identification.

Because principal component decomposition is itself non-resistant to outliers, we apply a functional version of Croux & Ruiz-Gazen's (2003) robust principal component analysis which uses a projection pursuit technique. This method was described and used in Hyndman & Ullah (2007).

31.2 Functional bagplot

The functional bagplot is based on the bivariate bagplot of Rousseeuw et al. (1999) applied to the first two (robust) principal component scores.

The bagplot is constructed on the basis of the halfspace location depth denoted by $d(\boldsymbol{\theta}, \boldsymbol{z})$ of some point $\boldsymbol{\theta} \in R^2$ relative to the bivariate data cloud $\{\boldsymbol{z}_i; i = 1, \ldots, n\}$. The depth region D_k is the set of all $\boldsymbol{\theta}$ with $d(\boldsymbol{\theta}, \boldsymbol{z}) \geq k$. Since the depth measurements are convex polygons, we have $D_{k+1} \subset D_k$. This concept is somewhat similar to the notion of a ball used in Ferraty and Vieu (2006). For a fixed center, the regions grow as the radius increases.

Thus, the data points are ranked according to their depth. The bivariate bagplot displays the median point (the deepest location), along with the selected percentages of convex hulls. Any point beyond the highest percentage of the convex hulls is considered as an outlier. Each point in the scores bagplot corresponds to a curve in the functional bagplot. The functional bagplot also displays the median curve which is the deepest location, the 95% confidence intervals for the median, and the 50% and 95% of surrounding curves ranking by depth. Any curve beyond the 95% convex hull is flagged as a functional outlier.

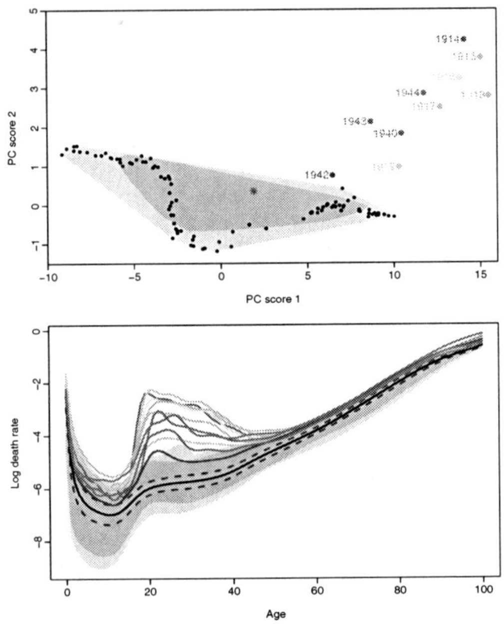

Fig. 31.2 The scores bagplot and functional bagplot for the French male mortality data.

An example is shown in Figure 31.2 using the French male mortality data. The red asterisk marks the median of the bivariate scores and corresponds to the solid black functional observation in the right panel. The dotted blue lines give 95% confidence intervals for the median curve. In the left panel, the dark grey regions show the 50% convex hull and the light grey regions show the 95% convex hull. These correspond directly with the regions of similar shading in the functional plot on the right. Points outside these regions are identified as outliers. The different colours for these outliers enable the individual curves on the right to be matched to the scores on the left.

31.3 Functional HDR boxplot

The functional highest density region (HDR) boxplot is based on the bivariate HDR boxplot of Hyndman (1996) applied to the first two (robust) principal component scores.

The HDR boxplot is constructed using the Parzen-Rosenblatt bivariate kernel density estimate $\hat{f}(\boldsymbol{w}; a, b)$. For a bivariate random sample $\{\boldsymbol{z}_i; i = 1, \ldots, n\}$, drawn from a density f, the product kernel density estimate is defined by (Scott, 1992)

$$\hat{f}(\boldsymbol{w}; a, b) = \frac{1}{nab} \sum_{i=1}^{n} K\left(\frac{w_1 - z_{i,1}}{a}\right) K\left(\frac{w_2 - z_{i,2}}{b}\right), \tag{31.1}$$

where $\boldsymbol{w} = (w_1, w_2)'$, K is a symmetric univariate kernel function such that $\int K(u)du = 1$ and (a, b) is a bivariate bandwidth parameter such that $a > 0$, $b > 0$, $a \to 0$ and $b \to 0$ as $n \to \infty$. The contribution of data point $\boldsymbol{z}_i$ to the estimate at some point $\boldsymbol{w}$ depends on how distant $\boldsymbol{z}_i$ and $\boldsymbol{w}$ are.

A highest density region is defined as

$$R_\alpha = \{\boldsymbol{z} : \hat{f}(\boldsymbol{z}; a, b) \geq f_\alpha\},$$

where f_α is such that $\int_{R_\alpha} \hat{f}(\boldsymbol{z}; a, b)d\boldsymbol{z} = 1 - \alpha$. That is, it is the region with probability coverage $1 - \alpha$ where every point within the region has higher density estimate than every point outside the region.

The beauty of ranking by the HDR is its ability to show multimodality in the bivariate data. The HDR boxplot displays the mode, defined as $\sup_{\boldsymbol{z}} \hat{f}(\boldsymbol{z}; a, b)$, along with the 50% HDR and the 95% HDR. All points not included in the 95% HDR are shown as outliers. The functional HDR boxplot is a one-to-one mapping of the scores HDR bivariate boxplot.

An example is shown in Figure 31.3 using the French male mortality data. The black circle (left panel) marks the mode of the bivariate scores and corresponds to the solid black functional observation in the right panel. In the left panel, the dark grey regions show the 50% HDR and the light grey regions show the 95% HDR. These correspond directly with the regions of similar shading in the functional plot on the right. Points outside these regions are identified as outliers. The different colours for these outliers enable the individual curves on the right to be matched to the scores on the left.

31.4 Comparison

The following table presents the outlier detection results from the proposed methods along with the functional depth measure of Febrero et al. (2007).

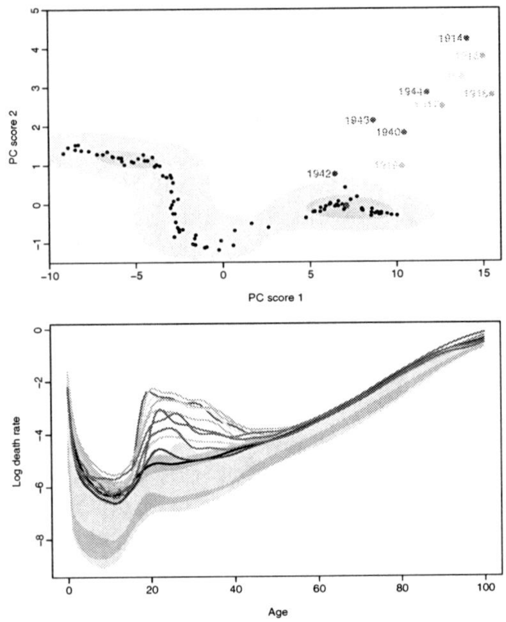

Fig. 31.3 The scores HDR boxplot, and functional HDR boxplot for the French male mortality data.

Method	Outlier (Year)
Functional depth	1915
Functional bagplot	1914–1919, 1940, 1943–1945
Functional HDR boxplot	1914–1919, 1940, 1943–1945

Table 31.1 Outlier detection performance between the proposed approach and the functional depth measure approach.

In this case, the functional depth measure approach performs the worst among all methods. In contrast, all of the apparent outliers in Figure 31.1 have been detected by both the functional bagplot and functional HDR boxplot methods.

Of the two new methods, we prefer the functional HDR boxplot as it also provides an additional advantage in that it can identify unusual "inliers" that full in sparse regions of the sample space.

R code for constructing the functional bagplot and HDR boxplot are available upon request from the first author.

References

[1] Croux, C. & Ruiz-Gazen, A.: High breakdown estimators for principal components: the projection-pursuit approach revisited. Journal of Multivariate Analysis. **95**, 206–226 (2003).

[2] Febrero, M., Galeano, P. & González-Manteiga, W.: A functional analysis of NOx levels: location and scale estimation and outlier detection. Computational Statistics. **23**(3), 411–427 (2007).

[3] Ferraty, F., & Vieu, P.: Nonparametric Functional Data Analysis. Springer. (2006).

[4] Filzmoser, P., Maronna, R. & Werner, M.: Outlier identification in high dimensions. Computational Statistics & Data Analysis. **52**, 1694–1711 (2008).

[5] Hall, P.G., Lee, Y. & Park, B.: A method for projecting functional data onto a low-dimensional space. Journal of Computational and Graphical Statistics. **16**, 799–812 (2007).

[6] Human Mortality Database, University of California, Berkeley (USA), and Max Planck Institute for Demographical Research (Germany). Viewed 15/4/07, available online at <www.mortality.org> or <www.humanmortality.de>. (2007).

[7] Hyndman, R.J.: Computing and graphing highest density regions. The American Statistician. **50**(2), 120–126 (1996).

[8] Hyndman, R.J. & Ullah, Md.S.: Robust forecasting of mortality and fertility rates: a functional data approach. Computational Statistics & Data Analysis. **51**, 4942–4956 (2007).

[9] López-Pintado, S & Romo, J.: Depth-based inference for functional data. Computational Statistics & Data Analysis. **51**, 4957–4968 (2007).

[10] Rousseeuw, P., Ruts, I. & Tukey, J.: The bagplot: a bivariate boxplot. The American Statistician. **53**(4), 382–387 (1999).

[11] Scott, D. W.: Multivariate density estimation: theory, practice, and visualization. John Wiley and Sons: new York. (1992).

Chapter 32
Marketing Applications of Functional Data Analysis

Gareth James, Ashish Sood and Gerard Tellis

Abstract The Bass (1969) model has been a standard for analyzing and predicting the market penetration of new products. The authors demonstrate the insights to be gained and predictive performance of Functional Data Analysis (FDA), on the market penetration of 760 categories drawn from 21 products and 70 countries. The authors compare a Functional Regression approach to several models including the Classic Bass model.

32.1 Introduction

Firms are introducing new products at an increasingly rapid rate. At the same time, the globalization of markets has increased the speed at which new products diffuse across countries, mature, and die off (Chandrasekaran and Tellis 2008). These two forces have increased the importance of the accurate prediction of the market penetration of an evolving new product. While research on modeling sales of new products in marketing has been quite insightful (Chandrasekaran and Tellis 2007; Peres, Mueller and Mahajan 2008), it is limited in a few respects. First, most studies rely primarily, if not exclusively, on the Bass model. Second, prior research, especially those based on the Bass model, need data past the peak sales or penetration for stable es-

Gareth James
Marshall School of Business, University of Southern California, Los Angeles, California 90089, USA, e-mail: gareth@usc.edu

Ashish Sood
Goizueta School of Business, Emory University, Atlanta, GA 30322, USA, e-mail: ashish_sood@bus.emory.edu

Gerard Tellis
Marshall School of Business, University of Southern California, Los Angeles, California 90089, USA, e-mail: tellis@usc.edu

timates and meaningful predictions. Third, prior research has not indicated how the wealth of accumulated penetration histories across countries and categories can be best integrated for good prediction of penetration of an evolving new product. For example, a vital unanswered question is whether a new product's penetration can be best predicted from past penetration of a) similar products in the same country, b) the same product in similar countries, c) the same product itself in the same country, or d) some combination of these three histories.

The current study attempts to address these limitations. In particular, it makes four contributions to the literature. First, we illustrate the potential advantages of using Functional Data Analysis (FDA) techniques for the analysis of penetration curves (Ramsay and Silverman, 2005). Second, we demonstrate how information about the historical evolution of new products in other categories and countries can be integrated to predict the evolution of penetration of a new product. Third, we compare the predictive performance of the Bass model versus an FDA approach, and some naïve models. Fourth, we indicate whether information about prior countries, other categories, the target product itself, or a combination of all three is most important in predicting the penetration of an evolving new product.

One important aspect of the current study is that it uses data about market penetration from most of 21 products across 70 countries, for a total of 760 categories (product x country combinations). The data include both developed and developing countries from Europe, Asia, Africa, Australasia, and North and South America. In scope, this study exceeds the sample used in prior studies. Yet the approach achieves our goals in a computationally efficient and substantively instructive manner. Another important aspect of the study is that it uses Functional Data Analysis to analyze these data. Over the last decade FDA has become a very important emerging field in statistics, although it is not well known in the marketing literature. FDA provides a set of techniques that can improve the prediction of future items of interest especially in cases where prior longitudinal data is available for the same products, data is available from histories of similar products, or complete data is not available for some years. The central paradigm of FDA is to treat each function or curve as the unit of observation. We apply the FDA approach by treating the yearly cumulative penetration data of each category as 760 curves or functions. By taking this approach we can extend several standard statistical methods for use on the curves themselves.

For instance, we use functional principal components analysis (PCA) to identify the patterns of shapes in the penetration curves. Doing so enables a meaningful understanding of the variations among the curves. An additional benefit of the principal component analysis is that it provides a parsimonious, finite dimensional representation for each curve. In turn this allows us to perform functional regression by treating the functional principal component scores as the predictors and future characteristics of the curves, such as future penetration or time to takeoff, as the response. We show that this approach to prediction is more accurate than the traditional approach of using information from only one curve. It also provides a deeper understanding of the evolutions of the penetration curves. Finally, we perform functional clustering by grouping the curves into clusters with similar patterns of evolution in penetration. The groups that we form show strong clustering among certain products and provide further insights into the patterns of evolution in penetration. In particular plotting the principal component scores allows us to visually assess the level of clustering among different products for all 760 curves simultaneously. Such a visual representation would be impossible using the original curves.

32.2 Methodology

In this section we describe the three different FDA techniques that we applied to the penetration data. We first describe our functional principal components approach. We then utilize the functional principal component scores to perform functional regression for predicting various future characteristics of the various products. Finally, the PCA scores are used to perform functional cluster analysis and hence identify groupings among curves.

Most FDA techniques assume that the curves have been observed at all time points but in practice this is rarely the case. Since we have regularly spaced observations for each curve we opt to use a simple smoothing spline approach to generate a continuous smooth curve from our discrete observations. To compute the functional principal components we divide the time period $t = 1$ to $t = T$ into p equally spaced points and evaluate $X_1(t), \ldots, X_n(t)$ at each of these time points. Finally, we perform standard PCA on this p dimensional data. The resulting principal component vectors provide accurate approximations to the functional principal components at each of the p grid points and likewise the principal component scores represent the functional PCA scores.

We use functional regression to predict several items of interest, such as future marginal penetration level in any given year or the year of takeoff. Let $X_i(t)$ be the smooth spline representation of the ith curve observed over time such as the first five years of cumulative penetration for a given category. Let Y_i represent a related item to be predicted, such as the marginal penetration in year six. Functional regression establishes a relationship between predictor, $X_i(t)$, and the item to be predicted, Y_i, as follows:

$$Y_i = f(X_i(t)) + \varepsilon_i, \quad i = 1, \ldots, n. \tag{32.1}$$

Equation (32.1) is difficult to work with directly because $X_i(t)$ is infinite dimensional. However, for any function f there exists a corresponding function g such that $f(X(t)) = g(e_1, e_2, \ldots)$ where e_1, e_2 etc. are the principal component scores of $X_i(t)$. We use this equivalence to perform functional regression with the functional principal component scores as the predictors. The simplest choice for g would be a linear function. We opt to use the more powerful model produced by assuming that g is an additive but non-linear, function (Hastie and Tibshirani, 1990). In this case, Equation (32.1) becomes

$$Y_i = \beta_0 + \sum_{j=1}^{D} g_j(e_{ij}) + \varepsilon_i \qquad (32.2)$$

where the g_j's are non-linear functions that are estimated as part of the fitting procedure and D is chosen so that the first D principal components explain most of the variability in $X(t)$. One advantage of using Equation (32.2) to implement a functional regression is that once the e_{ij}'s have been computed via the functional PCA, we can then use standard additive regression software to relate Y_i to the principal component scores. We can also extend Equation (32.2) by adding covariates that contain information about the curves beyond the principal components, such as product or country characteristics or marketing variables.

We use functional clustering for the purpose of better understanding the penetration patterns in the data. In particular, we wish to identify groups of similar curves and relate them to observed characteristics of these curves such as the product and country. We use the principal components described above to reduce the potentially large number of dimensions of variability and cluster all the curves in the sample. We apply the standard k-means clustering approach (MacQueen 1967) to the D-dimensional principal component scores, e_i, to cluster all the curves in the sample. We use the "jump" approach (Sugar and James 2003) to select the optimal number of clusters, k. Sugar and James (2003) show through the use of information theory and simulations that this approach provides an accurate estimate of the true number of clusters in the data. Once we compute the cluster centers, we assign each curve to its closest cluster mean curve. We then project the centers back into the original curve space and examine the shape of a typical curve from each cluster.

32.3 Results

We find that two functional principal components explain almost all the variability in the penetration curves. The first component measures the final penetration level by year ten while the second component records the pattern that the product took to get to that point.

We compare two versions of the functional regression approach with five more standard methodologies, including the Bass Model. Eight different response variables were also tested. For almost all combinations of method and response variable the functional regression approach produced superior predictions with most results being statistically significant.

The functional clustering suggested six different groupings in the penetration curves, corresponding to different rates, and different patterns, of growth. There are also clear dependencies between products and clusters with, for example, electronics goods tending to group in the high growth clusters and white goods favoring the low growth clusters. Patterns are also clear among geographic regions.

References

[1] Bass, F. M.:(1969), A new product growth for model consumer durables. Management Science. **15**, 215-227

[2] Chandrasekaran, Deepa and Gerard J Tellis: Diffusion of New Products: A Critical Review of Models, Drivers, and Findings, Review of Marketing, 39-80 (2007).

[3] Chandrasekaran, Deepa and Gerard J Tellis: Global Takeoff of New Products: Culture, Economics or Vanishing Differences forthcoming,. Marketing Science. (2008).

[4] Hastie, T. J. and Tibshirani, R. J.: Generalized Additive Models. Chapman and Hall. (1990).

[5] MacQueen, J. B.: Some Methods for classification and Analysis of Multivariate observations, Proceedings of 5-th Berkeley Symposium on Mathematical Statistics and Probability, Berkeley, University of California Press, **1**, 281-297 (1967).

[6] Peres, Renana, Eitan Mueller and Vijay Mahajan: "Review of Research on Diffusion. forthcoming, Marketing Science. (2007).

[7] Ramsay, J.O. and Silverman, B.W.: Functional Data Analysis. Springer (2nd Edition). (2005).

[8] Sugar, C. A. and James, G. M.: Finding the number of clusters in a data set: An information theoretic approach, Journal of the American Statistical Association. **98**, 750-763 (2003).

Chapter 33
Nonparametric Estimation in Functional Linear Model

Jan Johannes

Abstract We consider the problem of estimating the slope parameter in functional linear regression, where scalar responses $Y_1, \dots, Y_n$ are modeled in dependence of random functions $X_1, \dots, X_n$. In the case of second order stationary random functions and as well in the non stationary case estimators of the functional slope parameter and its derivatives are constructed based on a regularized inversion of the estimated covariance operator. In this paper the rate of convergence of the estimator is derived assuming that the slope parameter belongs to the well-known Sobolev space of periodic functions and that the covariance operator is finitely, infinitely or in some general form smoothing.

33.1 Introduction

Functional linear models are becoming very important in a diverse range of disciplines, including medicine, linguistics and chemometrics (see for instance Ramsay and Silverman (2005) and Ferraty and Vieu (2006), for several case studies). Roughly speaking, in all these applications the dependence of a response variable Y on the variation of an explanatory random function X is modeled by

$$Y = \int_0^1 \beta(t)X(t)dt + \varepsilon \tag{33.1}$$

for some error term ε. One objective is then to estimate nonparametrically the slope function β and its derivatives based on a sample of (Y, X).

Jan Johannes
Universität Heidelberg, Institut für Angewandte Mathematik, Im Neuenheimer Feld, 294, 69120 Heidelberg, Germany, e-mail: `johannes@statlab.uni-heidelberg.de`

In this paper we suppose that the random function X is taking its values almost surely in $L^2[0,1]$, which is endowed with the usual norm $\|\cdot\|$, and has a finite second moment, i.e., $\mathbb{E}\|X\|^2 < \infty$. In order to simplify notations we assume that the mean function of X is zero. Moreover, the random function X and the error term ε are independent, where ε has mean zero and a finite second moment, i.e, $\mathbb{E}\varepsilon^2 < \infty$. This situation has been considered, for example, in Cardot, Ferraty and Sarda (2003) or Stadtmuller (2005). Then multiplying both sides in (33.1) by X and taking the expectation leads to

$$g(s) := \mathbb{E}[YX(s)] = \int_0^1 \text{cov}(X(t), X(s))\beta(t)dt =: [T_{\text{cov}}\beta](s), \quad s \in [0,1], \tag{33.2}$$

where g belongs to $L^2[0,1]$. We assume that there exists a unique solution $\beta \in L^2[0,1]$ of equation (33.2), i.e., g belongs to the range $\mathcal{R}(T_{\text{cov}})$ of T_{cov} and T_{cov} is injective. However, as usual in the context of inverse problems all the results below can also straightforward be obtained for the unique least-square solution with minimal norm, which exists if and only if g is contained in the direct sum of $\mathcal{R}(T_{\text{cov}})$ and its orthogonal complement $\mathcal{R}(T_{\text{cov}})^{\perp}$ (for a definition and detailed discussion in the context of inverse problems see chapter 2.1 in EHN (2000), while in the special case of a functional linear model we refer to Cardot, Ferraty and Sarda (2003)). The normal equation (33.2) is the continuous equivalent of normal equations in the multivariate linear model. Estimation of β is thus linked with the inversion of the covariance operator T_{cov} of X defined in (33.2), which due to the finite second moment of X is a Hilbert-Schmidt operator. Thereby, unlike in the finite dimensional case, a continuous inverse for T_{cov} does not exist as long as the range of the operator T_{cov} is an infinite dimensional subspace of $L^2[0,1]$. This corresponds to the setup of ill-posed inverse problems (with the additional difficulty that T_{cov} is unknown and, hence has to be estimated). In the literature several approaches are proposed in order to circumvent this instability issue. Essentially, all of them replace in equation (33.2) the operator T_{cov} by a regularized version having a continuous inverse. A popular example is based on a functional principal components regression (c.f. Bosq (2000), Cardot, Mas and Sarda (2007) or Muller and Stadtmuller (2005)), this method is also called Spectral cut-off in the numerical analysis literature (Tautenhahn (1996)). An other example is the Tikhonov regularization (c.f. Hall and Horowitz (2007)), where the regularized solution β_α is defined as unique minimizer of the Tikhonov functional $F_\alpha(\beta) = \|T_{\text{cov}}\beta - g\|^2 + \alpha\|\beta\|^2$ for some strictly positive α. A regularization through a penalized least squares approach after projection onto some basis (such as splines) is also considered in Ramsay and Dalzell (1991), Eilers and Marx (1996) or Cardot, Ferraty and Sarda (2003). However, there is a large number of alternative regularization schemes in the numerical analysis literature available like the generalized Tikhonov regularization, Landweber iteration or the ν-Methods to name but a few (c.f. Tautenhahn (1996)). The common aspect of all these regularization schemes is that an additional regularization parameter α (for example, the parameter determining the weight of the penalty in the Tikhonov functional) is introduced. The risk of the resulting regularized estimator can then be decomposed, roughly speaking, into a function of the risk of the nonparametric estimators of g and T_{cov} plus an additional bias term which is a function of the regularization parameter α (for a detailed discussion in the context of inverse problems see Johannes *et al.* (2007)). In this paper we assume that β belongs to the Sobolev space of periodic functions $W_p[0,1]$ (defined below). The relationship between the range of the covariance operator T_{cov} and the Sobolev spaces is then essentially determining the functional form of the bias term. For example, if T_{cov} is finitely smoothing, i.e., the range of the T_{cov} equals $W_a[0,1]$ for some $a > 0$, then the bias is a polynomial of the parameter α. On the other hand, if T_{cov} is infinitely smoothing, i.e., the range of $|\log(T_{\text{cov}})|^{-1}$ equals $W_a[0,1]$ for some $a > 0$, then the bias is a logarithm of the parameter α. The theory behind these rates can be unified using an index function κ (c.f. Nair *et al.*(2005)), which 'links' the range of T_{cov} and the Sobolev spaces.

The paper is organized in the following way. In Section 2 we define the estimator of β when X is second order stationary as well as when X is not second order stationary. We investigate the asymptotic behavior of the estimator of β in case of a second order stationary X and a not second order stationary X in Section 3 and Section 4, respectively. All proofs of the results in this paper can be found in Johannes (2007).

33.2 Definition of the estimator of β

Let $W_p[0,1]$ denote the Sobolev space of periodic functions, which is defined for integer $m \in \mathbb{N}$ by $W_m[0,1] = \{f \in H_m[0,1] : f^{(j)}(0) = f^{(j)}(1), j = 0,\dots,m-1\}$, where
$H_m[0,1] = \{f : [0,1] \to R : f^{(m-1)}$ is absolutely continuous, $f^{(m)} \in L^2[0,1]\}$ are Sobolev spaces, and for real values p, $W_p[0,1]$ is defined by interpolation and duality. Considering the trigonometric basis

$$\phi_1 :\equiv 1,\ \phi_{2k}(s) := \sqrt{2}\cos(2\pi k s),\ \phi_{2k+1}(s) := \sqrt{2}\sin(2\pi k s), s \in [0,1], \tag{33.3}$$

$k = 1,2,\dots$ and the unbounded sequence

$$\gamma_1 := 1 \text{ and } \gamma_{2k} := \gamma_{2k+1} := 2k, \quad k = 1,2,\dots, \tag{33.4}$$

the Sobolev space of periodic functions can equivalently defined by (c.f. Neubauer (1988), Mair and Ruymgaart (1996) or Tsybakov (2004))

$$W_p[0,1] = \Big\{ f \in L^2[0,1] : \|f\|_p^2 := \sum_{j=1}^{\infty} \gamma_j^{2p} |\langle f, \phi_j\rangle|^2 < \infty \Big\}. \tag{33.5}$$

Estimation of β when X is second order stationary

If we suppose there exists a positive definite function $c : [-1,1] \to \mathbb{R}$ such that $\operatorname{cov}(X(t), X(s)) = c(t-s)$, $s,t \in [0,1]$, i.e., X is second order stationary. Then the eigenfunctions of T_{cov} are given by the trigonometric basis defined in (33.3). Moreover, only the eigenvalues $\{\lambda_1, \lambda_2, \dots\}$ of T_{cov} depend on the unknown covariance function $c(\cdot)$ and, hence have to be estimated. Let $(Y_1, X_1), \dots, (Y_n, X_n)$ be an i.i.d. sample of (Y, X), which we use to construct estimators $\widehat{g}$ and $\widehat{\lambda}_m$, $m \in \mathbb{N}$, of g and λ_m, $m \in \mathbb{N}$, respectively. The estimator $\widetilde{\beta_q}$ of β is then defined by introducing a threshold $\alpha > 0$, that is for $q \geq 0$ we consider

$$\widetilde{\beta_q} := \sum_{m=1}^{\infty} \frac{\widehat{g}_m}{\widehat{\lambda}_m} \cdot \phi_m \cdot \mathbf{1}\{\widehat{\lambda}_m^2 \geq \alpha \cdot \gamma_m^{2q}\}, \text{ with } \widehat{g}_m := \langle \widehat{g}, \phi_m \rangle, \ m = 1, 2, \dots, \tag{33.6}$$

where the threshold $\alpha = \alpha(n)$ has to tend to zero as the sample size n increases. The eigenvalues of T_{cov} satisfy $\lambda_m = \mathbb{E}\langle X, \phi_m \rangle^2$, for all $m \in \mathbb{N}$, which motivates the following unbiased estimators

$$\widehat{\lambda}_m = \frac{1}{n} \sum_{i=1}^{n} \langle X_i, \phi_m \rangle^2, \quad m = 1, 2, \dots . \tag{33.7}$$

Moreover, due to Parseval's formula we have $\sum_{m=1}^{\infty} \widehat{\lambda}_m = \frac{1}{n} \sum_{i=1}^{n} \|X_i\|^2$ which is almost surely finite. Thereby, the sum in (33.6) contains only a finite number of nonzero summands. On the other hand, an unbiased estimator of g is given by

$$\widehat{g}(s) := \frac{1}{n} \sum_{i=1}^{n} Y_i X_i(s), \quad s \in [0,1]. \tag{33.8}$$

Estimation of β when X is not second order stationary

We consider a generalized Tikhonov regularization using a Sobolev penalty in order to overcome the saturation effect of the classical Tikhonov regularization (for a detailed discussion in the context of an instrumental regression we refer to Florens *et al.* (2007)). Given an i.i.d. sample $(Y_1, X_1), \dots, (Y_n, X_n)$ of (Y, X) we consider again the estimator $\widehat{g}$ of g defined in (33.8). In addition we estimate the covariance function $\operatorname{cov}(t,s) := \operatorname{cov}(X(t), X(s))$ by its empirical counterpart, that is

$$\widehat{\operatorname{cov}}(t,s) := \frac{1}{n} \sum_{i=1}^{n} X_i(t) X_i(s), \quad t,s \in [0,1]. \tag{33.9}$$

Then the estimator $\widehat{T}_{\text{cov}}$ of the covariance operator T_{cov} is defined by

$$[\widehat{T}_{\text{cov}}f](s) := \int_0^1 \widehat{\text{cov}}(t,s)f(t)dt, \; s \in [0,1], \quad f \in L^2[0,1]. \tag{33.10}$$

The regularized estimator $\widehat{\beta}_q := \widehat{\beta}_{q,k}$ of β based on a generalized Tikhonov regularization of order k is obtained by solving iteratively the k minimization problems

$$\widehat{\beta}_{q,j+1} := \mathop{\arg\min}_{\beta \in W_q[0,1]} \{\|\widehat{T}_{\text{cov}}\beta - \widehat{g}\|^2 + \alpha\|\widehat{\beta}_{q,j} - \beta\|_q^2\}, \quad j = 0,\dots,k-1, \quad \widehat{\beta}_{q,0} := 0. \tag{33.11}$$

By definition $\widehat{\beta}_q$ belongs to $W_q[0,1]$ and the classical Tikhonov regularization is covered with $k=1$ and $q=0$. A numerical implementation needs a further discretization step such as a projection onto the first m trigonometric basis functions. However we ignore this step in the following presentation, since the obtained approximation can be chosen arbitrary close to the minimizer $\widehat{\beta}_q$ (depending only on the computational cost we are willing to pay).

33.3 Risk bound when X is second order stationary

We shall measure the performance of the estimator $\widetilde{\beta}_q$ defined in (33.6) by the W_q-risk, that is $\mathbb{E}\|\widetilde{\beta}_q - \beta\|_q^2$, provided $\beta \in W_p[0,1]$ for some $p > q \geq 0$. For an integer k the Sobolev norm $\|g\|_k$ is equivalent to $\|g\| + \|g^{(k)}\|$ with k-th weak derivative $g^{(k)}$ of g. Thereby, the W_k-risk reflects the performance of $\widetilde{\beta}_k$ and $\widetilde{\beta}_k^{(k)}$ as estimators of β and $\beta^{(k)}$, respectively.

Theorem 5 *Suppose that X is second order stationary and that $\beta \in W_p[0,1]$, $p > 0$. Let T_{cov} be finitely smoothing, that is $\mathcal{R}(T_{\text{cov}}) = W_a[0,1]$ for some $a > 0$. Consider for $0 \leq q < p$ the estimator $\widetilde{\beta}_q$ defined in (33.6). If $\mathbb{E}\|X\|^{4(p+a)/(a+q)} < \infty$, then with $\alpha = c \cdot n^{-(a+q)/(p+a)}$, $c > 0$ we obtain $\mathbb{E}\|\widetilde{\beta}_q - \beta\|_q^2 = O(n^{-(p-q)/(p+a)})$ as $n \to \infty$.*

If the operator T_{cov} is infinitely smoothing, that is $\mathcal{R}(|\log T_{\text{cov}}|^{-1}) = W_a[0,1]$ for some $a > 0$. Then it can be shown that under suitable regularity conditions on the parameter α and the moments of the random function X the W_q-risk of the estimator $\widetilde{\beta}_q$ is of order $O((\log n)^{-(p-q)/a})$. Moreover, the finitely and infinitely smoothing case can be unified, using an index function $k : (0,1] \to \mathbb{R}$, which is assumed to be a continuous, strictly increasing and concave function with $\kappa(0+) = 0$ (c.f. Nair *et al.* (2005)). Denote by Φ and ω the inverse functions of κ and $\omega^{-1}(t) := t\Phi(t)$, respectively. If the covariance operator T_{cov} is general smoothing, that is

$$d \leq \gamma_m^{2(q-p)} \cdot \kappa(c \cdot \lambda_m^2/\gamma_m^{2q}) \leq D, \quad m = 1,2,\dots, \quad \text{for some } d, D, c > 0.$$

Then under suitable regularity conditions on the parameter α and the moments of the random function X the W_q-risk of the estimator $\widetilde{\beta}_q$ is of order $O(\omega(1/n))$. It is remarkable that the function ω provides the same order than the modulus of continuity of the inverse operation of T_{cov} (for a detailed discussion in the context of a deconvolution problem we refer to Johannes (2007)).

33.4 Risk bound when X is not second order stationary

Theorem 6 *Suppose that X is not second order stationary and that $\beta \in W_p[0,1]$, $p > 0$. Let T_{cov} be finitely smoothing, that is $\mathcal{R}(T_{\text{cov}}) = W_a[0,1]$ for some $a > 0$. Consider for $0 \leq q < p$ the estimator $\widehat{\beta}_q$ defined in (33.11) with $m \geq (p-q)/(a+q) \vee 1$. If $\mathbb{E}\|X^m\|^2 < \infty$ and $\alpha = c \cdot n^{-(a+q)/(p+a)}$, $c > 0$, then $\mathbb{E}\|\widehat{\beta}_q - \beta\|_q^2 = O(n^{-(p-q)/(p+a)})$.*

If the operator T_{cov} is infinitely smoothing, i.e., $\mathcal{R}(|\log T_{\text{cov}}|^{-1}) = W_a[0,1]$, $a > 0$. Then under suitable regularity conditions on the parameter α and the moments of the random function X the W_q-risk of the estimator $\widehat{\beta}_q$ is of order $O((\log n)^{-(p-q)/a})$. Moreover, if the covariance operator T_{cov} is general smoothing, that is

$$d\|f\|_{q-p} \leq \|\kappa^{1/2}(B^{-\frac{q}{2}}T_{\text{cov}}^2 B^{-\frac{q}{2}})f\|^2 \leq D\|f\|_{q-p} \quad \text{for all } f \in L^2[0,1]$$

and some $d, D > 0$, where $B : W_2[0,1] \to L^2[0,1]$ with $Bf := -f''$ (c.f. Johannes *et al.* (2007)). Then under suitable regularity conditions on the parameter α and the moments of the random function X the W_q-risk of the estimator $\widehat{\beta}_q$ is of order $O(\omega(1/n))$.

References

[1] Bosq, D.: Linear Processes in Function Spaces. Lecture Notes in Statistics. Springer-Verlag. (2000).

[2] Cardot, H. and Johannes, J.: Nonparametric estimation in functional linear models. (2007). Preprint. University Heidelberg.

[3] Cardot, H. and Mas, A. and Sarda, P.: CLT in Functional Linear Regression Models. Prob. Theory and Rel. Fields. To appear (2007).

[4] Cardot, H. and Ferraty, F. and Sarda, P.: Spline Estimators for the Functional Linear Model. Statistica Sinica. **13**, 571-591 (2003).

[5] Eilers, P. H. and Marx, B. D.: Flexible smoothing with B-splines and penalties. Statistical Science. **11**, 89-102 (1996).

[6] Ferraty, F. and Vieu, P.: Nonparametric Functional Data Analysis: Methods, Theory, Applications and Implementations. Springer-Verlag, London. (2006).

[7] Florens, J. P. and Johannes, J. and Van Bellegem, S.: Identification and estimation by penalization in Nonparametric Instrumental Regression. Submitted (2007).

[8] Hall, P. and Horowitz, J. L.: Methodology and convergence rates for functional linear regression. Ann. Stat. To appear (2007).

[9] Johannes, J.: Nonparametric estimation in functional linear models. (2007). Preprint. University Heidelberg.

[10] Johannes, Jan.: Deconvolution with unknown error distribution. Submitted (2007).

[11] Johannes, J. and Van Bellegem, S. and Vanhems, Anne.: A unified approach to solve ill-posed inverse problems in econometrics. Submitted (2007).

[12] Mair, Bernard A. and Ruymgaart, Frits H.: Statistical inverse estimation in Hilbert scales. SIAM J. Appl. Math. **5** (56), 1424-1444 (1996).

[13] Müller, H.-G. and Stadtmüller, U.: Generalized Functional Linear Models. Ann. Stat. **33**, 774-805 (2005).

[14] Nair, M.T. and Pereverzev, S. V. and Tautenhahn, U.: Regularization in Hilbert scales under general smoothing conditions. Inverse Problems. **21**, 1851-1869 (2005).

[15] Neubauer, Andreas.: When do Sobolev spaces form a Hilbert scale? Proc. Amer. Math. Soc. **2** (103), 557-562 (1988).

[16] Ramsay, J. O. and Dalzell, C. J.: Some tools for Functional Data Analysis. Journal of the Royal Statistical Society, Series B. **53**, 539-572 (1991).

[17] Ramsay, J.O. and Silverman, B.W.: Functional Data Analysis. Springer, New York. Second Ed. (2005).

[18] Tsybakov, Alexandre B.: Introduction à l'estimation non-paramétrique (Introduction to nonparametric estimation). Springer, Paris. (2004).

Chapter 34
Presmoothing in Functional Linear Regression

Adela Martínez-Calvo

Abstract We consider the functional linear model with scalar response Y and explanatory variable X valued in a functional space. Functional Principal Components Analysis (FPCA) have been used to estimate the model parameter in recent literature. We propose to modify this methodology by presmoothing either X or Y. For these new estimates, consistency is stated and their efficiency by comparison with the FPCA approach are studied.

34.1 Introduction

Nowadays the progress of computing and measure tools allows us to have access to data that can be observed in a fine grid. In these cases, we can see the data as a discretized version of a functional variable. For this reason the classical regression models have been adapted to the functional context where the response variable Y and/or the explanatory variable X are valued in a functional space (see Ramsay and Silverman, 2005, for a *parametric* state of art and Ferraty and Vieu, 2006, for a *nonparametric* one). In particular, many authors have studied the functional linear model with scalar response and they have proposed techniques for estimating the model parameter, for example using basis function systems as B-splines (Cardot *et al.*, 2003, Ramsay and Silverman, 2005, Crambes *et al.*, 2007). Another well-known approach is based on FPCA and has been developed and analysed in many papers (Cardot *et al.*, 1999, Cardot *et al.*, 2003, Cai and Hall, 2006, Hall and Hosseini-Nasab, 2006, Hall and Horowitz, 2007).

In order to make everything formal, let $(E, \langle\cdot,\cdot\rangle)$ be a real separable Hilbert space (let us denote $\|\cdot\| = \langle\cdot,\cdot\rangle^{1/2}$ its associated norm), and let $\|\cdot\|_{E'}$ be the

Adela Martínez-Calvo
University of Santiago de Compostela, Spain, e-mail: adelamc@usc.es

norm defined as $\|T\|_{E'} = \left(\sum_{k=1}^{\infty}(Te_k)^2\right)^{1/2}$ $\forall T \in E'$, where E' is the dual space of E and $\{e_k\}_{k=1}^{\infty}$ is an orthonormal basis in E.

Let us consider the functional linear model given by

$$Y = m(X) + \varepsilon = \langle X, \theta \rangle + \varepsilon,$$

where Y is a real random variable, X is a random variable valued in E (such that $\mathbb{E}(X) = 0$ and $\mathbb{E}(\|X\|^2) < \infty$), θ is a square integrable parameter valued in E and ε is a real random variable that verifies $\mathbb{E}(\varepsilon) = 0$, $\mathbb{E}(\varepsilon^2) = \sigma^2$ and $\mathbb{E}(\varepsilon X) = 0$. We define the second moment operator Γ (let $\{(\lambda_j, v_j)\}_{j=1}^{\infty}$ be its eigenvalues and eigenfunctions, assuming that $\lambda_1 > \lambda_2 > \ldots$) and the cross second moment operator Δ as

$$\Gamma x = \mathbb{E}(X \otimes_E X(x)) = \mathbb{E}(\langle X, x\rangle X), \qquad \Delta x = \mathbb{E}(X \otimes_{E'} Y(x)) = \mathbb{E}(\langle X, x\rangle Y),$$

$\forall x \in E$. If $\sum_{j=1}^{\infty}(\frac{\Delta v_j}{\lambda_j})^2 < \infty$, Cardot *et al.* (2003) show that the optimization problem

$$\min_{\beta \in E} \mathbb{E}[(Y - \langle \beta, X\rangle)^2] \tag{34.1}$$

has an unique solution θ that satisfies $\langle \Gamma x, \theta\rangle = \Delta x$, $\forall x \in E$. In particular, when $x = v_j$, $\lambda_j\langle v_j, \theta\rangle = \Delta v_j$ for $j = 1, 2, \ldots$. Therefore, the solution of (34.1) can be expressed as

$$\theta = \sum_{k=1}^{\infty}\langle \theta, v_j\rangle v_j = \sum_{k=1}^{\infty}\frac{\Delta v_j}{\lambda_j}v_j.$$

This expansion for θ leads them to $\hat{\theta}_{K_n} = \sum_{j=1}^{K_n}\frac{\Delta_n \hat{v}_j}{\hat{\lambda}_j}\hat{v}_j$, where $\Delta_n = \frac{1}{n}\sum_{i=1}^{n} X_i \otimes_{E'} Y_i$ and $\{(\hat{\lambda}_j, \hat{v}_j)\}_{j=1}^{\infty}$ are the eigenvalues and the eigenfunctions of $\Gamma_n = \frac{1}{n}\sum_{i=1}^{n} X_i \otimes_E X_i$ (assuming that $\hat{\lambda}_1 \geq \hat{\lambda}_2 \geq \ldots$). This estimator converges almost surely (see Cardot *et al.*, 1999) and it is easy to show that $\hat{\theta}_{K_n}$ is the truncated version of the function θ_n that satisfies

$$\langle \Gamma_n x, \theta_n\rangle = \Delta_n x, \qquad \forall x \in E. \tag{34.2}$$

Besides elementary calculations allow us to write its conditional mean-squared error as

$$\mathbb{E}(\|\hat{\theta}_{K_n} - \theta\|^2 | X_1, \ldots, X_n) = \frac{\sigma^2}{n}\sum_{j=1}^{K_n}\frac{1}{\hat{\lambda}_j} + R_{K_n}, \tag{34.3}$$

where $R_{K_n} = \sum_{j > K_n}\langle \theta, \hat{v}_j\rangle^2$.

In this paper, we have revisited this estimator in order to improve its behavior in terms of the conditional mean-squared error (34.3) by presmoothing either X or Y. This is what we call efficiency: among two estimators, the more efficient is the one leading to the smaller conditional mean-squared error.

34.2 Back to the multivariate case

The reason why we have decided to introduce presmoothing techniques comes from the eighties when Faraldo-Roca and González-Manteiga (1985) state the efficiency of linear regression estimates obtained by preliminary nonparametric estimation in the real context (see Cristóbal-Cristóbal *et al.*, 1987, for an extension to the multivariate case). We have realized that those estimates can be seen as a presmoothing of the covariable X. In order to see this, let us suppose that $E = \mathbb{R}^p$. Following the steps taken in Faraldo-Roca and González-Manteiga (1985), we can consider the multivariate Nadaraya-Watson estimator $\hat{m}_H(x) = \frac{\sum_{k=1}^n Y_k K(H^{-1/2}(X_k - x))}{\sum_{k=1}^n K(H^{-1/2}(X_k - x))}$, in which H is a $p \times p$ symmetric positive definite matrix and K is a symmetrical kernel function and solve the problem

$$\min_{\beta \in \mathbb{R}^p} \int (\hat{m}_H(x) - x^t \beta)^2 d\mu_n(x) \tag{34.4}$$

where μ_n is a weighting function.

If we choose $\mu_n(x) = \int_{-\infty}^x \hat{f}_H(t)dt$ where $\hat{f}_H(x) = \frac{1}{n|H|^{1/2}} \sum_{i=1}^n K(H^{-1/2}(X_i - x))$ is the nonparametric estimate of the density function f of X, and K verifies that $\int zz^t K(z)dz = \mu_2(K)I$, it can be shown that θ_n is solution of (34.4) if and only if it satisfies

$$(\Gamma_n + \mu_2(K)H)\,\theta_n = \Delta_n, \tag{34.5}$$

where $\Gamma_n = \frac{1}{n}\sum_{i=1}^n X_i X_i^t$ and $\Delta_n = \frac{1}{n}\sum_{i=1}^n X_i Y_i$. This expression is a perturbation of (34.2) and allows us to define $\hat{\theta}^{H,X} = \sum_{j=1}^p \frac{\Delta_n^t \hat{v}_j^H}{\hat{\lambda}_j^H} \hat{v}_j^H$, where $\{(\hat{\lambda}_j^H, \hat{v}_j^H)\}_{j=1}^p$ are the eigenvalues and the eigenvectors of $\Gamma_n + \mu_2(K)H$. It is important to emphasize that $\hat{\theta}^{H,X}$ is also solution of

$$\min_{\beta \in \mathbb{R}^p} \mathbb{E}_{\hat{F^*}_H}((Y - (X+Z)^t\beta)^2),$$

where Z is a random vector independent of X with density $g(z) = |H|^{-1/2} K(H^{-1/2}(z))$, and $\hat{F^*}_H(x,y,z) = \frac{1}{n}\sum_{i=1}^n 1_{\{Y_i \le y\}} 1_{\{X_i \le x\}} \int_{-\infty}^z g(t)dt$, and therefore it can be seen as a "presmoothing" of X. Moreover, our study of $\hat{\theta}^{H,X}$ allows us to say that this estimate can be more efficient than the least-squares estimates.

In the other hand, when $\mu_n(x) = \frac{1}{n}\sum_{i=1}^n 1_{(-\infty,x]}(X_i)$, we obtain a new solution θ_n of (34.4) that satisfies

$$\Gamma_n \theta_n = \Delta_n^H, \tag{34.6}$$

in which $\Delta_n^H = \frac{1}{n}\sum_{i=1}^n X_i \hat{m}_H(X_i)$ is a "smooth" version of Δ_n. We can introduce the natural estimator $\hat{\theta}^{H,Y} = \sum_{j=1}^p \frac{(\Delta_n^H)^t \hat{v}_j}{\hat{\lambda}_j}\hat{v}_j$. Again (34.6) is a perturbation of (34.2) and here it is clear that we have smoothed Y by means of the nonparametric estimator $\hat{m}_H$. However, in this case we cannot insure the efficiency of $\hat{\theta}^{H,Y}$.

These reasonings have encouraged us to study in the following sections what happens when E is an arbitrary real separable Hilbert space and how presmoothing influences on the conditional mean-squared error.

34.3 Presmoothing Y

Bearing in mind (34.6), let us look for θ_n that verifies $\langle \Gamma_n x, \theta_n \rangle = \Delta_n^h x$ for all $x \in E$, in which $\Delta_n^h = \frac{1}{n}\sum_{i=1}^n X_i \otimes_{E'} \hat{m}_h(X_i)$ is a smooth version of $\Delta_n = \frac{1}{n}\sum_{i=1}^n X_i \otimes_{E'} Y_i$, and $\hat{m}_h(\cdot) = \frac{\sum_{k=1}^n Y_k K(h^{-1}d(\cdot,X_k))}{\sum_{k=1}^n K(h^{-1}d(\cdot,X_k))}$ is the nonparametric estimator proposed in Ferraty and Vieu (2006) and studied in Ferraty *et al.* (2007), where K is an asymmetrical kernel function, h is a strictly positive real and d is a semimetric on the space E. Hence let us define the estimator $\hat{\theta}_{K_n}^{h,Y}$ of θ as

$$\hat{\theta}_{K_n}^{h,Y} = \sum_{j=1}^{K_n} \frac{\Delta_n^h \hat{v}_j}{\hat{\lambda}_j}\hat{v}_j,$$

where $\{(\hat{\lambda}_j, \hat{v}_j)\}_{j=1}^\infty$ are the eigenvalues and the eigenfunctions of $\Gamma_n = \frac{1}{n}\sum_{i=1}^n X_i \otimes_E X_i$. The estimator $\hat{\theta}_{K_n}^{h,Y}$ converges and we have obtained its conditional mean-squared error.

Theorem 34.1. *If $m(x) = \langle x, \theta \rangle$ and $m_{K_n}^{h,Y}(x) = \langle x, \hat{\theta}_{K_n}^{h,Y} \rangle$, under some assumptions*

$$\left\| \hat{m}_{K_n}^{h,Y} - m \right\|_{E'} \to 0 \qquad a.s.$$

Let us introduce the *small ball probability function* $\varphi_{X_0}(h) = P(d(X_0, X) \leq h)$ and let us assume that it can be expressed as $\varphi_{X_0}(h) \approx \phi(h) f(X_0)$ for "small" h.

Theorem 34.2. *Under some assumptions,*

$$\mathbb{E}(\|\hat{\theta}_{K_n}^{h,Y} - \theta\|^2 | X_1, \ldots, X_n) \approx \frac{\sigma^2}{n\phi(h)} \sum_{j=1}^{K_n} \frac{A}{\hat{\lambda}_j} + h^2 \sum_{j=1}^{K_n} \frac{B}{\hat{\lambda}_j^2} + R_{K_n} \tag{34.7}$$

where A and B are positive constants and R_{K_n} is defined as in formula (34.3).

34.4 Presmoothing X

First presmoothing. By analogy with (34.5), we can look for a function θ_n in E such that

$$\langle (\Gamma_n + \Gamma_Z)x, \theta_n \rangle = \Delta_n x, \qquad \forall x \in E, \tag{34.8}$$

where Γ_Z is the second moment operator of a centered process Z independent of X and, if $\{(\hat{\lambda}_j^Z, \hat{v}_j^Z)\}$ are the eigenvalues and eigenfunctions of $\Gamma_n^Z = \Gamma_n + \Gamma_Z$, we can propose

$$\hat{\theta}_{K_n}^Z = \sum_{j=1}^{K_n} \frac{\Delta_n \hat{v}_j^Z}{\hat{\lambda}_j^Z} \hat{v}_j^Z .$$

For this estimator we have the following results.

Theorem 34.3. *If $m(x) = \langle x, \theta \rangle$ and $m_{K_n}^Z(x) = \langle x, \hat{\theta}_{K_n}^Z \rangle$, under some assumptions*

$$\left\| \hat{m}_{K_n}^Z - m \right\|_{E'} \to 0 \qquad a.s.$$

Theorem 34.4. *If $R_{K_n}^Z = \sum_{j>K_n} \langle \theta, \hat{v}_j^Z \rangle^2$, under some assumptions*

$$\mathbb{E}(\|\hat{\theta}_{K_n}^Z - \theta\|^2 | X_1, \ldots, X_n) = \frac{\sigma^2}{n} \sum_{j=1}^{K_n} \frac{\langle \Gamma_n \hat{v}_j^Z, \hat{v}_j^Z \rangle}{\left(\hat{\lambda}_j^Z\right)^2} +$$

$$\sum_{j=1}^{K_n} \left(\langle \theta, \frac{\Gamma_n \hat{v}_j^Z}{\hat{\lambda}_j^Z} - \hat{v}_j^Z \rangle \right)^2 + R_{K_n}^Z .$$

In order to simplify the previous expression, let us suppose that $\Gamma_Z = \alpha I$. In this case $\{(\hat{\lambda}_j^Z, \hat{v}_j^Z)\} \equiv \{(\hat{\lambda}_j + \alpha, \hat{v}_j)\}$ and $\hat{\theta}_{K_n}^Z$ can be written as $\hat{\theta}_{K_n}^{\alpha,X} = \sum_{j=1}^{K_n} \frac{\Delta_n \hat{v}_j}{\hat{\lambda}_j + \alpha} \hat{v}_j$.

Corollary 34.1. *Under some assumptions,*

$$\mathbb{E}(\|\hat{\theta}_{K_n}^{\alpha,X} - \theta\|^2 | X_1, \ldots, X_n) \approx \tag{34.9}$$

$$\mathbb{E}(\|\hat{\theta}_{K_n} - \theta\|^2 | X_1, \ldots, X_n) - \frac{2\alpha\sigma^2}{n} \sum_{j=1}^{K_n} \frac{1}{\hat{\lambda}_j^2} + \alpha^2 \sum_{j=1}^{K_n} \frac{\langle \theta, \hat{v}_j \rangle^2}{\hat{\lambda}_j^2} .$$

Second presmoothing. Given that we have turned the presmoothing in X into a perturbation of the empirical second moment operator Γ_n (remind

(34.8)) and their associated eigenvalues and eigenfunctions, we have considered other kind of smoothing FPCA such as those proposed by Pezzulli and Silverman (1993) and Silverman (1996).

Pezzulli and Silverman (1993) study the properties of $\{(\tilde{\lambda}_j, \tilde{v}_j)\}_{j=1}^{\infty}$ solution of the generalized eigenproblem $(\Gamma_n - \alpha Q)\,\tilde{v} = \tilde{\lambda}\tilde{v}$ where α is a "small" positive real and Q is a symmetric nonnegative operator. The standard technique of asymptotic expansions allows us to write $\tilde{v}_j = (1 - \alpha\Pi_j Q + o(\alpha))\,\hat{v}_j$ and $\tilde{\lambda}_j = \hat{\lambda}_j - \alpha\rho_j + o(\alpha)$ for each j, where $\Pi_j = \sum_{k\neq j}(\hat{\lambda}_j - \hat{\lambda}_k)^{-1}P_k$ (P_k is the projection onto the subspace of E spanned by $\hat{v}_k$) and $\rho_j = \langle \hat{v}_j, Q\hat{v}_j\rangle$. Defining $\tilde{\theta}_{K_n}^{PS} = \sum_{j=1}^{K_n} \frac{\Delta_n \tilde{v}_j}{\tilde{\lambda}_j}\tilde{v}_j$ and following the steps given for obtain Theorem 34.4, we can link the conditional mean-squared errors of $\tilde{\theta}_{K_n}^{PS}$ and $\hat{\theta}_{K_n}$.

Theorem 34.5. *Under some assumptions,*

$$\mathbb{E}(\|\tilde{\theta}_{K_n}^{PS} - \theta\|^2|X_1,\ldots,X_n) \approx \mathbb{E}(\|\hat{\theta}_{K_n} - \theta\|^2|X_1,\ldots,X_n) + \frac{2\alpha\sigma^2}{n}\sum_{j=1}^{K_n}\frac{\rho_j}{\hat{\lambda}_j^2}$$
$$+\alpha^2\sum_{j=1}^{K_n}\frac{\langle\theta, Q\hat{v}_j\rangle^2}{\hat{\lambda}_j^2} - 2\alpha\sum_{j>K_n}\langle\theta,\hat{v}_j\rangle\langle\theta,\Pi_j Q\hat{v}_j\rangle. \tag{34.10}$$

Let us consider the smoothed FPCA proposed by Silverman (1996) who works with the eigenproblem $\Gamma_n\tilde{v} = \tilde{\lambda}\,(I + \alpha Q)\,\tilde{v}$ (and with different normalization conditions from Pezzulli and Silverman's ones). In this case, we can write $\tilde{v}_j = \left(1 - \alpha\left(\frac{\rho_j}{2} + \hat{\lambda}_j\Pi_j Q\right) + o(\alpha)\right)\hat{v}_j$ and $\tilde{\lambda}_j = \hat{\lambda}_j\,(1 - \alpha\rho_j + o(\alpha))$, and define the estimator $\tilde{\theta}_{K_n}^{S} = \sum_{j=1}^{K_n}\frac{\Delta_n\tilde{v}_j}{\tilde{\lambda}_j}\tilde{v}_j$.

Theorem 34.6. *Under some assumptions,*

$$\mathbb{E}(\|\tilde{\theta}_{K_n}^{S} - \theta\|^2|X_1,\ldots,X_n) \approx \mathbb{E}(\|\hat{\theta}_{K_n} - \theta\|^2|X_1,\ldots,X_n) + \frac{\alpha\sigma^2}{n}\sum_{j=1}^{K_n}\frac{\rho_j}{\hat{\lambda}_j^2}$$
$$+\alpha^2\sum_{j=1}^{K_n}\langle\theta, Q\hat{v}_j\rangle^2 - \alpha\left(\sum_{j>K_n}\rho_j\langle\theta,\hat{v}_j\rangle^2 + 2\sum_{j>K_n}\hat{\lambda}_j\langle\theta,\hat{v}_j\rangle\langle\theta,\Pi_j Q\hat{v}_j\rangle\right). \tag{34.11}$$

34.5 Comments about efficiency

In this paper, we have considered different estimators for the linear model parameter θ and we have obtained the expressions of their conditional mean-squared error. Looking at (34.3), (34.7), (34.9), (34.10) and (34.11), we can deduce that, in general, presmoothing of Y fails in comparison with $\hat{\theta}_{K_n}$, while presmoothing of X can be more efficient that $\hat{\theta}_{K_n}$ with an adequate choice of smoothing parameter α (depending on the selection of Q for $\tilde{\theta}^{PS}_{K_n}$ and $\tilde{\theta}^{S}_{K_n}$). On the other hand, simulations have confirmed these conclusions. In addition, these expressions allow us to drive the smoothing parameter selection to obtain efficiency.

References

[1] Cai, T.T. and Hall, P.: Prediction in functional linear regression. Annals of Statistics **34**, 2159–2179 (2006).

[2] Cardot, H., Ferraty, F. and Sarda, P.: Functional linear model. Statistics and Probability Letters **45** (1), 11–22 (1999).

[3] Cardot, H., Ferraty, F. and Sarda, P.: Spline estimators for the functional linear model. Statistica Sinica **13** (3), 571–591 (2003).

[4] Crambes, C., Kneip, A. and Sarda, P.: Smoothing splines estimators for functional linear regression. submitted to Annals of Statistics.

[5] Cristóbal-Cristóbal, J. A., Faraldo-Roca, P. and González-Manteiga, W.: A class of linear regression parameter estimators constructed by nonparametric estimation. Annals of Statistics **15** (2), 603–609 (1987).

[6] Faraldo-Roca, P. and González-Manteiga, W.: On efficiency of a new class of linear regression estimates obtained by preliminary non-parametric estimation. New Perspectives in Theoretical and Applied Statistics (New York) (M. Puri, ed.), Wiley. 229–242 (1985).

[7] Ferraty, F., Mas, A. and Vieu, P.: Nonparametric regression on functional data : inference and practical aspects. Australian and New Zealand Journal of Statistics **49** (3), 267–286 (2007).

[8] Ferraty, F. and Vieu, P: Nonparametric functional data analysis : theory and practice. Springer, New York. (2006).

[9] Hall, P and Horowitz, J.L.: Methodology and convergence rates for functional linear regression. Annals of Statistics **35** (1), 70–91 (2007).

[10] Hall, P. and Hosseini-Nasab, M.: On properties of functional principal components analysis. Journal of the Royal Statistical Society: Series B. **68** (1), 109–126 (2006).

[11] Pezzulli, S. and Silverman, B.W.: Some properties of smoothed principal components analysis for functional data. Computational Statistics. **8**, 1–16 (1993).

[12] Ramsay, J.O. and Silverman, B.W.: Functional data analysis. Springer, New York. (2005).

[13] Silverman, B.W.: Smoothed functional principal components analysis by choice of norm. Annals of Statistics **24** (1), 1–24 (1996).

Chapter 35
Probability Density Functions of the Empirical Wavelet Coefficients of Multidimensional Poisson Intensities

José Carlos Simon de Miranda

Abstract We determine the probability density functions of the empirical wavelet coefficient estimator $\hat{\beta}_\eta = \int \psi_\eta dN$ in the wavelet series expansion $\hat{p} = \sum \hat{\beta}_\eta \psi_\eta$ of non homogeneous multidimensional Poisson processes intensity functions.

35.1 Introduction

Estimation of non homogeneous Poisson intensities is a research subject of both theoretical and practical interest given the importance of Poisson processes in point processes theory as well as its use in a large number of practical applications. Parametric, semi parametric, non parametric and bayesian methods have been used to estimate Poisson intensities. We cite Timmermann and Novak (1998) to Miranda (2006). In our works Miranda (2005), Miranda (2003) and Miranda and Morettin (2006) we have studied the wavelet estimator $\hat{p} = \sum \hat{\beta}_\eta \psi_\eta$ of p, the intensity of a general Point process. In this article we specialize to non homogeneous Poisson processes. This restriction on the set of probability structures is strong enough to let us obtain the probability density function of the wavelet coefficient estimators $\hat{\beta}_\eta$ and yet not too strong as to forbid its practical and theoretical use as seen above. Mathematical and statistical aspects of wavelet theory can be found in Meyer (1993) to Donoho *et al.* (1996).

This article is organized as follows. In section 2 we present some basics and notations, in section 3 we state and prove the main results and in section 4 we make some remarks.

José Carlos Simon de Miranda
University of São Paulo, São Paulo, São Paulo, Brazil, e-mail: simon@ime.usp.br

35.2 Some basics and notations

Let N be a point process on $\mathbb{R}^d$, with unknown intensity p. Let $\{\psi_{j,i}|i,j \in \mathbf{Z}\}$ be an orthonormal wavelet basis of $L^2(\mathbb{R})$ of the form $\psi_{j,i}(t) = 2^{j/2}\psi(2^j t - i)$ or $\psi_{j,i}(t) = 2^{j/2}\psi(2^j(t-t_1)+t_1-iT)$ for some mother wavelet ψ obtained, if necessary by the composition of a standard wavelet with an affine transformation, such that its support is $[t_1, t_2]$ with $T = t_2 - t_1$. Here i corresponds to translations and j to dilations. Let ϕ be the father wavelet corresponding to ψ. Similarly, let $\{\phi_{\ell i,k}, \psi_{j,i} : i,k \in \mathbf{Z},\ j \geq \ell i, j,\ \ell i \in \mathbf{Z}\}$ be an orthonormal wavelet basis that contains all the scales beyond some fixed extended integer ℓi. It is extremely pleasant to adopt the following notation. Let ${}_d\mathbf{Z} = \{z \in \mathbf{Z} : z \geq d\}, d \in \mathbf{Z} \cup \{-\infty\}$ and define $Ze(\ell i) = \begin{cases} \mathbf{Z} \cup ({}_{\ell i}\mathbf{Z} \times \mathbf{Z}) \text{ if } \ell i \in \mathbf{Z}, \\ \mathbf{Z}^2 \text{ if } \ell i = -\infty. \end{cases}$

Let us use Greek letters for indexes in $Ze(\ell i)$ and we shall write $\psi_\eta = \phi_{\ell i,\eta}$ if and only if $\eta \in \mathbf{Z}$ and $\psi_\eta = \psi_{j,i}$ if and only if $\eta = (j,i) \in \mathbf{Z}^2$. Thus, the wavelet expansions $f(t) = \sum_{i\in\mathbf{Z}}\sum_{j\in\mathbf{Z}} \delta_{ji}\psi_{j,i}(t)$ and $f(t) = \sum_{k\in\mathbf{Z}} \gamma_k \phi_{\ell i,k}(t) + \sum_{i\in\mathbf{Z}}\sum_{j\in{}_{\ell i}\mathbf{Z}} \delta_{ji}\psi_{j,i}(t)$ will be simply written $f = \sum_{\eta\in Ze(\ell i)} \alpha_\eta \psi_\eta$, for α_η given by $\int_{-\infty}^{\infty} f\psi_\eta dt = \int_{\mathbb{R}}(\sum_\xi \alpha_\xi\psi_\xi)\psi_\eta dt = \sum_\xi \int_{\mathbb{R}} \alpha_\xi\psi_\xi\psi_\eta dt = \sum_\xi \alpha_\xi < \psi_\xi, \psi_\eta >= \alpha_\eta$. Let for all n, $1 \leq n \leq d$, $\{\psi_{n,j,i}|i,j \in \mathbf{Z}\}$, $\psi_{n,j,i}(t) = 2^{j/2}\psi_n(2^j t - i)$ or $\psi_{n,j,i}(t) = 2^{j/2}\psi_n(2^j(t - a_n) + a_n - iT_n)$ and $\{\phi_{n,\ell i,k}, \psi_{n,j,i} : i,k \in \mathbf{Z},\ j \geq \ell i_n, j,\ \ell i_n \in \mathbf{Z} \cup \{-\infty\}\}$ be orthonormal wavelet bases of $L^2(\mathbb{R})$ as above where $\operatorname{supp}\psi_n = [a_n, b_n]$ and $T_n = b_n - a_n$.

These bases are simply written as $\{\psi_{n,\eta_n}|\eta_n \in Ze(\ell i_n)\}$ and they are, under restriction, also orthonormal bases of $L^2[a_n, b_n]$, $1 \leq n \leq d$. Taking tensor products we form the orthonormal basis $\{\tilde\psi_{\tilde\eta}|\tilde\psi_{\tilde\eta} = \otimes_{n=1}^d \psi_{n,\eta_n}, \tilde\eta = (\eta_1, \ldots, \eta_d) \in \prod_{n=1}^d Ze(\ell i_n)\}$ of $L^2(\mathbb{R}^d)$ and also, under restriction, of $L^2(\prod_{n=1}^d [a_n, b_n])$. Denote $\prod_{n=1}^d Ze(\ell i_n)$ by $\mathcal{Z}e(\tilde{\ell i})$; $\tilde{\ell i} = (\ell i_1, ..., \ell i_d)$. From now on we will drop the tilde and use simple notation for vectors in $\mathbb{R}^d$, tensor product wavelets and d-tuples in $\mathcal{Z}e(\tilde{\ell i})$. In this way if $f \in L^2(\mathbb{R}^d)$ we have $f = \sum_{\eta\in\mathcal{Z}e(\ell i)} \alpha_\eta\psi_\eta$ with $\alpha_\eta = \int_{\mathbb{R}^d} f\psi_\eta \mathrm{d}\ell$.

Frequently we want to obtain the restriction of p to $\prod_{n=1}^d [a_n, b_n] = [a,b] = \mathcal{O}$, an observation region, based on the points of a trajectory of the process that are contained in this $\mathbb{R}^d$ interval.

From now on we assume that p is locally square integrable. Therefore for the wavelet expansion of p restricted to bounded $\mathbb{R}^d$ interval observation regions, we have

$$p = \sum_\eta \beta_\eta \psi_\eta, \tag{35.1}$$

with

$$\beta_\eta = \int_{\mathbb{R}^d} p\psi_\eta d\ell. \tag{35.2}$$

The main estimation purpose is to obtain p through the expansion (1) and for this we need to estimate the wavelet coefficients β_η given by (2). The unbiased estimator we use is $\hat{\beta}_\eta = \int \psi_\eta dN$.

We use $O_F = (0,\ldots,0) \in \mathcal{Z}e(\ell i)$, $O_M = ((\ell i_1,0),\ldots,(\ell i_d,0)) \in \mathcal{Z}e(\ell i)$. We write for $\eta \in Ze(\ell i)$, $j(\eta) = \ell i$ if $\eta \in \mathbf{Z}$ and $j(\eta) = j$ if $\eta = (j,i)$. Also, if $\eta \in \mathcal{Z}e(\ell i)$, $j(\eta) = (j(\eta_1),\ldots,j(\eta_m))$ and $|j(\eta)| = \sum_{\ell=1}^m j(\eta_i)$.

35.3 Main results

In this section we present the central results of this paper. Theorem 1 tells us how to obtain the probability density function of the empirical wavelet coefficient $\hat{\beta}_\eta$. Note that this function depends on both the wavelet ψ_η and the Poisson intensity $p(x)$. Corollary 1 presents the series expansion of the characteristic function of $\hat{\beta}_\eta$ and Corollary 2 gives formulas for the first four centered moments of $\hat{\beta}_\eta$ as well as its asymmetry and kurtosis coefficients. Theorem 2 in the analog of Theorem 1 for the specific case of Haar wavelets.

Theorem 35.1. *Let N be a Poisson process on $\mathbb{R}^d$ with intensity function $p : \mathbb{R}^d \to \mathbb{R}_+$. Suppose the wavelet ψ_η is compactly supported and continuous. Then $f_\eta : \mathbb{R} \to \mathbb{R}_+$, the probability density function of $\hat{\beta}_\eta = \int_{\mathbb{R}^d} \psi_\eta dN$, is given by the principal value*

$$f_\eta(y) = \frac{1}{2\pi}\int_{\mathbb{R}} exp\left(\int_{supp\psi_\eta} p(x)\left(cos\left(w\psi_\eta(x)\right) - 1\right)dx\right)$$
$$cos\left(\int_{supp\psi_\eta} p(x) sin\left(w\psi_\eta(x)\right)dx - wy\right)dw.$$

From the proof of theorem 1, the characteristic function of $\hat{\beta}_\eta$ is given by the following:

Corollary 35.1. *Under theorem's 1 hypothesis we have*

$$\mathrm{E}(e^{iw\hat{\beta}_\eta}) = 1+$$
$$\sum_{n=1}^{\infty}\sum_{(\sum_{m=1}^{\infty} i_m = n)}\left\{\frac{(iw)^{\sum_{m=1}^{\infty} m i_m}}{\prod_{m=1}^{\infty}(i_m!(m!)^{i_m})}\prod_{m=1}^{\infty}\left(\int_{supp\psi_\eta} p(x)\psi_\eta^m(x)dx\right)^{i_m}\right\}$$

Since we have the series expansion of the characteristic function of $\hat{\beta}_\eta$ its moments are easily obtained. The variance, asymmetry and kurtosis of the wavelet coefficient distributions is the subject of the following:

Corollary 35.2. *Under theorem's 1 hypothesis, we have*

$$\beta_\eta = \mathrm{E}(\hat{\beta}_\eta) = \int \psi_\eta p d\ell, \ var(\hat{\beta}_\eta) = \int \psi_\eta^2 p d\ell,$$

$$\mu_3(\hat{\beta}_\eta) = \int \psi_\eta^3 p d\ell \ and \ \mu_4(\hat{\beta}_\eta) = \int \psi_\eta^4 p d\ell + 3(\int \psi_\eta^2 p d\ell)^2$$

so that the coefficients of asymmetry α_3 and kurtosis α_4 are given by:

$$\alpha_3(\eta) = \frac{\int \psi_\eta^3 p d\ell}{(\int \psi_\eta^2 p d\ell)^{(3/2)}} \ and \ \alpha_4(\eta) = 3 + \frac{\int \psi_\eta^4 p d\ell}{(\int \psi_\eta^2 p d\ell)^2}.$$

One of the most important and used wavelet families is the Haar family. This is a consequence of the extremely simple forms of its scale function and mother wavelet that makes it computationally easier to use Haar wavelets instead of other more elaborated ones. However, Haar wavelets are not continuous and theorem 1 does not apply to them. In this way, we present the following:

Theorem 35.2. *Let N be a Poisson process on $\mathbb{R}^d$ with intensity function $p : \mathbb{R}^d \to \mathbb{R}_+$. Denote $\psi_\eta^+ = (|\psi_\eta| + \psi_\eta)/2$ and $\psi_\eta^- = (|\psi_\eta| - \psi_\eta)/2$. Suppose the wavelet family used is Haar, that is, the wavelets in this family are tensor products of one dimensional Haar wavelets only. Then $\hat{\beta}_\eta \sim ||\psi_\eta||_\infty(X^+ - X^-)$, where X^+ and X^- are independent Poisson random variables with means $\lambda_\eta^+ = \int_{supp\psi_\eta^+} p d\ell$ and $\lambda_\eta^- = \int_{supp\psi_\eta^-} p d\ell$. The probability function of $\hat{\beta}_\eta$, $f_\eta : ||\psi_\eta||_\infty \mathbf{Z} \to \mathbb{R}_+$, is given by:*

$$f_\eta(||\psi_\eta||_\infty z) = exp(-\int_{supp\psi_\eta} p d\ell) \sum_{k \geq max\{0,z\}} \frac{(\lambda_\eta^+)^k (\lambda_\eta^-)^{k-z}}{k!(k-z)!}.$$

35.4 Final remarks

We remark that if the intensity may be regarded as constant on $supp\psi_\eta$ then we can write the following approximations:

$$\alpha_3(\eta) = \frac{\int \psi_\eta^3 p d\ell}{(\int \psi_\eta^2 p d\ell)^{(3/2)}} \cong \frac{2^{|j(\eta)|/2}}{p^{\frac{1}{2}}} \int \psi_{z(\eta)}^3 d\ell$$

and

$$\alpha_4(\eta) = 3 + \frac{\int \psi_\eta^4 p d\ell}{(\int \psi_\eta^2 p d\ell)^2} \cong 3 + \frac{2^{|j(\eta)|}}{p} \int \psi_{z(\eta)}^4 d\ell,$$

where $\psi_{z(\eta)}$ is any re-scaled wavelet that corresponds to ψ_η such that $j(z(\eta)) = 0 \in \mathbf{Z}^d$. Since, for all $\eta \in \mathcal{Z}e(\ell i)$ $\int \psi_{z(\eta)}^3 d\ell$ and $\int \psi_{z(\eta)}^4 d\ell$ are limited by constants, we observe that for continuous intensities the kurtosis coefficient will increase without bound as $|j(\eta)|$ goes to infinity; and the same will happen to the absolute value of the asymmetry coefficient in case the wavelet has non vanishing integral of its third power. Note that one can have all $\alpha_3(\eta)$'s equal to zero if the multidimensional wavelet basis is formed by tensor products of one dimensional wavelets such that the integral of the third power of each of these wavelets is zero.

It is also worth noting that in case we have n independent replications of the Poisson process, i.e. we have n independent trajectories of the process, we can form the estimators $\tilde{\beta}_\eta = \frac{1}{n}\sum_{i=1}^{n}\hat{\beta}_\eta(i)$, $\tilde{p} = \frac{1}{n}\sum_{i=1}^{n}\hat{p}(i)$, where $\hat{\beta}_\eta(i)$ and $\hat{p}(i)$ are the estimated wavelet coefficient and intensity obtained from the ith observation. These estimators inherit the unbiasedness of $\hat{\beta}_\eta(i)$ and $\hat{p}(i)$. Moreover, $\tilde{\beta}_\eta$ also presents the desired feature of asymptotical normality as a consequence of the finiteness of the first and second moments of the wavelet coefficient estimators $\hat{\beta}_\eta(i)$ that guarantees the central limit theorem can be applied to the independent sum $\tilde{\beta}_\eta$. As a matter of fact the asymptotic normality of $\tilde{\beta}_\eta$ is not restricted to Poisson process setting; in Miranda and Morettin (2005), Miranda (2003) and Miranda and Morettin (2006) we have also shown that the finiteness requirements mentioned above are also valid for a larger class of point processes so that they will also exhibit this feature in case of independent replications.

Acknowledgements The author thanks our Lord and Saviour Jesus Christ. This work was partially supported by FAPESP grant 03/10105-2.

References

[1] Timmermann, K. E. and Novak, R. D.: Multiscale Bayesian Estimation of Poisson Intensities. IEEE. 85-90 (1998).

[2] Novak, R. D. and Kolaczyk, E. D.: A Multiscale MAP Estimation Method for Poisson Inverse Problems. IEEE 1682-1686 (1998).

[3] Heikkinen, J. and Arjas, E.: Non-Parametric Bayesian Estimation of a Spatial Poisson Intensity. Scand J Statist. **25** 435-450 (1998).

[4] Novak, R. D. and Kolaczyk, E. D.: Multiscale Maximum Penalized Likelihood Estimators. IEEE. **156**, (2002).

[5] Müller, P. and Vidakovic, B.: Bayesian Inference with Wavelets: Density Estimation Journal of computational and Graphical Statistics. **7** (4), 456-468 (1998).

[6] Barber, S., Nason, G. P. and Silverman, B. W.: Posterior Probability Intervals for Wavelet Thresholding, J. R. Statist. Soc. B. **64**, part2, 189-205 (2002).

[7] Kolaczyk, E. D. and Novak, R. D.: Multiscale Likelihood Analysis and Complexity Penalized Estimation The Annals of Statistics. **32** (2), 2004, 500-527 (2004).

[8] Lam, W. M. and Wornell, G. W.: Multiscale Representation and Estimation of Fractal Point Processes IEEE Transactions on Signal Processing. **43** (11), 2606-2617 (1995).
[9] Winter, A., Maître, H., Cambou, N. and Legrand, E.: Object Detection Using A Multiscale Probability Model IEEE. 269-272 (1996).
[10] Figueiredo, M. A. T. and Novak, R. O.: Wavelet-Based Image Estimation: An Empirical Bayes Approach Using Jeffrey's Noninformative Prior. IEEE Transaction on Image Processing. **10** (9), September, 1322–1331 (2001).
[11] Miranda, J.C.S. and Morettin, P.A.: Estimation of the Density of Point Processes on $\mathbb{R}^m$ via Wavelets, Technical Report - Department of Mathematics -IME-USP. No. **09**, June, 2005.
[12] Miranda, J.C.S.: Sobre a estimação da intensidade dos processos pontuais via ondaletas. São Paulo. 92 p. Tese de Doutorado. Instituto de Matemática e Estatística da Universidade de São Paulo. (2003).
[13] Miranda, J.C.S. and Morettin, P.A.: On the Estimation of the Intensity of Point Processes on via Wavelets, Technical Report - Department of Statistics - IME-USP. **6**, (2006).
[14] De Miranda, J.C.S.: Adaptive Maximum Probability Estimation of Multidimensional Poisson Processes Intensity Function. Technical Report - Department of Mathematics -IME-USP. **01**, March. (2006).
[15] Meyer, Y.: Wavelets and Operators, Cambridge Studies in Advanced Mathematics, **37**, April. (1993).
[16] Daubechies, I.: Ten Lectures on Wavelets, Philadelphia, P.A. Society for Industrial and Applied Mathematics (CBMS - NSF Regional Conference Series in Applied Mathematics) **61** (1992).
[17] Donoho, D. L., Johnstone, I. M.: Ideal Spatial Adaptation by Wavelet Shrinkage Biometrika. **81** (3), 425-455 (1994).
[18] Donoho, D. L., Johnstone, I. M. Kerkyacharian, G. and Picard, D. : Wavelet Shrinkage: Asymptopia. J. R. Statist. Soc. B. **57** (2), 301-369 (1995).
[19] Donoho, D. L., Johnstone, I. M. I. M., Kerkyacharian, G. and D. Picard: Density Estimation by Wavelet Thresholding. The Annals of Statistics. **24** (2), 508-539 (1996).

Chapter 36
A Cokriging Method for Spatial Functional Data with Applications in Oceanology

Pascal Monestiez and David Nerini

Abstract We propose a method based on a functional linear model which takes into account the spatial dependencies between sampled functions. The problem of estimating a function when spatial samples are available is turned to a standard cokriging problem for suitable choices of the regression function. This work is illustrated with environmental data in Antarctic where marine mammals operate as samplers. In the framework of second order stationarity, the application points out some difficulties when estimating the structure of spatial covariance between observations.

36.1 Introduction

One of the hottest challenge for Functional Data Analysis (Ramsay and Silverman, 2005) is the development of statistical methods adapted for spatially connected curves. Even if theoretical works are currently scarce in this area (Dabo-Niang and Yao, 2007), many applications raise the need to include spatial relations between functional variables into statistical analysis especially in environmental sciences (Meiring, 2007). In oceanography, surveys provide vertical profiles of temperature, salinity or other variables that are spatially dependent and sampled along the depth. Most of the time, the analysis of such data involves geostatistical methods (Wackernagel, 2003). In the best case, the vertical dimension is included as a third spatial dimension and analysis is

Pascal Monestiez
INRA, Unité de Biostatistique et Processus Spatiaux Domaine Saint Paul, Site Agroparc, 84914 AVIGNON Cedex, France, e-mail: Pascal.Monestiez@avignon.inra.fr

David Nerini
Laboratoire de Microbiologie, Géochimie et Ecologie Marines UMR 6117 CNRS, Centre d'Océanologie de Marseille Case 901, Campus de Luminy, 13288 MARSEILLE Cedex, France, e-mail: david.nerini@univmed.fr

achieved with standard kriging. However, this approach is often problematic due to strong and complex anisotropy and to non-stationarity along the vertical dimension. An alternative way is to discretize the curves and to modelize them using multivariate geostatistics as in Goulard and Voltz (1992). The latter approach also suffers from several drawbacks : data analysis does not include the functional form of the variable along the profile and computation is rapidely limited when profiles are recorded on a fine grid. Thus, the aim of this work is to propose a method which extend the coregionalization approach taking into account the functional nature of the data. We show that, in the framework of a functional linear model with spatial dependancy, the estimation of the regressor reduces to a multivariate cokriging problem, for suitable choice of the regression function. The method is illustrated with data analysis of temperature profiles in Antarctic.

36.2 Spatial linear model

Let us consider a collection of curves $E = \{y_i, i = 1, ..., N\}$ sampled at N random spatial locations $\mathbf{x}_i$ over a domain $\mathcal{D}$. Each $y_i(t)$ is an unique observation of $Y_i(t)$, random function at location $\mathbf{x}_i$ where argument t varies in a compact interval τ of $\mathbb{R}$. The functions Y_i take values in a Hilbert space $\mathcal{H}$ with an associated norm denoted by $\|\cdot\|$. If we suppose second order stationnarity, the spatial covariance function C_{ij} between Y_i at location $\mathbf{x}_i$ and Y_j at location $\mathbf{x}_j = \mathbf{x}_i + \mathbf{h}$ reads

$$\begin{cases} \mathbb{E}(Y_i) = \mu, \forall \mathbf{x}_\mathrm{i} \in \mathcal{D} \\ C_{ij} = \mathbb{E}\left[(Y_i - \mu) \otimes (Y_j - \mu)\right] \end{cases} .$$

The mean of each function is equal to the same function $\mu(t)$ at any point of the domain. The spatial covariance only depends on the vector $\mathbf{h}$ connecting the functional variable pair and is invariant for any translation of the pair into the domain. Following the work of Cardot *et al.* (1999), we seek to estimate Y_0, the curve at unknown location $\mathbf{x}_0$, with the linear model

$$\widehat{Y}_0(t) = \sum_{i=1}^{N} \int_\tau \beta_i(s,t) Y_i(s)\, ds \tag{36.1}$$

such that the weighting functions $\beta_i(s,t)$ are chosen to minimize

$$\mathbb{E}\left\|\widehat{Y}_0 - Y_0\right\|^2 .$$

We turn the problem of estimating the $\beta_i(s,t)$ by considering the case where the Y_i's are expressed in terms of a linear combination of K known basis functions $\phi_1, ..., \phi_K$

$$Y_i(t) = \sum_{k=1}^{K} \alpha_k(\mathbf{x}_i)\,\phi_k(t) = \boldsymbol{\alpha}_i'\boldsymbol{\phi}(t)$$

where $\boldsymbol{\alpha}_i = (\alpha_1(\mathbf{x}_i), ..., \alpha_K(\mathbf{x}_i))'$ is the vector of coefficients at location $\mathbf{x}_i$ and $\boldsymbol{\phi}(t) = (\phi_1(t), ..., \phi_K(t))'$ the vector of basis functions. In the same way, the bivariate function $\beta_i(s,t)$ is expanded in the ϕ-basis such that

$$\begin{aligned}\beta_i(s,t) &= \sum_{k=1}^{K}\sum_{l=1}^{K} b_{kl}^i \phi_k(s)\,\phi_l(t)\\ &= \boldsymbol{\phi}'(s)\,\mathbf{B}_i\boldsymbol{\phi}(t)\end{aligned}$$

for each (s,t) belonging to $\tau \times \tau$, where $\mathbf{B}_i$ is the $K \times K$ matrix of coefficients b_{kl}^i at location $\mathbf{x}_i$. Replacing these expressions in (36.1) gives

$$\widehat{Y}_0(t) = \sum_{i=1}^{N} \boldsymbol{\alpha}_i'\mathbf{W}\mathbf{B}_i\boldsymbol{\phi}(t).$$

For a suitable choice of an orthonormal basis, the matrix

$$\mathbf{W} = \int \boldsymbol{\phi}(t)\,\boldsymbol{\phi}'(t)$$

of the inner products of the ϕ-basis is identity and

$$\widehat{Y}_0(t) = \sum_{i=1}^{N}\sum_{k,l=1}^{K} b_{kl}^i \alpha_l(\mathbf{x}_i)\,\phi_k(t).$$

The computation of $\widehat{Y}_0$ expressed in terms of linear combination of known functions, is equivalent to an ordinary cokriging on coefficients of sampled curves in the isotopic case *i. e.* when all coefficients are available at all sampling points.

36.3 Cokriging on coefficients

The stationarity hypothesis of the random functions Y_i expressed into the ϕ-basis becomes

$$\begin{cases} \mathbb{E}(\boldsymbol{\alpha}_i) = \mathbf{a},\ \forall \mathbf{x}_i \in \mathcal{D} \\ \mathbf{C}_{ij} = \mathbb{E}\left[(\boldsymbol{\alpha}_i - \mathbf{a})(\boldsymbol{\alpha}_j - \mathbf{a})'\right] \end{cases}$$

where the mean $\mathbf{a}$ is a $K-$vector of coefficients, $\mathbf{C}_{ij}$ the $K \times K$ cross covariance matrix with entries $cov(\alpha_k(\mathbf{x}_i), \alpha_l(\mathbf{x}_j))$, $k,l = 1,...,K$. Here again, the cross-covariance only depends on distance between locations $\mathbf{x}_i$ and $\mathbf{x}_j$. The cokriging estimator of $\boldsymbol{\alpha}_0$ at location $\mathbf{x}_0$ is defined as

$$\widehat{\boldsymbol{\alpha}}_0 = \sum_{i=1}^{N} \mathbf{B}_i^{'} \boldsymbol{\alpha}_i \tag{36.2}$$

which minimizes

$$trace\left(var\left(\widehat{\boldsymbol{\alpha}}_0 - \boldsymbol{\alpha}_0\right)\right). \tag{36.3}$$

Condition of unbiasedness

$$\mathbb{E}\left(\widehat{\boldsymbol{\alpha}}_0 - \boldsymbol{\alpha}_0\right) = \mathbf{0}$$

is satisfied by choosing weights that fulfill the constraints

$$\sum_{i=1}^{N} b_{kl}^{i} = \delta_{kl} = \begin{cases} 1 \text{ if } \alpha_{\mathrm{l}} = \alpha_{\mathrm{k}} \\ 0 \text{ otherwise} \end{cases}.$$

The computation of the K coefficients of $\widehat{\boldsymbol{\alpha}}_0$ is achieved by global constrained minimization of (36.3). As usual in geostatistics, the quality of estimations relies on the choice of a suitable model for the multivariate covariance functions for which positive semi-definite properties must be well checked. As shown in the following, by fitting a Linear Model of Coregionalization (LMC), the choice of an obtainable model of spatial covariance structure at different scales is conducted from a multivariate nested variogram fit (Goulard *et al.*, 1992).

36.4 Dealing with real data

The above method is illustrated with data in oceanography where marine mammals operate as samplers. As a matter of fact, the southern Antarctic ocean is probably one of the less accessible area on Earth. This place plays a key role in heat exchanges between ocean and atmosphere and very few oceanographic data are available. Since the 90's, scientists are interested in the possibility to explore unkown parts of Antarctic ocean using elephant seals equipped with Argos transmitters including pressure, temperature and salinity sensors. The elephant seal becomes a valuable auxiliary for operational oceanography and enables scientists to study the hydrology of the southern ocean and the animal behaviour (Bailleul *et al.*, 2007).

Figure 1 displays temperature profiles sampled in Antarctic ocean by an elephant seal travelling from Kerguelen Islands to the Antarctic continental shelf. Temperature profiles are recorded at each dive at different discrete locations in space and depth. Typically, these data require to be fitted in order to form a set of functional profiles. We choose to express each function into

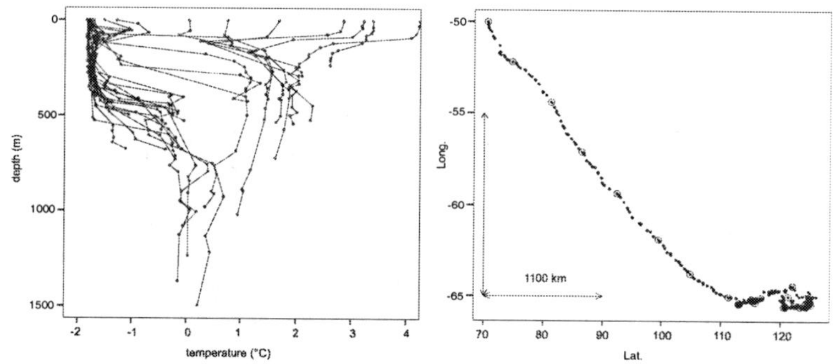

Fig. 36.1 Example of sampled temperature (řC) profiles by elephant seal and spatial location of displayed profiles along the trajectory (white circles). The cruise duration is 4.5 monthes. During the travel, the animal crosses and samples different water structures.

an orthonormal polynomial basis where coefficients have been estimated taking into account the variability of the sampling devices. The set of estimated coefficient vectors $\{\boldsymbol{\alpha}_i, i = 1, ..., N\}$ constitutes the sample E and estimations of a functional profile may be achieved at any location $\mathbf{x}_0$ along the trajectory, using the estimator in (36.2). Modelization of covariance structures between coefficients is realized through the fit of a multivariate nested variogram model

$$\boldsymbol{\Gamma}(\mathbf{h}) = \sum_{u=1}^{S} \mathbf{P}_u \gamma_u (\mathbf{h})$$

where S is the number of chosen structures, the $\gamma_u(\mathbf{h})$ are normalized variograms and $\mathbf{P}_u$ are positive semi-definite matrices. This coregionalization approach with two different scales provides a acceptable estimation of the field of temperature profiles ranging from 0 to 600 m (Fig. 2).

The fitted nested covariograms of Fig. 2 point out the main difficulty of the method. It is a hard task to find the set of variogram models γ_u, selected among the family of parametric models used in geostatistics (spherical, exponential, ...), taking care to keep S reasonably small. In our case, the range of each nested models is fixed by the practitioner (20 km and 100km) so as to provide the most graphically satisfactory fit. The matrices $\mathbf{P}_u$ are then fitted by a least squares algorithm (Wackernagel, 2003).

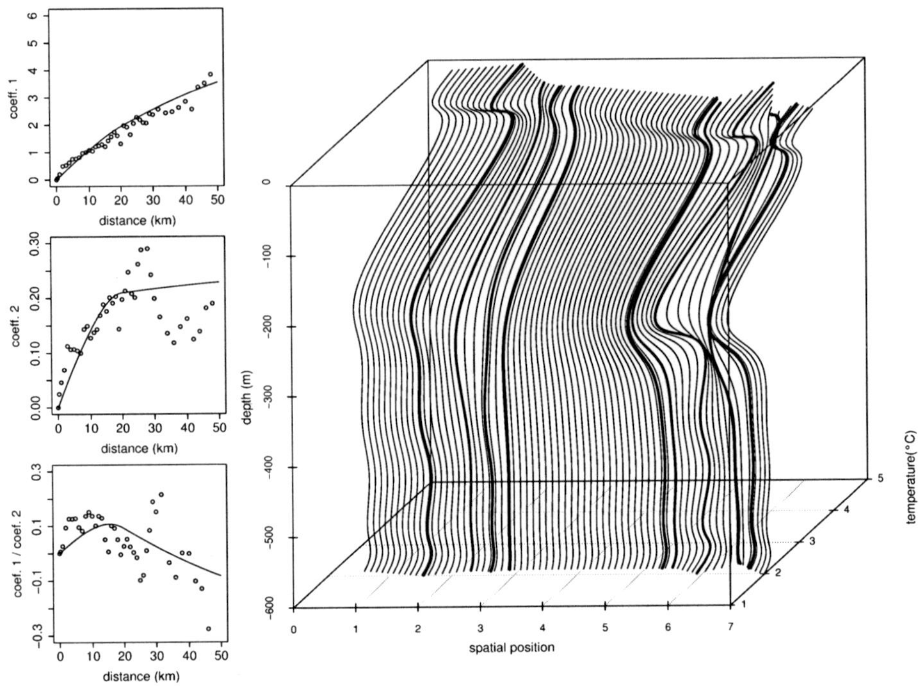

Fig. 36.2 Example of least squares fit of a multivariate nested variogram model on 3 polynomial coefficients and associated estimated temperature profiles. The covariance structure is decomposed in two scales (20 km and 100 km) which provides an obtainable model for estimating the coefficients at different locations along the trajectory of the elephant seal. Bolded curves are observed profiles. Thin curves show the predicted curves with cokriging

References

[1] Bailleul F., Charrassin J-B, Ezraty R, Girard-Ardhuin F., McMahon C. R. , Field I. C. and C. Guinet C: Southern elephant seals from Kerguelen Islands confronted by Antarctic Sea ice. Changes in movements and in diving behaviour. Deep Sea Research Part II: Topical Studies in Oceanography **54**, 343-355 (2007).

[2] Cardot H., Ferraty F. and P. Sarda: Functional Linear Model. Statistics and Probability Letters **45**, 11-22 (1999).

[3] Dabo-Niang S. and A. -F. Yao: Kernel regression estimation for continuous spatial processes. Mathematical Methods for Statistics **16**, 298-317 (2007).

[4] Goulard M. and M. Voltz: Linear coregionalization model : tools for estimation and choice of multivariate variograms. Mathematical Geology **24**, 269-286 (1992).

[5] Meiring W.: Oscillations and Time Trends in Stratospheric Ozone Levels : A Functional Data Analysis Approach. J. Am. Stat. Ass. **102**, 788-802 (2007).

[6] Ramsay J. O. and B. W. Silverman: Functional data analysis. Springer, New-York. (2005).

[7] Wackernagel H.: Multivariate geostatistics : an introduction with applications. Springer, New-York. (2003).

Chapter 37
On the Effect of Curve Alignment and Functional PCA

Juhyun Park

Abstract When dealing with multiple curves as functional data, it is a common practice to apply functional PCA to summarise and characterise random variation in finite dimension. Often functional data however exhibits additional time variability that distorts the assumed common structure. This is recognized as the problem of curve registration. While the registration step is routinely employed, this is considered as a preprocessing step prior to any serious analysis. Consequently, the effect of alignment is mostly ignored in subsequent analyses and is not well understood. We revisit the issue by particularly focusing on the effect of time variability on the FPCA and illustrate the phenomena from a borrowed perturbation viewpoint. The analysis further suggests an iterative estimating procedure to optimise FPCA.

37.1 Introduction

Repeated measurements in the form of curves are increasingly common in various scientific applications including biomedicine and physical sciences (Ramsay and Silverman, 2002, 2005). Individual measurements are taken at consecutive time points (index set) and repeatedly observed for different subjects. Usually the sample of curves is assumed to have some homogeneous structure in the functional shape, while allowed for individual variability. It is desirable that the additional variability is summarised with a few components which are able to extract most variability and which are easy to interpret (Park et al. 2007).

Functional PCA utilises the well-known Karhunen-Loève expansion to provide an optimal representation of the function with a small number of

---

Juhyun Park
Department of Mathematics and Statistics, Lancaster University, Lancaster, U.K., e-mail: juhyun.park@lancaster.ac.uk

common components. This is based on the assumption that the underlying random function shares the common mean and covariance function. To fix the idea, consider a stochastic process $X \in L^2(\mathbf{T})$ with compact support $\mathbf{T} = [0, T]$, with the mean function $\mu(t)$ and the covariance function $\gamma(s,t) = Cov(X(s), X(t))$. Assume that $\int_{\mathbf{T}} E[X(t)^2] < \infty$. Let $\lambda_1 \geq \lambda_2 \geq \dots$ be the ordered eigenvalues of the covariance operator defined through γ with the corresponding eigenfunctions $\phi_1, \phi_2, \dots$. We assume that $\sum_k \lambda_k < \infty$. Then

$$X(t) = \mu(t) + \sum_k \xi_k \phi_k(t) , \qquad (37.1)$$

where $E[\xi] = 0$ and $E[\xi_j \xi_k] = \lambda_j I(j = k)$.

With a sample of curves available, these quantities are replaced by their estimates and a finite number of components are usually considered sufficient to extract *significant* observed variation. Theoretical properties of estimators are studied in Dauxois et al. (1982), Rice and Silverman (1991), Kneip (1994) and Hall et al. (2006).

Often functional data exhibits additional time variability, which is mainly dealt with in pre-processing step, by aligning curves to eliminate the time variability prior to any serious analysis. This is known as registration problem and there are several methods developed. Basically when the functions exhibit identifiable features, curves can be aligned to match those features, which is known as landmark registration (Gasser and Kneip, 1995). This works well as long as features are correctly identified. Several other methods have been developed to automate the procedure when the features are less prominent. An overview can be found in Ramsay and Silverman (2005).

Although the issue has been rightly acknowledged, because most analysis treats registration as a preprocessing step, its carry-on effects on later analysis was not well studied. A recent work of Kneip and Ramsay (2007) address a similar problem and propose a new procedure to combine registration to fit functional PCA models, extending the covex averaging idea of registration (Liu and Müller, 2004).

Instead we focus on quantifying our misconduct. What happens then if registration was not carried out or was made improperly? The obvious problem arises when estimating global mean structure. Generally, how does the time variability propagate through to functional PCA analysis? Some issues with interpretability in functioanl PCA may also be attributed to the improper registration. We concentrate on relations of eigenvalues and eigenfunctions between unregistered and registered curves, in the sense that we do not want our registrations step to be *perfect* but we would like to be able to *correct* the residual difference from our imperfect analysis later.

Assume that the observed variable is $\tilde{X}(t) = X(\eta(t))$ for a monotone transformation $\eta(t)$ with $E[\eta(t)] = t$ for $t \in \mathbf{T}$. Suppose that we proceed to functional PCA without correcting η at the earlier stage to obtain $\tilde{\lambda}$ and $\tilde{\phi}$. How much do we lose by ignoring η?

We may start with the representation in (37.1) as

$$\tilde{X}(t) = \mu(\eta(t)) + \sum_k \xi_k \phi_k(\eta(t)) .$$

Now $E[\tilde{X}(t)] = \mu(\eta(t))$ but note that the series is not any longer orthonormal decomposition. Write $\tilde{\gamma}(s,t) = Cov(\tilde{X}(s), \tilde{X}(t))$. Then

$$\tilde{\gamma}(s,t) = \gamma(s,t) + \tilde{\gamma}(s,t) - \gamma(s,t) .$$

With some Taylor appromixation argument, it may be shown that $\tilde{\gamma}(s,t) - \gamma(s,t) = \varepsilon v(s,t)$ for some ε and v, then, under some regularity conditions and for small ε, we would have

$$\tilde{\lambda}_k = \lambda_k + \varepsilon < \phi_k, V\phi > + O(\varepsilon^2) ,$$
$$\tilde{\phi}_k \propto \phi + \varepsilon \sum_{l \neq k} \frac{< \phi_k, V\phi_l >}{\lambda_k - \lambda_l} + O(\varepsilon^2) ,$$

where V denotes the corresponding operator for v. A similar derivation is made in Hall et al. (2006) to quantify sampling variability. We extend the idea to include time variability. Our interest is to recover λ and ϕ from $\tilde{\lambda}$ and $\tilde{\phi}$ using a sample of curves and a registration. Our estimators will be obtained from the estimators of unregistered curves with some correction made based on a registration. The precision of registration will be reflected on that of V and thus the correction terms in general. Based on these relations some properties of estimators will be studied and illustrated.

References

[1] Dauxois, J. Pousse, A. and Romain, Y.: Asymptotic theory for the principal component analysis of a vector random function: some applications to statistical inference. Journal of Multivariate Analysis, **12**, 136-154 (1982).

[2] Liu, X. and Müller, H. G.: Functional convex averaging and synchronization for time-warped random curves. Journal of the American Statistical Association, **99**, 687-699 (2004).

[3] Gasser, T. and Kneip, A.: Searching for structure in curve samples. Journal of the American Statistical Association, **90**, 1179-1188 (1995).

[4] Hall, P, Müller, H. G. and Wang, J. L.: Properties of principal componenet methods for functional and longitudinal data analysis. Annals of Statistics, **34**, 1493-1517 (2006).

[5] Kneip, A.: Nonparametric estimation of common regressors for similar curve data. Annals of Statistics, **22**, 1386-1427 (1994).

[6] Kneip, A. and Ramsay, J. O.: Combining registration and fitting for functional models. technical report. (2007).

[7] Park, J. Gasser, T. and Rousson, V.: Structural components in functional data. technical report. (2007).

[8] Ramsay, J. O. and Silverman, B. W.: Applied functional data analysis, New York: Springer. (2002).

[9] Ramsay, J. O. and Silverman, B. W.: Functional data analysis, New York: Springer. (2005).

[10] Rice, J. W. and Silverman, B. W.: Estimating the mean and the covariance structure nonparametrically when the data are curves. Journal of Royal Statistical Society, B, **53**, 233-243 (1991).

Chapter 38
K-sample Subsampling

Dimitris Politis and Joseph Romano

Abstract The problem of subsampling in two-sample and K-sample settings is addressed where both the data and the statistics of interest take values in general spaces. We show the asymptotic validity of subsampling confidence intervals and hypothesis tests in the case of independent samples, and give a comparison to the bootstrap in the K-sample setting.

38.1 Introduction

Subsampling is a statistical method that is most generally valid for nonparametric inference in a large-sample setting. The applications of subsampling are numerous starting from i.i.d. data and regression, and continuing to time series, random fields, marked point processes, etc.; see Politis, Romano and Wolf (1999) for a review and list of references.

Interestingly, the two-sample and K-sample settings have not been explored yet in the subsampling literature; we attempt to fill this gap here. So, consider K independent datasets: $\underline{X}^{(1)}, \ldots, \underline{X}^{(K)}$ where $\underline{X}^{(k)} = (X_1^{(k)}, \ldots, X_{n_k}^{(k)})$ for $k = 1, \ldots, K$. The random variables $X_j^{(k)}$ take values in an arbitrary space[1] $\mathbf{S}$; typically, $\mathbf{S}$ would be $\mathbf{R}^d$ for some d, but $\mathbf{S}$ can very well be a function space. Although the dataset $\underline{X}^{(k)}$ is independent of $\underline{X}^{(k')}$ for $k \neq k'$, there may exist some dependence *within* a dataset. For

Dimitris Politis
University of California—San Diego, USA, e-mail: dpolitis@ucsd.edu

Joseph Romano
Stanford University, USA, e-mail: romano@stanford.edu

[1] Actually, one can let S vary with k as well, but we do not pursue this here for lack of space.

conciseness, we will focus on the case of independence within samples here; the general case will be treated in a follow-up bigger exposition.

Thus, in the sequel we will assume that, for any k, $X_1^{(k)}, \ldots, X_{n_k}^{(k)}$ are i.i.d. An example in the i.i.d. case is the usual two-sample set-up in biostatistics where d 'features' (body characteristics, gene expressions, etc.) are measured on a group of patients, and then again measured on a control group. The probability law associated with such a K-sample experiment is specified by $P = (P_1, \ldots, P_K)$, where P_k is the underlying probability of the kth sample; more formally, the joint distribution of all the observations is the product measure $\prod_{k=1}^K P_k^{n_k}$. The goal is inference (confidence regions, hypothesis tests, etc.) regarding some parameter $\theta = \theta(P)$ that takes values in a general normed linear space $\mathbf{B}$ with norm denoted by $||\cdot||$. Denote $\mathbf{n} = (n_1, \ldots, n_K)$, and let $\hat{\theta}_{\mathbf{n}} = \hat{\theta}_{\mathbf{n}}(\underline{X}^{(1)}, \ldots, \underline{X}^{(K)})$ be an estimator of θ. It will be assumed that $\hat{\theta}_{\mathbf{n}}$ is consistent as $\min_k n_k \to \infty$. In general, one could also consider the case where the number of samples K tends to ∞ as well.

Let $g : \mathbf{B} \to \mathbf{R}$ be a continuous function, and let $J_{\mathbf{n}}(P)$ denote the sampling distribution of the "root" $g[\tau_{\mathbf{n}}(\hat{\theta}_{\mathbf{n}} - \theta(P))]$ under P, with corresponding cumulative distribution function

$$J_{\mathbf{n}}(x, P) = Prob_P\{g[\tau_{\mathbf{n}}(\hat{\theta}_{\mathbf{n}} - \theta(P))] \leq x\} \tag{38.1}$$

where $\tau_{\mathbf{n}}$ is a normalizing sequence; in particular, $\tau_{\mathbf{n}}$ is to be thought of as a fixed function of $\mathbf{n}$ such that $\tau_{\mathbf{n}} \to \infty$ when $\min_k n_k \to \infty$. As an example, $g(\cdot)$ might be a continuous function of the norm $||\cdot||$ or a projection operator.

As in the one-sample case, the basic assumption that is required for subsampling to work is existence of a *bona fide* large-sample distribution, i.e.,

Assumption 38.1.1 *There exists a nondegenerate limiting law $J(P)$ such that $J_{\mathbf{n}}(P)$ converges weakly to $J(P)$ as $\min_k n_k \to \infty$.*

The α–quantile of $J(P)$ will be denoted by $J^{-1}(\alpha, P) = \inf\{x : J(x, P) \geq \alpha\}$. In addition to Assumption 38.1.1, we will use the following mild assumption.

Assumption 38.1.2 *As $\min_k n_k \to \infty$, $\tau_{\mathbf{b}}||\hat{\theta}_{\mathbf{n}} - \theta(P)|| = o_P(1)$.*

Assumptions 38.1.1 and 38.1.2 are implied by the following assumption, as long as $\tau_{\mathbf{b}}/\tau_{\mathbf{n}} \to 0$.

Assumption 38.1.3 *As $\min_k n_k \to \infty$, the distribution of $\tau_{\mathbf{n}}(\hat{\theta}_{\mathbf{n}} - \theta(P))$ under P converges weakly to some distribution (on the Borel σ-field of the normed linear space $\mathbf{B}$).*

Here, weak convergence is understood to be taken in the modern sense of Hoffmann-Jorgensen; see Section 1.3 of van der Vaart and Wellner (1996). That Assumption 38.1.3 implies both Assumptions 38.1.1 and 38.1.2 follows by the Continuous Mapping Theorem; see Theorem 1.3.6 of van der Vaart and Wellner (1996).

38.2 Subsampling hypothesis tests in K samples

For $k = 1, \ldots, K$, let $\mathcal{S}_k$ denote the set of all size b_k (unordered) subsets of the dataset $\{X_1^{(k)}, \ldots, X_{n_k}^{(k)}\}$ where b_k is an integer in $[1, n_k]$. Note that the set $\mathcal{S}_k$ contains $Q_k = \binom{n_k}{b_k}$ elements that are enumerated as $S_1^{(k)}, S_2^{(k)}, \ldots, S_{Q_k}^{(k)}$. A K–fold subsample is then constructed by choosing one element from each super-set $\mathcal{S}_k$ for $k = 1, \ldots, K$. Thus, a typical K–fold subsample has the form: $S_{i_1}^{(1)}, S_{i_2}^{(2)}, \ldots, S_{i_K}^{(K)}$, where i_k is an integer in $[1, Q_k]$ for $k = 1, \ldots, K$. It is apparent that the number of possible K–fold subsamples is $Q = \prod_{k=1}^K Q_k$. So a subsample value of the general statistic $\hat{\theta}_{\mathbf{n}}$ is

$$\hat{\theta}_{\mathbf{i},\mathbf{b}} = \hat{\theta}_{\mathbf{b}}(S_{i_1}^{(1)}, \ldots, S_{i_K}^{(K)}) \tag{38.2}$$

where $\mathbf{b} = (b_1, \ldots, b_K)$ and $\mathbf{i} = (i_1, \ldots, i_K)$. The subsampling distribution approximation to $J_{\mathbf{n}}(P)$ is defined by

$$L_{\mathbf{n},\mathbf{b}}(x) = \frac{1}{Q} \sum_{i_1=1}^{Q_1} \sum_{i_2=1}^{Q_2} \cdots \sum_{i_K=1}^{Q_K} 1\{g[\tau_{\mathbf{b}}(\hat{\theta}_{\mathbf{i},\mathbf{b}} - \hat{\theta}_{\mathbf{n}})] \leq x\}. \tag{38.3}$$

The distribution $L_{\mathbf{n},\mathbf{b}}(x)$ is useful for the construction of subsampling confidence sets as discussed in Section 38.3. For hypothesis testing, however, we instead let

$$G_{\mathbf{n},\mathbf{b}}(x) = \frac{1}{Q} \sum_{i_1=1}^{Q_1} \sum_{i_2=1}^{Q_2} \cdots \sum_{i_K=1}^{Q_K} 1\{g[\tau_{\mathbf{b}}(\hat{\theta}_{\mathbf{i},\mathbf{b}})] \leq x\}, \tag{38.4}$$

and consider the general problem of testing a null hypothesis H_0 that $P = (P_1, \ldots, P_k) \in \mathbf{P}_0$ against H_1 that $P \in \mathbf{P}_1$. The goal is to construct an asymptotically valid null distribution based on some test statistic of the form $g(\tau_{\mathbf{n}}\hat{\theta}_{\mathbf{n}})$, whose distribution under P is defined to be $G_{\mathbf{n}}(P)$ (with c.d.f. $G_n(\cdot, P)$). The subsampling critical value is obtained as the $1-\alpha$ quantile of $G_{\mathbf{n},\mathbf{b}}(\cdot)$, denoted $g_{\mathbf{n},\mathbf{b}}(1-\alpha)$. We will make use of the following assumption.

Assumption 38.2.1 *If $P \in \mathbf{P}_0$, there exists a nondegenerate limiting law $G(P)$ such that $G_{\mathbf{n}}(P)$ converges weakly to $G(P)$ as* $\min_k n_k \to \infty$.

Let $G(\cdot, P)$ denote the c.d.f. corresponding to $G(P)$. Let $G^{-1}(1-\alpha, P)$ denote a $1-\alpha$ quantile of $G(P)$. The following result gives the consistency of the procedure under H_0, and under a sequence of contiguous alternatives; for the definition contiguity see Section 12.3 of Lehmann and Romano (2007). One could also obtain a simple consistency result under fixed alternatives.

Theorem 38.1. *Suppose Assumption 38.2.1 holds. Also, assume that, for each* $k = 1, \ldots, K$*, we have* $b_k/n_k \to 0$*, and* $b_k \to \infty$ *as* $\min_k n_k \to \infty$*.*
(i) Assume $P \in \mathbf{P}_0$*. If* $G(\cdot, P)$ *is continuous at its* $1-\alpha$ *quantile* $G^{-1}(1-\alpha, P)$*, then*

$$g_{\mathbf{n},\mathbf{b}} \xrightarrow{P} G^{-1}(1-\alpha, P) \tag{38.5}$$

and

$$Prob_P\{g(\tau_{\mathbf{n}}\hat{\theta}_{\mathbf{n}}) > g_{\mathbf{n},\mathbf{b}}(1-\alpha)\} \to \alpha \quad \text{as } \min_k n_k \to \infty \,. \tag{38.6}$$

(ii) Suppose, that for some $P = (P_1, \ldots, P_k) \in \mathbf{P}_0$*,* $P_{k,n_k}^{n_k}$ *is contiguous to* $P_k^{n_k}$ *for* $k = 1, \ldots K$*. Then, under such a contiguous sequence,* $g(\tau_{\mathbf{n}}\hat{\theta}_{\mathbf{n}})$ *is tight. Moreover, if it converges in distribution to some random variable* T *and* $G(\cdot, P)$ *is continuous at* $G^{-1}(1-\alpha, P)$*, then the limiting power of the test against such a sequence is* $P\{T > G^{-1}(1-\alpha, P)\}$*.*

Proof: To prove (i), let x be a continuity point of $G(\cdot, P)$. We claim

$$G_{\mathbf{n},\mathbf{b}}(x) \xrightarrow{P} G(x, P) \,. \tag{38.7}$$

To see why, note that $E[G_{\mathbf{n},\mathbf{b}}(x)] = G_{\mathbf{b}}(x, P) \to G(x, P)$. So, by Chebychev's inequality, to show (38.7), it suffices to show $Var[G_{\mathbf{n},\mathbf{b}}(x)] \to 0$. To do this, let $d = d_{\mathbf{n}}$ be the greatest integer $\leq \min_k(n_k/b_k)$. Then, for $j = 1, \ldots, d$, let $\bar{\theta}_{j,\mathbf{b}}$ be equal to the statistic $\hat{\theta}_{\mathbf{b}}$ evaluated at the data set where the observations from the kth sample are $(X^{(k)}_{b_k(j-1)+1}, X^{(k)}_{b_k(j-1)+2}, \ldots, X^{(k)}_{b_k(j-1)+b_k})$. Then, set

$$\bar{G}_{\mathbf{n},\mathbf{b}}(x) = d^{-1} \sum_{j=1}^{d} 1\{g(\tau_{\mathbf{b}}\bar{\theta}_{j,\mathbf{b}}) \leq x\} \,.$$

By construction, $\bar{G}_{\mathbf{n},\mathbf{b}}(x)$ is an average of i.i.d. 0–1 random variables with expectation $G_{\mathbf{b}}(x, P)$ and variance that is bounded above by $1/(4d_{\mathbf{n}}) \to 0$. But, $G_{\mathbf{n},\mathbf{b}}(x)$ has smaller variance than $\bar{G}_{\mathbf{n},\mathbf{b}}(x)$. This last statement follows by a sufficiency argument from the Rao-Blackwell Theorem; indeed,

$$G_{\mathbf{n},\mathbf{b}}(x) = E[\bar{G}_{\mathbf{n},\mathbf{b}}(x) | \hat{P}^{(k)}_{n_k}, \; k = 1, \ldots, K] \,,$$

where $\hat{P}^{(k)}_{n_k}$ is the empirical measure in the kth sample. Since these empirical measures are sufficient, it follows that

$$Var(G_{n,b}(x)) \leq Var(\bar{G}_{n,b}(x)) \to 0 \,.$$

Thus, (38.7) holds. Then, (38.5) follows by Lemma 11.2.1(ii) of Lehmann and Romano (2005). Application of Slutsky's Theorem yields (38.6).

To prove (ii), we know that $g_{\mathbf{n},\mathbf{b}} \xrightarrow{P} G^{-1}(1-\alpha, P)$ under P. Contiguity forces the same convergence under the sequence of contiguous alternatives. The result follows by Slutsky's Theorem. ◇

38.3 Subsampling confidence sets in K samples

Let $c_{\mathbf{n},\mathbf{b}}(1-\alpha) = \inf\{x : L_{\mathbf{n},\mathbf{b}}(x) \geq 1-\alpha\}$ where $L_{\mathbf{n},\mathbf{b}}(x)$ was defined in (38.3).

Theorem 38.2. *Assume Assumptions 38.1.1 and 38.1.2, where g is assumed uniformly continuous. Also assume that, for each $k = 1, \ldots, K$, we have $b_k/n_k \to 0$, $\tau_{\mathbf{b}}/\tau_{\mathbf{n}} \to 0$, and $b_k \to \infty$ as $\min_k n_k \to \infty$.*
(i) Then, $L_{\mathbf{n},\mathbf{b}}(x) \xrightarrow{P} J(x, P)$ for all continuity points x of $J(\cdot, P)$.
(ii) If $J(\cdot, P)$ is continuous at $J^{-1}(1-\alpha, P)$, then the event

$$\{g[\tau_{\mathbf{n}}(\hat{\theta}_{\mathbf{n}} - \theta(P))] \leq c_{\mathbf{n},\mathbf{b}}(1-\alpha)\} \tag{38.8}$$

has asymptotic probability equal to $1-\alpha$; therefore, the confidence set $\{\theta : \ g[\tau_{\mathbf{n}}(\hat{\theta}_{\mathbf{n}} - \theta)] \leq c_{\mathbf{n},\mathbf{b}}(1-\alpha)\}$ has asymptotic coverage probability $1-\alpha$.

Proof: Assume without loss of generality that $\theta(P) = 0$ (in which case $J_{\mathbf{n}}(P) = G_{\mathbf{n}}(P)$). Let x be a continuity point of $J(\cdot, P)$. First, we claim that

$$L_{\mathbf{n},\mathbf{b}}(x) - G_{\mathbf{n},\mathbf{b}}(x) \xrightarrow{P} 0 \ . \tag{38.9}$$

Given $\varepsilon > 0$, there exists $\delta > 0$, so that $|g(x) - g(x')| < \varepsilon$ if $\|x - x'\| < \delta$. But then, $|g[\tau_{\mathbf{b}}(\hat{\theta}_{\mathbf{i},\mathbf{b}} - \hat{\theta}_{\mathbf{n}})] - g(\tau_{\mathbf{b}}\hat{\theta}_{\mathbf{i},\mathbf{b}})| < \varepsilon$ if $\|\tau_{\mathbf{b}}\hat{\theta}_{\mathbf{n}}\| < \delta$; this latter event has probability tending to one. It follows that, for any fixed $\varepsilon > 0$,

$$G_{\mathbf{n},\mathbf{b}}(x-\varepsilon) \leq L_{\mathbf{n},\mathbf{b}}(x) \leq G_{\mathbf{n},\mathbf{b}}(x+\varepsilon)$$

with probability tending to one. But, the behavior of $G_{\mathbf{n},\mathbf{b}}(x)$ was given in Theorem 38.1. Letting $\varepsilon \to 0$ through continuity points of $J(\cdot, P)$ yields (38.9) and (i). Part (ii) follows from Slutsky's Theorem. ◇

Remark. The uniform continuity assumption for g can be weakened to continuity if Assumptions 38.1.1 and 38.1.2 are replaced by Assumption 38.1.3. However, the proof is much more complicated and relies on a K-sample version of Theorem 7.2.1 of Politis, Romano and Wolf (1999).

In general, we may also try to approximate the distribution of a studentized root of the form $g(\tau_{\mathbf{n}}[\hat{\theta}_{\mathbf{n}} - \theta(P)]/\hat{\sigma}_{\mathbf{n}}$, where $\hat{\sigma}_{\mathbf{n}}$ is some estimator which tends in probability to some finite nonzero constant $\sigma(P)$. The subsampling approximation to this distribution is

$$L^+_{\mathbf{n},\mathbf{b}}(x) = \frac{1}{Q}\sum_{i_1=1}^{Q_1}\sum_{i_2=1}^{Q_2}\cdots\sum_{i_K=1}^{Q_K} 1\{g[\tau_{\mathbf{b}}(\hat\theta_{\mathbf{i},\mathbf{b}} - \hat\theta_{\mathbf{n}})]/\hat\sigma_{\mathbf{i},\mathbf{b}} \le x\}, \tag{38.10}$$

where $\hat\sigma_{\mathbf{i},\mathbf{b}}$ is the estimator $\hat\sigma_{\mathbf{b}}$ computed from the $\mathbf{i}$th subsampled data set. Also let $c^+_{\mathbf{n},\mathbf{b}}(1-\alpha) = \inf\{x : L^+_{\mathbf{n},\mathbf{b}}(x) \ge 1-\alpha\}$.

Theorem 38.3. *Assume Assumptions 38.1.1 and 38.1.2, where g is assumed uniformly continuous. Let $\hat\sigma_{\mathbf{n}}$ satisfy $\hat\sigma_{\mathbf{n}} \xrightarrow{P} \sigma(P) > 0$. Also assume that, for each $k = 1,\ldots,K$, we have $b_k/n_k \to 0$, $\tau_{\mathbf{b}}/\tau_{\mathbf{n}} \to 0$, and $b_k \to \infty$ as $\min_k n_k \to \infty$.*
(i) Then, $L_{\mathbf{n},\mathbf{b}}(x) \xrightarrow{P} J(x\cdot\sigma(P),P)$ if $J(\cdot,P)$ is continuous at $x\sigma(P)$.
(ii) If $J(\cdot,P)$ is continuous at $J^{-1}(1-\alpha,P)/\sigma(P)$, then the event

$$\{g[\tau_{\mathbf{n}}(\hat\theta_{\mathbf{n}} - \theta(P))]/\hat\sigma_{\mathbf{n}} \le c^+_{\mathbf{n},\mathbf{b}}(1-\alpha)\} \tag{38.11}$$

has asymptotic probability equal to $1-\alpha$.

38.4 Random subsamples and the K–sample bootstrap

For large values of n_k and b_k, $Q = \prod_k \binom{n_k}{b_k}$ can be a prohibitively large number; considering *all* possible subsamples may be impractical and, thus, we may resort to Monte Carlo. To define the algorithm for generating random subsamples of sizes $b_1,\ldots,b_K$ respectively, recall that subsampling in the i.i.d. single-sample case is tantamount to sampling *without* replacement from the original dataset; see e.g. Politis et al. (1999, Ch. 2.3). Thus, for $m = 1,\ldots,M$, we can generate the mth joint subsample as $\underline{X}^{(1)}_m, \underline{X}^{(2)}_m, \ldots, \underline{X}^{(K)}_m$ where $\underline{X}^{(k)}_m = \{X^{(k)}_{I_1},\ldots,X^{(k)}_{I_{b_k}}\}$, and $I_1,\ldots,I_{b_k}$ are b_k numbers drawn randomly *without* replacement from the index set $\{1,2,\ldots,n_k\}$. Note that the random indices drawn to generate $\underline{X}^{(k)}_m$ are independent to those drawn to generate $\underline{X}^{(k')}_m$ for $k \ne k'$.

Thus, a randomly chosen subsample value of the statistic $\hat\theta_{\mathbf{n}}$ is given by $\hat\theta_{m,\mathbf{b}} = \hat\theta_{\mathbf{b}}(\underline{X}^{(1)}_m,\ldots,\underline{X}^{(K)}_m)$, with corresponding subsampling distribution defined as

$$\tilde L_{\mathbf{n},\mathbf{b}}(x) = \frac{1}{M}\sum_{m=1}^{M} 1\{g[\tau_{\mathbf{b}}(\hat\theta_{m,\mathbf{b}} - \hat\theta_{\mathbf{n}})] \le x\}. \tag{38.12}$$

The following corollary shows that $\tilde L_{\mathbf{n},\mathbf{b}}(x)$ and its $1-\alpha$ quantile $\tilde c_{\mathbf{n},\mathbf{b}}(1-\alpha)$ can be used for the construction of large-sample confidence regions for θ; its proof is analogous to the proof of Corollary 2.1 of Politis and Romano (1994).

Corollary 38.1. *Assume the conditions of Theorem 38.2. As $M \to \infty$, parts (i) and (ii) of Theorem 38.2 remain valid with $\tilde L_{\mathbf{n},\mathbf{b}}(x)$ and $\tilde c_{\mathbf{n},\mathbf{b}}(1-\alpha)$ instead of $L_{\mathbf{n},\mathbf{b}}(x)$ and $c_{\mathbf{n},\mathbf{b}}(1-\alpha)$.*

Similarly, hypothesis testing can be conducted using the notion of random subsamples. To describe it, let $\tilde{g}_{\mathbf{n},\mathbf{b}}(1-\alpha) = \inf\{x : \tilde{G}_{\mathbf{n},\mathbf{b}}(x) \geq 1-\alpha\}$ where

$$\tilde{G}_{\mathbf{n},\mathbf{b}}(x) = \frac{1}{M} \sum_{m=1}^{M} 1\{g[\tau_{\mathbf{b}}(\hat{\theta}_{m,\mathbf{b}})] \leq x\}. \tag{38.13}$$

Corollary 38.2. *Assume the conditions of Theorem 38.1. As $M \to \infty$, parts (i) and (ii) of Theorem 38.1 remain valid with $\tilde{G}_{\mathbf{n},\mathbf{b}}(x)$ and $\tilde{g}_{\mathbf{n},\mathbf{b}}(1-\alpha)$ instead of $G_{\mathbf{n},\mathbf{b}}(x)$ and $g_{\mathbf{n},\mathbf{b}}(1-\alpha)$.*

The bootstrap in two-sample settings is often used in practical work; see Hall and Martin (1988) or van der Vaart and Wellner (1996). In the i.i.d. setup, resampling and (random) subsampling are very closely related since, as mentioned, they are tantamount to sampling *with* vs. *without* replacement from the given i.i.d. sample. By contrast to subsampling, however, no general validity theorem is available for the bootstrap *unless* a smaller resample size is used; see Politis and Romano (1993).

Actually, the general validity of K-sample bootstrap that uses a resample size b_k for sample k follows from the general validity of subsampling as long as $b_k^2 << n_k$. To state it, let $J^*_{\mathbf{n},\mathbf{b}}(x)$ denote the bootstrap (pseudo-empirical) distribution of $g[\tau_{\mathbf{b}}(\hat{\theta}^*_{\mathbf{n},\mathbf{b}} - \hat{\theta}_{\mathbf{n}})]$ where $\hat{\theta}^*_{\mathbf{n},\mathbf{b}}$ is the statistic $\hat{\theta}_{\mathbf{b}}$ computed from the bootstrap data. Similarly, let $c^*_{\mathbf{n},\mathbf{b}}(1-\alpha) = \inf\{x : J^*_{\mathbf{n},\mathbf{b}}(x) \geq 1-\alpha\}$. The proof of the following corollary parallels the discussion in Section 2.3 of Politis et al. (1999).

Corollary 38.3. *Under the additional condition $b_k^2/n_k \to 0$ for all k, Theorem 38.2 is valid as stated with $J^*_{\mathbf{n},\mathbf{b}}(x)$ and $c^*_{\mathbf{n},\mathbf{b}}(1-\alpha)$ in place of $L_{\mathbf{n},\mathbf{b}}(x)$ and $c_{\mathbf{n},\mathbf{b}}(1-\alpha)$ respectively.*

References

[1] Hall, P. and Martin, M.: On the bootstrap and two-sample problems, Australian Journal of Statistics. **30A**, 179-192 (1988).

[2] Lehmann, E.L. and Romano, J.: Testing Statistical Hypotheses. 3rd edition, Springer, New York. (2005).

[3] Politis, D.N., and Romano, J.P.: Estimating the Distribution of a Studentized Statistic by Subsampling, in Bulletin of the International Statistical Institute, 49th Session. Firenze, August 25 - September 2, 1993, Book **2**, pp.315-316 (1993).

[4] Politis, D.N., and Romano, J.P.: Large sample confidence regions based on subsamples under minimal assumptions, Ann. Statist. **22**, 2031-2050 (1994)..

[5] Politis, D., Romano, J. and Wolf, M.: Subsampling. Springer, New York. (1999).

[6] van der Vaart, A. and Wellner, J.: Weak Convergence and Empirical Processes. Springer, New York. (1996).

Chapter 39
Inference for Stationary Processes Using Banded Covariance Matrices

Mohsen Pourahmadi and Wei Biao Wu

Abstract We consider prediction and estimation problems by banding covariance matrices of stationary processes. Under a novel short-range dependence condition for a class of nonlinear processes, it is shown that the banded covariance matrix estimates converge in operator norm to the true covariance matrix with reasonable rates of convergence. A sub-sampling approach is proposed to choose the banding parameter.

39.1 Introduction

Given a realization $X_1, \ldots, X_n$ of a stationary process $\{X_t\}$ with the autocovariance function $\gamma_k = \mathrm{cov}(X_0, X_k)$, estimation of the covariance matrix $\Sigma_n = (\gamma_{i-j})_{1 \le i,j \le n}$ is important in almost every aspect of prediction and statistical inference. A good covariance matrix estimate should necessarily be positive definite and be uniformly close to the true one (Hannan and Deistler, 1988, Sec. 5.3) so that one can invert the estimated covariance matrix to perform prediction and other inferential tasks. Assuming $\mu = E(X_t) = 0$ and $E(X_t^2) < \infty$, the autocovariances can be estimated by

$$\hat{\gamma}_k = \frac{1}{n} \sum_{i=1}^{n-|k|} X_i X_{i+|k|}, \quad k = 0, \pm 1, \ldots, \pm(n-1). \tag{39.1}$$

It is known that for fixed $k \in \mathbf{Z}$, under the ergodicity condition, $\hat{\gamma}_k \to \gamma_k$ in probability. However, this entry-wise convergence does not automatically

---

Mohsen Pourahmadi
Northern Illinois University, USA, e-mail: pourahm@math.niu.edu

Wei Biao Wu
University of Chicago, USA, e-mail: wbwu@galton.uchicago.edu

imply that the corresponding estimated matrix $\hat{\Sigma}_n = (\hat{\gamma}_{i-j})_{1\le i,j\le n}$ is a good estimate of Σ_n (Hannan and Deistler, 1988, Sec. 5.3). Indeed, though $\hat{\Sigma}_n$ is positive definite (see Chapter 5 in Pourahmadi (2001)), it is not uniformly close to the population covariance matrix Σ_n, in the sense that the largest eigenvalue or the operator norm of $\hat{\Sigma}_n - \Sigma_n$ does not converge to zero.

Our covariance matrix estimate, for an l a nonnegative integer, is of the form

$$\hat{\Sigma}_{n,l} = (\hat{\gamma}_{i-j}\mathbf{1}_{|i-j|\le l})_{1\le i,j\le n}. \tag{39.2}$$

It is a truncated version of $\hat{\Sigma}_n$, preserving the diagonal and the $2l$ main sub-diagonals; note that if $l \ge n-1$, then $\hat{\Sigma}_{n,l} = \hat{\Sigma}_n$. Following Bickel and Levina (2007), we call $\hat{\Sigma}_{n,l}$ *the banded covariance matrix estimate* and l its band parameter. The motivation for banding comes from the fact that for a large lag k, either γ_k is close to zero or that $\hat{\gamma}_k$ is an unreliable estimate of γ_k. Thus, prudent use of banding may bring considerable computational economy in the former case and statistical efficiency in the latter by keeping small or unreliable $\hat{\gamma}_k$ out of the calculations.

There are important differences between our setup and results here, and those in Bickel and Levina (2007) and Zeitouni and Anderson (2008) where the observations are iid random vectors and can be viewed as m rows of an $m \times n$ random matrix or a multivariate dataset. They considered the banded version of the *sample covariance matrix*, and obtained consistency results under some regularity condition when $\log n/m \to 0$. However, we work with only one ($m = 1$) realization or time series data and establish consistency by banding the *sample autocovariance matrix.* Also we impose very mild moment and dependence conditions on a class of nonlinear processes using a new concept of short-range dependence (Wu, 2005). The selection of band parameters of our covariance matrix estimates is an adaptation of a resampling and risk-minimization approach due to Bickel and Levina (2007). Its performance is assessed via simulations for linear and nonlinear processes and the results will be reported elsewhere.

39.2 The results

We first introduce some structural assumptions on the process $\{X_t\}$ and work within the framework of nonlinear stationary processes which includes the standard linear processes. Hannan and Deistler (1988) have considered certain linear ARMA processes and obtained the uniform bound $\|\hat{\Sigma}_{n,\ell} - \Sigma_n\|_\infty = O(\sqrt{\log\log n}/\sqrt{n})$, $\ell \le (\log n)^\alpha, \alpha < \infty$; see Theorem 5.3.2 therein. In this section, we obtain comparable results for nonlinear processes and allow a wider range of ℓ; see Theorem 39.2 below.

39.2.1 A class of nonlinear processes

Let ε_i, $i \in \mathbf{Z}$, be independent and identically distributed (iid) random variables. Assume that $\{X_i\}$ is a causal process of the form

$$X_i = g(\ldots, \varepsilon_{i-1}, \varepsilon_i), \tag{39.3}$$

where g is a measurable function such that X_i is a well-defined second-order process. Many stationary processes fall within the framework of (39.3) (see Tong (1990) and Wu (2005)). To introduce the dependence structure, let $(\varepsilon'_i)_{i\in\mathbf{Z}}$ be an independent copy of $(\varepsilon_i)_{i\in\mathbf{Z}}$ and $\xi_i = (\ldots, \varepsilon_{i-1}, \varepsilon_i)$. Following Wu (2005), for $\alpha > 0$ and $i \geq 0$, define the physical dependence measure

$$\delta_\alpha(i) = \|X_i - X'_i\|_\alpha, \text{ where } X'_i = g(\xi'_i) \text{ and} \tag{39.4}$$

$$\xi'_i = (\ldots, \varepsilon_{-1}, \varepsilon'_0, \varepsilon_1, \ldots, \varepsilon_{i-1}, \varepsilon_i).$$

Observe that $X'_i = g(\xi'_i)$ is a coupled version of $X_i = g(\xi_i)$ with ε_0 in the latter replaced by an iid copy ε'_0. The quantity $\delta_\alpha(i)$ measures the dependence of X_i on ε_0. As in Wu (2005), we say that $\{X_i\}$ is short-range dependent if

$$\Delta_\alpha = \sum_{i=0}^{\infty} \delta_\alpha(i) < \infty. \tag{39.5}$$

Namely the cumulative impact of ε_0 on future values of the process or $(X_i)_{i\geq 0}$ is finite, thus suggesting a short-range dependence. In many applications it is easy to work with $\delta_\alpha(i)$ which is directly related to the data generating mechanism of the underlying process as indicated in the next two examples.

Example 39.1. Let $X_j = K(\sum_{i=0}^{\infty} a_i \varepsilon_{j-i})$, where a_i are real coefficients with $\sum_{i=0}^{\infty} |a_i| < \infty$, ε_i are iid with $\varepsilon_i \in \mathcal{L}^\alpha, 2 < \alpha \leq 4$, and K is a Lipschitz continuous function. Easy calculation shows that $\delta_\alpha(i) = O(|a_i|)$, hence (39.5) holds.

Example 39.2. Let ε_i be iid random variables and set $X_i = R(X_{i-1}, \varepsilon_i)$, where R is a bivariate function such that the system admits a stationary solution. Many nonlinear time series models follow this framework. Wu and Shao (2004) showed that, under mild and natural conditions, one has $\delta_\alpha(i) = O(r^i)$, $0 < r < 1$. So (39.5) clearly follows.

39.2.2 Convergence of banded covariance estimators

First we show that $\hat{\sum}_n$ is not a consistent estimate of $\sum_n$ in the sense that the operator norm or the largest eigenvalue of $\hat{\sum}_n - \sum_n$ does not converge to zero. On the positive side, we are able to show the convergence to zero and obtain an explicit upper bound for $\rho(\hat{\Sigma}_{n,l} - \Sigma_n)$ in our Theorem 39.2. All the proofs will appear elsewhere.

Theorem 39.1. *Define the projection operator* $\mathcal{P}_k$, $k \in \mathbf{Z}$, *by*

$$\mathcal{P}_k Z = \mathrm{E}(Z|\xi_k) - \mathrm{E}(Z|\xi_{k-1}), \quad Z \in \mathcal{L}^1.$$

If the process $\{X_t\}$ *in (39.3) satisfies*

$$\sum_{i=0}^{\infty} \|\mathcal{P}_0 X_i\| < \infty, \tag{39.6}$$

with $\sigma = \|\sum_{i=0}^{\infty} \mathcal{P}_0 X_i\| > 0$, *then,* $\rho(\hat{\Sigma}_n - \Sigma_n) \not\to 0$ *in probability.*

It is very difficult to find the asymptotic distribution of the maximal eigenvalue $\rho(\hat{\Sigma}_n - \Sigma_n)$, even in the special case that X_i are iid. Recently, Bryc, Dembo and Jiang (2006) studied spectral measures of Toeplitz matrices with sub-diagonals being independent. In our case the matrix $\hat{\Sigma}_n - \Sigma_n$ is Toeplitz. However, the sub-diagonals are dependent. Hence the results by Bryc et al (2006) are not directly applicable. For other contributions for inconcsistency of largest eigenvalues of sample covariance matrices see Johnstone (2001) and El Karoui (2007) among others.

Lemma 39.1. *Assume that (39.5) holds for some* $2 < \alpha \le 4$ *and let* $q = \alpha/2, B_q = 18q^{3/2}(q-1)^{-1/2}$ *if* $q < 2$ *and* $B_q = 1$ *if* $q = 2$. *Then for any* $j \in \mathbf{Z}$,

$$\left\| \sum_{i=1}^{n} X_i X_{i+j} - n\gamma_j \right\|_q \le 2B_q n^{1/q} \|X_1\|_\alpha \Delta_\alpha. \tag{39.7}$$

Theorem 39.2. *Let* $2 < \alpha \le 4$ *and* $q = \alpha/2$. *Assume that (39.5) holds and* $0 \le l < n-1$. *Then*

$$\|\rho(\hat{\Sigma}_{n,l} - \Sigma_n)\|_q \le c_\alpha (l+1) n^{1/q-1} \|X_1\|_\alpha \Delta_\alpha + \frac{2}{n} \sum_{j=1}^{l} j|\gamma_j| \tag{39.8}$$

$$+2 \sum_{j=l+1}^{n} |\gamma_j|,$$

where $c_\alpha > 0$ *is a constant only depending on* α.

39.2.3 Band selection

The band selection problem is intuitively related to the order selection for fitting MA models to the data, and bandwidth selection in the nonparametric estimation of the spectral density function. A method motivated by our Theorem 39.2 suggests that l should satisfy:

$$l \to \infty,\ ln^{1/q-1} \to 0,\ \text{or}\ ln^{1/q-1} \asymp \sum_{j=l+1}^{\infty} |\gamma_j|. \tag{39.9}$$

As a data-driven choice of l, one could propose the following naive algorithm:

1. Choose l such that $\sum_{k=-l}^{l} \hat{\gamma}(k)$ is a "good" estimate of σ^2 or the spectral density of $\{X_t\}$ at zero.
2. Check whether $\Sigma_{n,l}$ is positive definite. If so, let $l^* = l$.
3. Otherwise, let $l^* = l - 1$ and go to Step 2.

The finer details for implementing this method is worked out in this section using the idea of resampling and risk-minimization, in a manner similar to that in Bickel and Levina (2007, Section 5). While they show that "nonoverlapped" splitting of the data works well for band selection in the multivariate data framework, our preliminary numerical experiments showed this scheme to be unsatisfactory for the time series data. Instead, the technique of subsampling (Politis, Romano and Wolf, 1999) which amounts to "overlapped" splitting of the data proved to be more suitable for time series data. Interestingly, some of the details for implementing subsampling fall within the conceptual framework of estimating the spectral density of $\{X_t\}$ at zero as in 1 above, which happens to be a familiar topic in the literature of time series analysis; for an excellent review see Politis and Romano (1995).

For linear processes, a natural way to select the band parameter ℓ in (39.2) is to minimize the risk

$$R(\ell) = E\|\hat{\Sigma}_{n,\ell} - \Sigma_n\|_{(1,1)}, \tag{39.10}$$

where for two $n \times n$ matrices A and B, $\|A - B\|_{(1,1)} = \max_i \sum_{j=1}^{n} |a_{ij} - b_{ij}|$ is the same norm used in Bickel and Levina (2007). Of course, the "oracle" ℓ is given by $\ell_0 = \arg\min_{\ell} R(\ell)$. The following subsampling scheme will be used to estimate the risk in (39.10) and hence ℓ_0. An asymptotic justification for it can be found by focusing on the estimation of the vector of parameters $\theta = (\gamma_0, \cdots, \gamma_K)', K \geq 1$, for a stationary process, and using Theorem 3.3.1 and the results related to Example 3.3.4 in Politis et al. (1999, pp. 83-85).

Given the stationary, centered time series data $X_1, X_2, \cdots, X_n$ of length n, the $\hat{\gamma}_k$ in (39.1) is usually computed for $k = 0, 1, \cdots, K$. The choice of K

is important in practice, since $\hat{\gamma}_k$ is not an accurate estimate of γ_k for k large. A useful guide which is part of the folklore of time series analysis is to use $K \leq n/4$, but the default value in the SAS software is $K = 24$ and in R it is $K = 10\log(10n)$. In our calculations here we fix it at $K = 30$, and when using subsampling to estimate (39.10), the unknown $\sum_n$ will be replaced by the $K \times K$ sample autocovariance matrix $\hat{\sum}_K$ as the "target" and the whole data $X_1, \cdots, X_n$ will be used to estimate its entries. The $\hat{\sum}_{n,\ell}$ will be replaced by the $K \times K$ banded matrix $\hat{\sum}_{b,\ell,\nu}$ whose entries are computed using the νth block (subseries) of length b, i.e. $\{X_\nu, \cdots, X_{\nu+b-1}\}, \nu = 1, \cdots, n-b+1$. Finally, (39.10) is estimated by

$$\hat{R}(\ell) = \frac{1}{n-b+1} \sum_{\nu=1}^{n-b+1} \|\hat{\Sigma}_{b,\ell,\nu} - \hat{\Sigma}_K\|_{(1,1)}, \tag{39.11}$$

and $\hat{\ell}$ is selected to minimize $\hat{R}(\cdot)$. Note that whereas ℓ_0 is the best choice in terms of the risk (39.10), $\hat{\ell}$ tries to adapt to the time series data at hand via (39.11). The optimal choice of the block size b plays a crucial role in selecting the band ℓ. As a general guide, Politis et al (1999, Chaps 3,9) show that for consistency in estimation of a parameter, the block size b must grow to infinity while $b/n \to 0$ with a rate like $n^{1/3}$. Note that this requirement is similar to (39.9) corresponding to the choice of $q = 3/2$ or $\alpha = 3$ in Theorem 39.2. For the computations here, we take $b > K$, and it is fixed at $b = 40$.

Only in a simulation setup where $\sum_n$ is known, it is possible and useful to compare $\hat{\ell}$ from above to the best band choice for the time series data, i.e.

$$\ell_1 = \arg\min_{\ell} \|\hat{\Sigma}_{K,\ell} - \Sigma_K\|_{(1,1)}, \tag{39.12}$$

where $\sum_K$ is the first $K \times K$ principal minor of $\sum_n$ and $\hat{\sum}_{K,\ell}$ is the ℓ-banded version of $\hat{\sum}_K$ in (39.11). Also, the losses of the $K \times K$ and $n \times n$ sample autocovariance matrices, i.e. $\|\hat{\sum}_K - \sum_K\|_{(1,1)}$ and $\|\hat{\sum}_n - \sum_n\|_{(1,1)}$, do serve as useful guides on the merits of these estimators and the relevance of (39.10)-(39.12) for band selection.

References

[1] Bickel, P. J. and Levina, E.: Regularized estimation of large covariance matrices. (2006).
www.stat.berkeley.edu/~bickel/techrep.pdf

[2] Bryc, W., Dembo, A. and and Jiang, T.: Spectral measure of large random Hankel, Markov and Toeplitz matrices. Ann. Probab. **34**, 138 (2006).

[3] El Karoui, N.: Tracy-Widom limit for the largest eigenvalue of a large class of complex sample covariance matrices. Ann. Probab. **35**, 663-714 (2007).

[4] Hannan, E.J. and Deistler, M.: The Statistical Theory of Linear Systems. Wiley, New York. (1988).

[5] Johnstone, I.M.: On the distribution of the largest eigenvalue in principal components analysis. Ann. Statist. **29**, 295-327 (2001).

[6] Politis, D.N., Roman, J.P. and Wolf, M.: Subsampling. Springer Series in Statistics, New York. (1999).

[7] Pourahmadi, M.: Foundations of Time Series Analysis and Prediction Theory. Wiley, New York. (2001).

[8] Tong, H.: Non-linear Time Series: A Dynamical System Approach. Oxford Scientific Publications. (1990).

[9] Wu, W. B.: Nonlinear system theory: Another look at dependence. Proceedings of the National Academy of Sciences USA. **102**, 14150-14154 (2005).

[10] Wu, W.B. and X. Shao: Limit Theorems for Iterated Random Functions. Journal of Applied Probability. **41**, 425-436 (2004).

[11] Zeitouni, O. and Anderson, G.W.: A CLT for regularized sample covariance matrices. Ann. of Statist. To appear. (2008).

Chapter 40
Automatic Local Spectral Envelope

Ori Rosen and David Stoffer

Abstract The concept of spectral envelope for the scaling and analysis of categorical time series in the frequency domain was developed in Stoffer et al. (1993) under the assumption of homogeneity. Here, we present a method for fitting a local spectral envelope for nonstationary sequences.

40.1 Introduction

The concept of spectral envelope for the scaling and analysis of categorical time series in the frequency domain was first introduced in Stoffer et al. (1993). There, we addressed the basic question of how to efficiently discover periodic components in categorical time series. This was accomplished as follows. Let X_t, $t = 0, \pm 1, \pm 2, \ldots$, be a categorical-valued time series with finite state-space $\mathcal{C} = \{c_1, c_2, \ldots, c_{k+1}\}$. Assume that X_t is stationary and $p_j = \Pr\{X_t = c_j\} > 0$ for $j = 1, 2, \ldots, k+1$. For $\boldsymbol{\beta} = (\beta_1, \beta_2, \ldots, \beta_k)' \in \mathbb{R}^k$, denote by $X_t(\boldsymbol{\beta})$ the real-valued stationary time series corresponding to the scaling that assigns the category c_j the numerical value β_j, for $j = 1, 2, \ldots, k$; the category c_{k+1}, is assigned the fixed value of zero (this is without loss of generality, as was shown in the article). Our goal was to find scalings $\boldsymbol{\beta}$ so that the spectral density of $X_t(\boldsymbol{\beta})$, say $f_X(\omega; \boldsymbol{\beta})$, is in some sense interesting, and to summarize the spectral information by what we called the spectral envelope. We chose $\boldsymbol{\beta}$ to maximize the power at each frequency relative to the total power,

---

Ori Rosen
Department of Mathematical Sciences, University of Texas, El Paso, USA, e-mail: `ori@math.utep.edu`

David Stoffer
Department of Statistics, University of Pittsburgh, USA, e-mail: `stoffer@pitt.edu`

$$\lambda(\omega) = \sup \frac{f_X(\omega; \boldsymbol{\beta})}{\sigma^2(\boldsymbol{\beta})}, \tag{40.1}$$

where the sup is over $\boldsymbol{\beta} \neq \mathbf{0}_k$, the $k \times 1$ vector of zeros, and $\sigma^2(\boldsymbol{\beta}) = \text{var}\{X_t(\boldsymbol{\beta})\}$.

It was useful to represent the categories in terms of the vectors $\boldsymbol{e}_1, \boldsymbol{e}_2, \ldots, \boldsymbol{e}_k$, where $\boldsymbol{e}_j$ represents the $k \times 1$ vector with a one in the j-th row, and zeros elsewhere; $\boldsymbol{e}_{k+1}$ is the $k \times 1$ zero vector. We then defined a k-dimensional stationary time series $\boldsymbol{Y}_t$ by $\boldsymbol{Y}_t = \boldsymbol{e}_j$ when $X_t = c_j$. If $\boldsymbol{Y}_t$ has a continuous spectral matrix $f_Y(\omega)$, then $X_t(\boldsymbol{\beta}) = \boldsymbol{\beta}'\boldsymbol{Y}_t$ implies that $f_X(\omega;\, \boldsymbol{\beta}) = \boldsymbol{\beta}' f_Y^{re}(\omega)\boldsymbol{\beta}$, where *re* denotes the real part [$f_Y(\omega)$ is skew-symmetric, so the imaginary part is annihilated by pre- and post-multiplication by a real-valued vector]. The optimality criterion can be expressed as

$$\lambda(\omega) = \sup \frac{\boldsymbol{\beta}' f_Y^{re}(\omega)\boldsymbol{\beta}}{\boldsymbol{\beta}' V \boldsymbol{\beta}}, \tag{40.2}$$

where V is the variance-covariance matrix of $\boldsymbol{Y}_t$.

We defined $\lambda(\omega)$ to be the *spectral envelope* of a stationary categorical time series and $\boldsymbol{\beta}(\omega)$ the *optimal scaling*. The name spectral envelope is appropriate because $\lambda(\omega)$ envelopes the standardized spectrum of any scaled process; that is, setting $\sigma^2(\boldsymbol{\beta}) = \text{var}\{X_t(\boldsymbol{\beta})\} = 1$, we have $f_X(\omega; \boldsymbol{\beta}) \leq \lambda(\omega)$, with equality if and only if $\boldsymbol{\beta}$ is proportional to $\boldsymbol{\beta}(\omega)$. Although information is lost when one restricts attention to the spectrum of $X_t(\boldsymbol{\beta})$, less information is lost when one considers the spectrum of $\boldsymbol{Y}_t$. Dealing directly with the spectral density $f_Y(\omega)$ itself is somewhat cumbersome because it is a function into the set of complex Hermitian matrices. Alternatively, one can view the spectral envelope as an easily understood, parsimonious tool for exploring the periodic nature of a categorical time series with a minimal loss of information. We mention that the spectral envelope methodology draws heavily from ideas developed in spectral domain principal component analysis (e.g. Brillinger, 2001, Ch. 9)

For the stationary case, estimation proceeds in an obvious way, first by obtaining an consistent estimate of $f_Y(\omega)$ in the usual way, and then by obtaining the largest eigenvalue of the estimate in the metric of the sample covariance matrix of the data $\boldsymbol{Y}_t$. More details and some examples of the theory and methodology can be found in Shumway & Stoffer (2006, Ch. 7); R programs for computing the spectral envelope are available on the website for the text (`http://www.stat.pitt.edu/stoffer/tsa2/`).

Recently, local methods based on the spectral envelope have been discussed by other researchers from various fields such as soil science, and signal processing (e.g., Wang and Johnson, 2002). We realized early on that the stationary assumption would have to be relaxed for the spectral envelope to be a truly useful tool. For example, a common problem in analyzing long DNA sequences is in identifying coding sequences (CDS) that are dispersed throughout the sequence and separated by regions of noncoding. Local behavior is encountered even within short subsequences of DNA. To address this problem of local behavior in categorical-valued time series in general, we developed a technique to use the spectral envelope methodology in conjunction with a dyadic tree-based adaptive segmentation (TBAS) method for analyzing locally stationary processes. These and related techniques were reported in Stoffer & Ombao (2001), Stoffer (2002), and Stoffer, Ombao & Tyler (2002). In these papers, we developed various *local* spectral envelope techniques; in particular we focused on a TBAS method that automatically divides a sequence into smaller stationary segments. Once the optimal segmentation was found, we extracted the pertinent spectral information from these segments. We provided numerous examples in these papers to exhibit the viability of the technique in detecting genes. The problem with the TBAS method was that it produced dyadic subdivisions of a sequence that were considered stationary. Hence, when the analysis was completed, one only had knowledge of an approximate location of a CDS (or some other interesting segments, such as repeat regions).

40.2 Basic approach

Our present focus is on developing a better method for estimating a local (in a generic sense) spectral envelope. The initial method we are proposing is based on fitting local splines. The first step was to develop a method to estimate a spectral matrix function of a stationary vector process via smoothing splines. This step was accomplished in Rosen and Stoffer (2007). The basic idea is as follows. We assumed that we have a sufficiently large number, n, of observations from a p-dimensional stationary time series, $\boldsymbol{x}_t$, whose $p \times p$ autocovariance matrix, $\Gamma(h) = \{\gamma_{jk}(h)\}$, satisfies $\sum_{h=-\infty}^{\infty} |\gamma_{jk}(h)| < \infty$ for all $j, k = 1, \ldots, p$. The $p \times p$ spectral density matrix is given by

$$f(\omega) = \sum_{h=-\infty}^{\infty} \Gamma(h) e^{-2\pi i \omega h}, \quad -1/2 \le \omega \le 1/2 \,,$$

where $f(\omega) = \{f_{jk}(\omega)\}$, for $j, k = 1, \ldots, p$. In addition, we assumed $f(\omega)$ is positive definite so that we may employ the Whittle likelihood

$$L(\boldsymbol{y}_0, \ldots, \boldsymbol{y}_{n-1}; f_0, \ldots, f_{n-1}) \doteq \prod_{k=0}^{n-1} \det(f_k)^{-1} \exp(-\boldsymbol{y}_k^* f_k^{-1} \boldsymbol{y}_k) \,, \qquad (40.3)$$

where $\boldsymbol{y}_k$ is the $p \times 1$ discrete Fourier transform of the data at frequency k/n,

$$\boldsymbol{y}_k = n^{-1/2} \sum_{t=1}^{n} \boldsymbol{x}_t \exp\{-2\pi i \tfrac{k}{n} t\},$$

and $f_k = f(k/n)$.

Our goal was to obtain smooth estimators of the elements of f as a function of ω while satisfying the constraint that f is positive definite. To this end, we expressed the inverse of the spectral matrix at frequency k/n as the modified complex Cholesky factorization

$$f_k^{-1} = T_k^* D_k^{-1} T_k \ , \tag{40.4}$$

where T_k is a complex unit lower triangular matrix, and D_k is a diagonal matrix. To be more specific,

$$T_k = \begin{pmatrix} 1 & & & & \\ -\theta_{21}^{(k)} & 1 & & & \\ -\theta_{31}^{(k)} & -\theta_{32}^{(k)} & 1 & & \\ \vdots & \vdots & & \ddots & \\ -\theta_{p1}^{(k)} & -\theta_{p2}^{(k)} & \dots & -\theta_{p,p-1}^{(k)} & 1 \end{pmatrix}$$

and $D_k = \text{diag}(\delta_{1k}^2, \dots, \delta_{pk}^2)$. Note that in general the $\theta_{il}^{(k)}$'s are complex-valued.

It is difficult to model the elements of the spectral matrix directly because of the constraint that the spectral matrix must be positive definite at each frequency, but in the factorization (40.4), the $\theta_{il}^{(k)}$'s are unconstrained and the δ_{jk}^2's are positive. Thus, it is much easier to model these parameters rather than the elements of the spectral matrix. Once T_k and D_k have been estimated, the resulting estimator of f_k is automatically positive definite.

To facilitate the estimation of the $\theta_{il}^{(k)}$'s and the δ_{jk}^2's and thereby the estimation of the spectral matrix, we used the likelihood (40.3) in combination with the factorization (40.4). We first rewrote the likelihood (40.3) as a function of the $\theta_{il}^{(k)}$'s and the δ_{jk}^2's. Let $N = [n/2]$, $\boldsymbol{\theta}_k$ be the $p(p-1)/2$-dimensional vector $(\theta_{21}^{(k)}, \theta_{31}^{(k)}, \theta_{32}^{(k)}, \dots, \theta_{p,p-1}^{(k)})'$, $\Theta = (\boldsymbol{\theta}_1, \dots, \boldsymbol{\theta}_N)$, $\Delta = \{\delta_{1k}^2, \dots, \delta_{pk}^2\}_{k=1}^N$ and $Y = (\boldsymbol{y}_1, \dots, \boldsymbol{y}_N)$. >From (40.3) and (40.4) it followed that the likelihood can be expressed as

$$L(Y; \Delta, \Theta) \propto \prod_{k=1}^{N} \prod_{j=1}^{p} \delta_{jk}^{-2} \exp\{(\boldsymbol{y}_k - Z_k\boldsymbol{\theta}_k)^* D_k^{-1} (\boldsymbol{y}_k - Z_k\boldsymbol{\theta}_k)\} \ , \tag{40.5}$$

where Z_k is a $p \times p(p-1)/2$ design matrix with elements that are either 0 or a component of $\boldsymbol{y}_k$. Note that in (40.5), the endpoint involving $\boldsymbol{y}_0$ is ignored. Next, we placed linear smoothing spline priors on the $\theta_{il}^{(k)}$'s and the δ_{jk}^2's. In our experience, linear smoothing splines were better suited to estimating the spectral matrix, as they can better accommodate narrowband peaks. In particular, each of the $\log \delta_{jk}^2$'s and the real and imaginary parts of each of the negative $\theta_{il}^{(k)}$'s are expressed as

$$\alpha_0 + \alpha_1 \omega_k + \sum_{s=1}^{N} \psi_s(\omega_k)\beta_s \,, \tag{40.6}$$

where $\omega_k = k/n$ and $\psi_s(\omega_k) = \sqrt{2}\cos\{(s-1)\pi\omega_k\}$. The $\psi_s(\cdot)$'s are the Demmler-Reinsch basis functions for linear smoothing splines (Eubank, 1999). Let X_β be the matrix whose columns are the basis functions $\psi_s(\cdot)$ evaluated at $\omega_1, \ldots, \omega_N$, and let X_α be a matrix whose columns are the vector of ones and $(\omega_1, \ldots, \omega_N)'$. Let $X = (X_\alpha \mid X_\beta)$ be the matrix formed by binding X_α and X_β columnwise, $\boldsymbol{\gamma}_j = (\boldsymbol{\alpha}_j', \boldsymbol{\beta}_j')'$, $\boldsymbol{\Delta}_j = (\delta_{j1}^2, \ldots, \delta_{jN}^2)'$ and $\boldsymbol{\theta}_{il} = (\theta_{il}^{(1)}, \ldots, \theta_{il}^{(N)})'$. Then

$$\log \boldsymbol{\Delta}_j = X\boldsymbol{\gamma}_j, \quad -\Re(\boldsymbol{\theta}_{il}) = X\boldsymbol{\gamma}_{il(re)}, \quad -\Im(\boldsymbol{\theta}_{il}) = X\boldsymbol{\gamma}_{il(im)} \,, \tag{40.7}$$

for $j = 1, \ldots, p$, $i = 2, \ldots, p$, and $l = 1, \ldots, i-1$, where $\Re(\cdot)$ and $\Im(\cdot)$ denote the real part and the imaginary part, respectively. Corresponding to (40.7), the priors on $\boldsymbol{\alpha}_j$, $\boldsymbol{\alpha}_{il(re)}$ and $\boldsymbol{\alpha}_{il(im)}$ are taken to be $N(\mathbf{0}, \sigma_\alpha^2 I_2)$, and those on $\boldsymbol{\beta}_j$, $\boldsymbol{\beta}_{il(re)}$ and $\boldsymbol{\beta}_{il(im)}$ are taken to be $N(\mathbf{0}, \tau_j^2 I_N)$, $N(\mathbf{0}, \tau_{il(re)}^2 I_N)$ and $N(\mathbf{0}, \tau_{il(im)}^2 I_N)$, respectively. With the $\theta_{il}^{(k)}$'s and the δ_{jk}^2's viewed as functions of ω, the parameters τ_j^2, $\tau_{il(re)}^2$ and $\tau_{il(im)}^2$ are smoothing parameters, governing the amount of smoothing of each of these functions. A zero value of a smoothing parameter corresponds to a linear fit, while a value tending to infinity results in an interpolating linear spline. The priors on the smoothing parameters are $p(\tau_j^2) \propto 1/\tau_j^2$, $p(\tau_{il(re)}^2) \propto 1/\tau_{il(re)}^2$ and $p(\tau_{il(im)}^2) \propto 1/\tau_{il(im)}^2$. We estimated the spectral matrix by its posterior mean using Markov chain Monte Carlo methods to perform the required multidimensional integration. Details about the sampling scheme and numerous examples, including an analysis of a DNA nucleotide sequence via the spectral envelope can be found in Rosen and Stoffer (2007).

The next step is to establish a method for fitting local univariate spectra and then to combine the results from the univariate local case and the multivariate stationary case to the local multivariate case. The stationary univariate case can be handled using the stationary multivariate approach previously described, but where there is no need to use the Cholesky decomposition; that is, in $\rho_F eq : whittle - -\rho_F eq : like$, there are no θs and there is only one δ_k^2 at frequency k/n.

Suppose we are given $\{X(t/N);\, t = 1, \ldots, N\}$ observations from a Dahlhaus-locally stationary process with spectrum $f(u, \omega)$, for $u \in (0, 1]$ and $\omega \in (-1/2, 1/2]$, let $\{X_s(t);\, t = 1, \ldots, N/S;\, s = 1, \ldots, S\}$ be a piecewise stationary process, where in any small non-overlapping segment $(\frac{s-1}{S}, \frac{s}{S}]$, the spectrum of $X_t(s)$ is $f(s, \omega)$. Under general smoothness conditions, a Dahlhaus-locally stationary processes can be well approximated by piecewise stationary processes, and we therefore focus on the estimation of $f(s, \omega)$ as an approximation to the Dahlhaus-spectrum $f(u, \omega)$. The choice of the number of segments, S, will be discussed after we present the model, but the basic idea is that the finer the partition of the unit time interval, the better the estimate of the time varying spectrum.

Given S, our goal is the estimation of $g_s(\omega) = \log f(s,\omega)$ from the data, $\{X_t(s); t = 1,\ldots,n; s = 1,\ldots,S\}$, where $n = N/S$. To this end, we will first calculate the periodogram corresponding to each segment. Let $\boldsymbol{y}_s = (y_{s,0},\ldots,y_{s,n/2})'$ be the log-periodogram for segment s, $s = 1,\ldots,S$, evaluated at the Fourier frequencies. We model these observations as

$$\boldsymbol{y}_s(\omega_k) = g_s(\omega_k) + \varepsilon_k$$

where $g_s(\omega)$ is the log of the spectral density for segment s, for $s = 1,\ldots S$, and the ε_k's for $k = 0,\ldots,[n/2]$ are independent, $\varepsilon_k \sim \log(\chi_2^2/2)$ for $k = 1,\ldots,[n/2]-1$, and $\varepsilon_k \sim \log(\chi_1^2)$ for $k = 0,[n/2]$.

Our basic approach is to model $g_s(\omega)$ as a mixture of an unknown but finite number of spectra so that

$$g_s(\omega) = \sum_{j=1}^{J} g_{js}(\omega)\Pr(j),$$

where J is the maximum number of components, $\Pr(j)$ is the prior probability that the mixture contains j components, and $g_{js}(\omega)$ is the log of the spectral density of a mixture of j components in segment s. For a given number of mixture components j and segment s we model $g_{js}(\omega)$ as

$$g_{js}(\omega) = \sum_{r=1}^{j} \pi_{rsj}\log(f_{rj}(\omega)),$$

where f_{rj} is the spectral density of the rth component and π_{rjs} is the unknown weight assigned to the rth component in segment s, with $\sum_{r=1}^{j}\pi_{rjs} = 1$. Note that the spectral density f_{rj} is common to all segments. The value of π_{rjs} represents the probability that in a mixture of j components, the data in segment s have spectral density f_{rj}. A key point to note is that these probabilities are parameterized to depend upon the segment s and are modeled using a multinomial logistic regression; we will discuss how they are specified next. This means that, although the component spectra are common to all segments, a time varying estimate of the spectral density is obtained by allowing the weights of the common spectra to change across segments.

It is important to note that the choice of the number of segments, S, is not crucial to our estimation process subject to certain constraints. In theory there are potentially as many segments as there are data points. However, practically, we need a minimum number of observations in each segment to estimate the spectral density and for the Whittle approximation to the likelihood to hold. In our preliminary experiments, it appears that using a

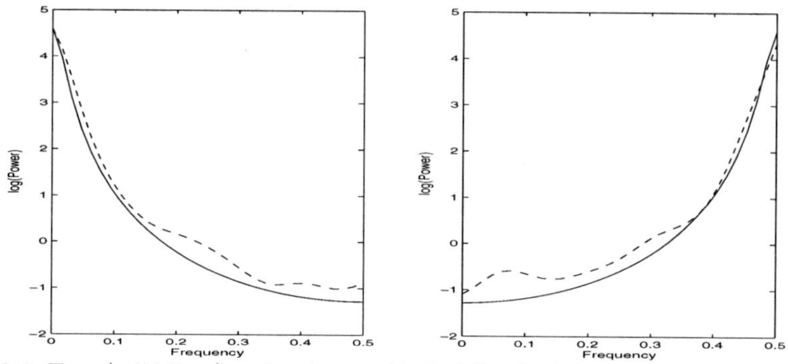

Fig. 40.1 True (solid lines) and estimates (dashed lines) of the log spectral density. The left panel shows $\log(f_{12})$ which is the log of the spectral density of the time series $x_t = 0.9x_{t-1} + \varepsilon_t$. The right panel shows $\log(f_{22})$, the log of the spectral density for $x_t = -0.9x_{t-1} + \varepsilon_t$.

minimum of 64 observations in each segment gives reliable results assuming that the true local spectra have well separated peaks.

Most, if not all, of local spectral techniques, including the adaptive techniques that rely on orthogonal libraries and use entropy-based basis algorithms such as the Best Basis Algorithm, use an arbitrary maximum level of segmentation of the unit interval to initialize the local analysis. In addition, most of these techniques use dyadic segmentation for ease. Although our method also requires picking a maximal segmentation through the choice of S, it is not crucial that the segmentation be dyadic. In some cases, such as for DNA, it may be more appropriate to consider triadic segmentation, which our method easily accommodates.

Finally, but perhaps most importantly, is that the parameters of the mixing function π_{rjs} in our model are of more importance than the number of segments because these parameters control the location and rate at which the time series moves from one stationary process to another.

To illustrate our method, consider a time series of length 1024 generated from the following piecewise stationary model

$$x_t = \begin{cases} 0.9x_{t-1} + \varepsilon_t \text{ if } & 1 \le t \le 450, \\ -0.9x_{t-1} + \varepsilon_t \text{ if } & 451 \le t \le 1024, \end{cases}$$

where $\varepsilon_t \sim N(0,1)$. For illustrative purposes, suppose we know that there are two components [i.e., $\Pr(j=2)=1$]. We first divide the time series into non-overlapping segments each containing 64 data points. This gives a total of 16 segments. Our estimate of the log spectra in segment s for $s = 1, \ldots 16$ is

$$g_{2s}(\omega) = \pi_{1s2} \log(f_{12}(\omega)) + (1 - \pi_{1s2}) \log(f_{22}(\omega)) \,. \tag{40.8}$$

Figure 40.1 shows the true (solid line) and estimated (dashed line) log spectral density for $x_t = 0.9x_{t-1} + \varepsilon_t$ (left panel) and for $x_t = -0.9x_{t-1} + \varepsilon_t$ (right panel), $\varepsilon_t \sim N(0,1)$. Figure 40.2 plots the estimated mixing function π_{12s} as a function of the segment. This figure shows that the probability that the data have spectral density f_{12} is close to 0.91 at the beginning of the time series. This probability decreases to 0.5 by the 7th segment and is approximately 0.03 by the end of the time series.

In the general setup, we will we express $\log(f_{rj})$ as

$$\log(f_{rj}(\omega_k)) = \alpha_{0rj} + h_{rj}(\omega_k)$$

and write $\boldsymbol{h}_{rj} = X\boldsymbol{\beta}_{rj}$. The priors on $\boldsymbol{\beta}_{rj}$, and on α_{0rj} for $r = 1, \ldots, j$ and $j = 1, \ldots, J$ and on τ^2_{rj} is similar to the discussion around $\rho_F eq : demmler -$ $-\rho_F eq : smoothing$; in addition, we impose an ordering on τ^2_{rj} for $r = 1, \ldots, j$ and $j = 1, \ldots, J$, so that for a given j, $\tau^2_{1j} >, \ldots, > \tau^2_{jj}$. This ensures that the likelihood is identified.

The mixing probabilities are expressed using the multinomial linear logit model so that

$$\pi_{rjs} = \frac{\exp(\boldsymbol{\delta}'_{rj}\mathbf{u}_s)}{\sum_{h=1}^{j} \exp(\boldsymbol{\delta}'_{hj}\mathbf{u}_s)} \tag{40.9}$$

with parameters $\boldsymbol{\delta}_{rj}$, $r = 1, \ldots, j$ and $j = 1, \ldots, J$. In (40.9), $\mathbf{u}_s = (1, u_s)'$, where the covariate u_s is taken as $u_s = s/S$, and $\boldsymbol{\delta}_{rj} = (\delta_{0rj}, \delta_{1rj})'$. For identifiability, $\boldsymbol{\delta}_{1j}$ is set to zero. Such logistic weights are also used in the mixtures-of-experts model (Jacobs et al., 1991). The priors on $\boldsymbol{\delta}_{rj}$ for $r = 1, \ldots, j$ and $j = 1, \ldots, J$ are bivariate normal with zero mean and variance $\sigma^2_\delta I_2$ and are assumed independent across all r and j. In all of our analyses, σ^2_δ was equal to 4. In addition, we impose the following dominance condition, which is a restriction on $\tilde{\boldsymbol{\delta}} = (\boldsymbol{\delta}_{11}, \ldots, \boldsymbol{\delta}_{JJ})$; let $\mathcal{U} = \{u_1, \ldots, u_S\}$. For $r = 1, \ldots, j$, let $m_{rj} = \max_{u \in \mathcal{U}} \pi_{rsj}(u)$ be the maximum of π_{rsj} over $\mathcal{U}$ and let $u_{(rj)}$ be the point at which the maximum is attained, i.e. $u_{(rj)} = \arg\max_{u \in \mathcal{U}} \pi_{rsj}(u)$. We assume that $m_{rj} > \pi_{isj}\{u_{(rj)}\}$ for all $i \neq j$. That is, each component is required to have a point in $\mathcal{U}$ at which the probability of that component exceeds the probabilities of all other components. This requirement is equivalent to placing a prior on $\tilde{\boldsymbol{\delta}}$ which puts zero probability on the values of $\tilde{\boldsymbol{\delta}}$ in the parameter space where this requirement is not met. The reason for imposing this dominance condition is to penalize for too many components in the model.

Finally we will assume *a priori* that the maximum number of components is J and that $\Pr(j = k) = 1/J$ for $k = 1, \ldots, J$.

For inference, we will estimate the log of the spectral density in segment s, for $s = 1, \ldots S$ by its posterior mean $E(g_s|\mathbf{y})$, with all unknown parameters integrated out and we use MCMC to perform the required multidimensional integration. The expectation $E\{g_s(\omega)|\mathbf{y}\}$ is defined as

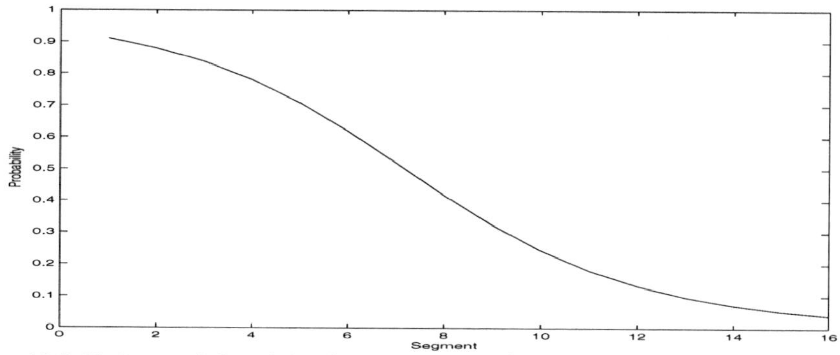

Fig. 40.2 Estimate of the mixing function π_{12s} in (40.8) based on 16 segments.

$$E\{g_s(\omega)|\mathbf{y}\} = \sum_{j=1}^{J} \int E\{g_s(\omega)|\mathbf{y}, \boldsymbol{\theta}_j, j\} p(\boldsymbol{\theta}_j|\mathbf{y}, j) d\boldsymbol{\theta}_j \Pr(j|\mathbf{y})$$

where $\boldsymbol{\theta}_j = (\boldsymbol{\alpha}_j', \boldsymbol{\beta}_j', \boldsymbol{\tau}_j', \boldsymbol{\delta}_j')'$, $\boldsymbol{\alpha}_j = (\alpha_{01}, \ldots, \alpha_{0j})'$, $\boldsymbol{\beta}_j = (\boldsymbol{\beta}_{1j}', \ldots, \boldsymbol{\beta}_{jj}')'$, $\boldsymbol{\tau}_j = (\tau_{1j}^2, \ldots, \tau_{jj}^2)'$ and $\boldsymbol{\delta}_j = (\boldsymbol{\delta}_{2j}', \ldots, \boldsymbol{\delta}_{jj}')$. This integral cannot be evaluated explicitly and we use MCMC simulation to estimate it. In addition to the point estimates, we construct $(1-\alpha)$-level pointwise credible intervals for the log spectra by obtaining the $\alpha/2$ and $1-\alpha/2$ percentiles of the MCMC fitted log spectra based on all the iterates after the burn-in period. Note that these credible intervals reflect the uncertainty surrounding not only our estimate of $\log f_{rj}$ but also our uncertainty surrounding the number of components J and the mixing probabilities π_{rsj}.

To simplify the simulation from the posterior distribution $p(\boldsymbol{\theta}_j, j|\mathbf{y})$, we introduce latent variables that are generated during the simulation. The first of these is the number of components j that are active at any point in the simulation. Then, given j, define the vector of indicator variables γ_{srj} for $s = 1, \ldots, S$ and $r = 1, \ldots, j$, where $\gamma_{srj} = 1$ if $\mathbf{y}_s$ originated from the rth component, and $\gamma_{srj} = 0$, otherwise.

The basic MCMC scheme for our model can be outlined in the following steps. The sampling scheme consists of two parts; a between-model move followed by a within-model move. The number of components j is first initialized, then conditional on this value, the other model parameters $\boldsymbol{\alpha}_j$, $\boldsymbol{\beta}_j$, $\boldsymbol{\tau}_j = (\tau_{1j}^2, \ldots, \tau_{jj}^2)'$ and $\boldsymbol{\delta}_j$ are initialized.

1. Between Model Move: A new value of j is proposed, and conditional on this value, parameter values for $\boldsymbol{\alpha}_j$, $\boldsymbol{\beta}_j$, $\boldsymbol{\tau}_j$ and $\boldsymbol{\delta}_j$ are proposed. These proposed values are then accepted or rejected using a Metropolis-Hastings step.
2. Within Model Move: Given the value of j, the parameters specific to a model of j components are then updated as follows.

a. Let $\boldsymbol{\beta}^*_{rj} = (\alpha_{rj}, \boldsymbol{\beta}'_{rj})'$, $r = 1, \ldots, j$, $\boldsymbol{\beta}^*_j = (\boldsymbol{\beta}^{*'}_{1j}, \ldots, \boldsymbol{\beta}^{*'}_{jj})'$ and $X^* = (\mathbf{1}, X)$. Generate $\boldsymbol{\beta}^*_j$ from $p(\boldsymbol{\beta}^*_j \mid \boldsymbol{\tau}_j, \boldsymbol{\gamma}_j, X^*, \tilde{\mathbf{y}})$ via Metropolis-Hastings steps, where $\boldsymbol{\gamma}_j = \{\gamma_{srj}\}$, for $r = 1, \ldots, j$ and $s = 1, \ldots, S$, are the component indicators, and $\tilde{\mathbf{y}} = (\mathbf{y}'_1, \ldots, \mathbf{y}'_S)'$.
b. Generate $\boldsymbol{\tau}_j$ from $p(\boldsymbol{\tau}_j \mid \boldsymbol{\beta}_j)$.
c. Generate $\boldsymbol{\delta}_j$ from $p(\boldsymbol{\delta}_j \mid \boldsymbol{\gamma}_j, U)$ via a Metropolis-Hastings step, where U is the matrix whose sth row is $\mathbf{u}'_s$, $s = 1, \ldots, S$.
d. Let $\gamma_{sj} = r$ if $\gamma_{srj} = 1$. Generate the component indicators from $p(\gamma_{sj} = r \mid \boldsymbol{\beta}^*_j, \boldsymbol{\delta}_j, X^*, \mathbf{y}_s)$.

The proposal densities for generating $\boldsymbol{\beta}^*_j$ and $\boldsymbol{\delta}_j$ are multivariate normal and so far, the resulting acceptance rates are around 30% and 80%, respectively.

As previously indicated, the next step is to combine the univariate nonstationary methods with the multivariate stationary method to obtain an estimator for the local spectral envelope.

References

[1] Brillinger, D.R.: Time Series: Data Analysis and Theory, 2nd ed. Philadelphia: SIAM. (2001).

[2] Eubank, R.L.: Nonparametric Regression and Spline Smoothing. Second Edition, Marcel Dekker, New York. (1999).

[3] Rosen, O. and Stoffer, D.S.: Automatic estimation of multivariate spectra via smoothing splines. Biometrika. **94**, 335-345 (2007).

[4] Shumway, R.H. and Stoffer, D.S.: Time Series Analysis and Its Applications: With R Examples, 2nd ed. New York: Springer. (2006).

[5] Stoffer, D.S.: Nonparametric Frequency Detection and Optimal Coding in Molecular Biology. In Modeling Uncertainty: An Examination of Stochastic Theory, Methods, and Applications. Moshe Dror, Pierre L'Ecuyer, Ferenc Szidarovszky (eds). Boston: Kluwer Academic Publishers. Chapter 7, pp 129-154 (2002).

[6] Stoffer, D.S. & Ombao H.: Evolutionary Spectral Envelope via Tree Based Adaptive Segmentation. In Proceedings of the 2nd International Symposium on the Frontiers of Time Series Modeling. T. Higuichi and G. Kitagawa (eds). Institute of Statistical Mathematics. (2001).

[7] Stoffer, D.S., H. Ombao & D.E. Tyler: Local Spectral Envelope: An Approach Using Dyadic Tree Based Adaptive Segmentation. Ann. Inst. Statist. Math. **54**, 201-223 (2002).

[8] Stoffer, D. S., Tyler, D. E. & McDougall, A. J.: Spectral analysis for categorical time series: Scaling and the spectral envelope. Biometrika. **80**, 611-622 (1993).

[9] Wang, W. and Johnson, D.H.: Computing linear transforms of symbolic signals. IEEE Trans. Signal Process. **50**, 628-634 (2002).

Chapter 41
Recent Advances in the Use of SVM for Functional Data Classification

Fabrice Rossi and Nathalie Villa

Abstract In the past years, several works were dealing with the use of Support Vector Machine (SVM) for classifying functional data. Here, we propose to give an overview of these works and to introduce a new result based on the use of smoothing conditions on the observed functions. The originality of this approach both lies in the fact that the consistency result allows to work with the derivatives of the function instead of the function itself but also that it is relative to the observed discretization and not to the entire knowledge of the functions.

41.1 Introduction

As the number of data coming from continuous recording has increased, the analysis of data taking the form of curves has also been developed. After the pioneering work of Deville (1974), Cardot *et al.* (1999), Ramsay and Silverman (1997) in the framework of linear models, various statistical methods have been adapted to what is now called *functional data analysis* (FDA): this is the case of nonparametric estimation, Ferraty and Vieu (2006), Ferraty and Vieu (2002), of neural networks Ferre and Villa (2006), Rossi and Conanguez (2005) or of k-nearest neighbors Biau*et al.* (2005), to name a few.

Fabrice Rossi
Projet AxIS, INRIA Rocquencourt, Domaine de Voluceau, Rocquencourt, B.P. 105, F-78153 Le Chesnay cedex, France, e-mail: fabrice.rossi@inria.fr

Nathalie Villa
Université de Toulouse, IMT (Institut de Mathématiques), 118 route de Narbonne, F-31062 Toulouse cedex 9, France and IUT de Perpignan, Département STID, Domaine Universitaire d'Auriac, F-11000 Carcassonne, France, e-mail: nathalie.villa@math.univ-toulouse.fr

SVM were introduced in the past years and they appear to be a competitive tool for solving binary classifications. One of their main interest is that they are less sensitive to the dimensionality of the predictor than other methods. Then, they are potentially an interesting approach in FDA. In his PhD thesis Lee (2004), Lee first uses the SVM for classifying curves: his approach was based on PCA pre-processing and was illustrated by several examples. Unfortunately, no consistency result was given. In Rossi and Villa (2006a, 2006b), the authors present various ways for dealing with binary classification of curves by the way of SVM: the first article presents a projection approach that is valid for any Hilbert space and the second one uses smoothness constraints by the way of a spline interpolation.

This article intends to summarize the past theoretical results obtained for classification of curves with SVM and to introduce a new consistency result with respect to the discretization of the observations. This approach is original as it allows to work on the derivatives of the observations which can be a relevant task for many kind of problems Ferraty and Vieu (2002), Rossi and Villa (2005), Dejean *et al.* (2007). In section 41.2, we recall the SVM algorithm and the existing consistency results in the multi-dimensional context. Then, section 41.3 presents the adaptation of this algorithm to the FDA context. To that aim, section 41.3.2 develops a consistency result by a projection method and section 41.3.3 a consistent method on derivatives which uses smoothing splines approximation of the predictors.

The proof of the results given in this paper as long as several applications on real data sets can be found in Rossi and Villa (2006b), Rossi and Villa (2007).

41.2 SVM classifiers

41.2.1 Definition

Vapnik (1998) introduces a theoretical context to model statistical learning and popularized Support Vector Machines (SVM), particularly in the framework of binary classification. To recall what is the principle of SVM, suppose that a training set of size n, $(z_i, y_i)_i$, of i.i.d. observations is given: (z_i) take their values in a space $\mathcal{X}$ and (y_i) in $\{-1, 1\}$. SVM are classifiers that belong to a family of semi-linear classifiers of the form $\phi_n(z) = \text{Sign}\left\{\langle w, \psi(z)\rangle_{\mathcal{F}} + b\right\}$ where $\psi : \mathcal{X} \to \mathcal{F}$ is a given nonlinear function from $\mathcal{X}$ to a Hilbert space $\mathcal{F}$, called *feature space*. Then, w and b are parameters that have to be learnt from the data set: they are chosen by an optimization problem that aims at maximizing the margin between the observations $(\psi(z_i))$ from both classes and the decision frontier. More precisely, they are the solution of:

$$\begin{aligned}(P_{C,\mathcal{F}})\ &\min_{w,b,\xi} \|w\|_{\mathcal{F}}^2 + C\sum_{i=1}^n \xi_i \\ &\text{such that } y_i(\langle w, \psi(z_i)\rangle_{\mathcal{F}} + b) \geq 1 - \xi_i,\ 1 \leq i \leq n, \\ &\qquad\qquad \xi_i \geq 0,\ 1 \leq i \leq n.\end{aligned}$$

The problem $(P_{C,\mathcal{F}})$ has a dual formulation that doesn't directly use the transformed data $\psi(z_i)$ but the inner product $\langle\psi(z_i), \psi(z_j)\rangle_{\mathcal{F}}$. Thus, the nonlinear transformation ψ and the feature space, $\mathcal{F}$ don't have to be explicitly known: they are implicitly used by defining the scalar product, $\langle\psi(z_i), \psi(z_j)\rangle_{\mathcal{F}}$ by the way of a *kernel trick*. A symmetric and positive definite kernel $K : \mathcal{X} \times \mathcal{X} \to \mathbb{R}$ is chosen: according to Moore-Aronszajn theorem (1), this ensures that there is a Hilbert space $\mathcal{F}$ and an application $\psi : \mathcal{X} \to \mathcal{F}$ such that $\langle\psi(x_i), \psi(x_j)\rangle_{\mathcal{F}} = K(x_i, x_j)$.

41.2.2 Universal consistency of SVM

SVM are known to have good generalization properties when $\mathcal{X}$ is a finite dimensional space. More precisely, Steinwart (2001-2002) show that d-dimensional SVM are universally consistent, under some hypothesis i.e., that $\lim_{n\to+\infty} L\phi_n = L^*$ where $L\phi_n$ is the probability of misclassification of the classifier ϕ_n, $L\phi_n = \mathbb{P}(\phi_n(Z) \neq Y)$, and L^* is the *Bayes error*, the optimal misclassification rate for the random pair (Z, Y) having same distribution as (z_i, y_i), $L^* = \inf_{\phi:\mathcal{X}\to\{-1,1\}} \mathbb{P}(\phi(Z) \neq Y)$.

This result is obtained with particular kernels: if $\mathcal{X}$ is a compact subset of $\mathbb{R}^d$, the kernel K used has to be *universal* i.e., the set $\{z \in \mathcal{X} \to \langle w, \psi(z)\rangle_{\mathcal{F}},\ w \in \mathcal{F}\}$ has to be dense in the set of continuous functions on $\mathcal{X}$. Secondly, for $\varepsilon > 0$, $\mathcal{N}(\varepsilon, K)$ is the *covering number* of the space $\mathcal{F}$ i.e., the minimum number of balls of radius ε that are needed to cover $\mathcal{F}$; consistency of SVM also requires that $\mathcal{N}(\varepsilon, K) = \mathcal{O}(\varepsilon^{-\nu_d})$ for a $\nu_d > 0$. Among others, Gaussian kernels, $K_\gamma^d(u, v) = exp\left(-\gamma\|u - v\|_{\mathcal{X}}^2\right)$, satisfy both assumptions with $\nu_d = 1/d$ (see Steinwart, 2002) but this can't be extended to the case where $\mathcal{X}$ has infinite dimension both because the covering number assumption is not fulfilled for usual kernels (as Gaussian kernel) and because assuming that the variable takes its values in a compact set is too much restrictive in infinite dimensional spaces.

In the following, $K_\gamma^d \in$ **(Acv)** will denote any kernel on $\mathbb{R}^d$ that satisfy these two conditions and that possibly can depend on a parameter γ. Moreover, if the calculation of $K_\gamma^d(u, v)$ is only based on the inner product of u and v in $\mathbb{R}^d$, such a kernel can be generalized into K_γ^∞ which is a kernel on L^2 that has the same form as K_γ^d except that the $\mathbb{R}^d$-inner product is replaced by the L^2-inner product.

41.3 Using SVM to classify functional data

As was explained above, the consistency result obtained for d-dimensional SVM can't be applied directly to the infinite dimensional case. Moreover, in FDA, the observations are not direct realizations of a random pair having a functional predictor: if (X,Y) is a random couple taking its values in $L^2 \times \{-1,1\}$, then i.i.d. realizations of (X,Y), (x_i,y_i), are not directly observed as (x_i) are only known through a discretization, $\mathbf{x}_i = (x_i(t))_{t\in\tau}$ where τ is a finite subset of $[0,1]$.

41.3.1 Kernels for functional data

To obtain consistency result for functional SVM, a pre-processing is required that takes into account the functional nature of X. Depending on the problem, two kinds of pre-processing are investigated in this paper:

- A *projection approach* (developed in Rossi and Villa (2006b)) where the pre-processing step is $\mathcal{P} : x \in \mathcal{H} \to \sum_{j=1}^{d} \langle x, e_j \rangle_{\mathcal{H}} e_j$ where $(e_j)_{j\geq 1}$ is a Hilbert basis of any Hilbert space, $\mathcal{H}$ which is the space where X is taking its values (e.g., a Fourier basis if $\mathcal{H} = L^2$, as stated above). In this approach, $\mathcal{P}(X)$ is a random variable taking its values in a d-dimensional space; then, as it is usual in FDA, a d-dimensional SVM can be computed on the d coordinates of the projection.
- A *differential approach* where a prior assumption on X is used: X is supposed to be "smooth" and, more formally, it is supposed to belong to the Sobolev space $\mathcal{H}^m = \{x \in L^2([0,1]) : \ D^m x \text{ exists (in a weak sense) and } D^m x \in L^2\}$. This Sobolev space is a Hilbert space with respect to the inner product $\langle u,v\rangle_{\mathcal{H}^m} = \int_0^1 u^{(m)}(t)v^{(m)}(t)dt + \sum_{j=1}^m B^j u B^j v$ where (B^j) denotes m boundary conditions that defines an infinite dimensional subspace of $\mathcal{H}^m$, $\mathcal{H}_1^m$, such that $\mathcal{H}^m = \mathcal{H}_0^m \oplus \mathcal{H}_1^m$ with $\mathcal{H}_0^m = \mathrm{Ker} D^m$ (see Kimerldorf and Wahba, 1971). Thus, in this approach, the pre-processing consists in using the derivatives of the original function: $\mathcal{P}^s(X) = (D^m X, (B^j X)_j)$.

The following sections are dedicated to the presentation of consistency results associated to these two approaches and to the description of their advantages and weaknesses.

41.3.2 Projection approach

The consistency of the projection approach depends on a validation procedure that aims at choosing optimal parameters of the model. Indeed, three parameters have to be chosen for using the SVM on the pre-processed data $(\mathcal{P}x_i)_i$: the best dimension of projection, d, the best regularization parameter, C, in $(P_{C,\mathcal{F}})$ and the best kernel among a finite set of kernels, $\mathcal{K}_d$. If $\mathcal{A}$ denotes a set of lists of parameters to explore, the choice of the optimal parameters, a^* in $\mathcal{A}$ has to be done by the validation procedure described in Algorithm 1.

Algorithm 1 Functional SVM by projection: a validation approach

1: **for** all $a \equiv d \in \mathbb{N}^*, K_\gamma^d \in \mathcal{K}_d, C \in [0; C_d]$ in $\mathcal{A}$ **do**
2: Split the data set into $\mathcal{B}_1 = (x_i, y_i)_{i=1,\ldots,l}$ and $\mathcal{B}_2 = (x_i, y_i)_{i=l\ldots,n}$.
3: Solve $(P_{C,\mathcal{F}})$ with $z_i = \mathcal{P}x_i$ for the chosen parameters a; the corresponding classifier will be denoted by ϕ_l^a.
4: **end for**
5: Choose $a^* = \arg\min_{a\in\mathcal{A}} \hat{L}_{n-l}\phi_l^a + \frac{\lambda_d}{\sqrt{n-l}}$ with $L_{n-l} = \frac{1}{n-l}\sum_{i=l+1}^n \mathbb{I}_{\{\phi_l^a(x_i)\neq y_i\}}$ and $\lambda_d \in \mathbb{R}$.
6: Finally, keep the classifier $\phi_n = \phi_l^{a^*}$.

A consistency result can be deduced from this procedure:

Theorem 41.1. *(17) Suppose that:*

Assumption on X: *X takes its value in a bounded subset of $\mathcal{X}$;*
Assumptions on $\mathcal{A}$: *for all $d \geq 1$, $\mathcal{K}_d$ is a finite set that contains a kernel $K_\gamma^d \in$* **(Acv)** *at least, $C_d > 1$ and $\sum_{d\geq 1} |\mathcal{K}_d| e^{-2\lambda_d^2} < +\infty$;*
Assumptions on the training and the validation sets: $\lim_{n\to+\infty} l = +\infty$, $\lim_{n\to+\infty} n - l = +\infty$ *and* $\lim_{n\to+\infty} \frac{l\log(n-l)}{n-l} = +\infty$.

Then, ϕ_n is universally consistent: $\lim_{n\to+\infty} L\phi_n = L^*$ where $L\phi_n = \mathbb{P}(\phi_n(X) \neq Y)$ and $L^* = \inf_{\phi:\mathcal{H}\to\{-1,1\}} \mathbb{P}(\phi(X) \neq Y)$.

Two applications of this approach in the context of voice recognition are given in Rossi and Villa (2006b). Moreover, Park *et al.* (2007) also uses this approach to classify gene expression data into functional groups but with a linear kernel.

41.3.3 Differentiation approach

The projection pre-processing shows interesting results on real data but is somehow restrictive: the form of the representation of X is constrained by an Hilbert basis and the derivatives of X, that are known to be relevant in some practical applications (such as spectrometric data), don't lead to a consistent result with this approach. Moreover, the problem of using a discretization of the observations isn't addressed.

41.3.3.1 Representing X

In the differential approach, x_i is expressed directly in function of its discretization: that allows to obtain its derivatives directly from $\mathbf{x}_i$. In Rossi and Villa (2006a), we investigated a method that is close to this one by relying on interpolating splines. But, as the observations of X can be noisy, smoothing splines can be usefull to provide more relevant representations of x_i.

Suppose that $(\tau_d)_d$ is a series of distinct discretization points such that $\tau_d \subset \tau_{d+1}$, then representing x_i by a smoothing spline, from its discretization $\mathbf{x}_i^d = (x_i(t))_{t\in\tau_d}$, consists in solving the optimization problem $x_i^{\lambda,d} = \arg\min_{h\in\mathcal{H}^m} \frac{1}{d}\sum_{t\in\tau_d}(x_i(t) - h(t)) + \lambda\int_0^1 (h^{(m)}(t))^2 dt$ (see Kimerldorf and Wahba (1971), Cox (1984), Ragozin (1983), Utreras (1988) for several consistency results of this approximation to the real x_i). The most interesting point of this approach is that it links the derivatives of the smoothing spline estimate with the discretization of the observation: it exists a matrix $\mathbf{M}_d$, symmetric and positive definite, such that

$$\langle \hat{x}_i^{\lambda,d}, \hat{x}_j^{\lambda,d}\rangle_{\mathcal{H}^m} = \mathbf{x}_i^T \mathbf{M}_d \mathbf{x}_j. \tag{41.1}$$

41.3.3.2 Differentiation kernel for consistent functional SVM

Therefore, using equation (41.1), a kernel on the derivatives of (x_i) can be defined that is directly computed from the discretizations $\mathbf{x}_i^d$. The following theorem links SVM computed on the derivatives of (x_i) with a more usual kernel affected by the matrix $\mathbf{M}_d$:

Theorem 41.2 (Consistency of differentiation SVM). *The SVM classifier on* $(z_i)_i = (D^m x_i^{\lambda,d}, (B^j x_i^{\lambda,d})_j)_i$ *obtained with kernel* $K_\gamma^\infty \otimes K_\gamma^m$ *is equivalent to the SVM classifier on* $(\mathbf{x}_i)_i$ *obtained with kernel* $K_\gamma^d \circ \mathbf{M}_d^{-1/2}$.

If this classifier is denoted by $\phi_{n,d}$*, and if*

Assumptions on the discretization points: *for all* d, $(B^j)_j$ *are linearly independent from* $\{h \to h(t)\}_{t\in\tau_d}$ *and, if* F *is the limit of* $F_d(\zeta) = \frac{1}{|\tau_d|}\sum_{t\in\tau_d} \mathbb{I}_{\{\zeta=t\}}$ *for the norm* $\|u-v\|_\infty = \sum_{t\in[0,1]}|u(t)-v(t)|$, *then* F *is* $\mathcal{C}^\infty$,

Assumption on X*:* $X[0,1]$ *is a bounded subset of* $\mathbb{R}$,

Assumptions on the kernel: $K_\gamma^d \in$ **(Acv)**,

Assumptions on the parameters: *if* $S_d = \|F_d - F\|_\infty$ *then* $\lim_{d\to+\infty}\lambda_d = 0$ *and* $\lim_{d\to+\infty} S_d\lambda_d^{-5/(4m)} = 0$ *and the regularization parameter* C *of the optimization problem* $(P_{C,\mathcal{X}})$ *is such that* $C_{n,d} = \mathcal{O}\left(n^{1-\beta_d}\right)$ *where* $0 < \beta_d < \nu_d$,

then, $\lim_{d\to+\infty}\lim_{n\to+\infty} L\phi_{n,d} = L^*$ *for* $L\phi_{n,d}$ *and* L^* *defined as in theorem 41.1.*

Remark 41.1. Assumptions on (τ_d) are fulfilled by $\tau_d = \left\{\frac{j}{2^d}\right\}_{j=0,\ldots,2^d}$, for example (see Ragozin, 1983).

References

[1] Aronszajn, N.: Theory of reproducing kernels. Transactions of the American Mathematical Society. **68** (3), 337–404 (1950).

[2] Biau, G., Bunea, F. and Wegkamp, M.: Functional classification in Hilbert spaces. IEEE Transactions on Information Theory. **51**, 2163–2172 (2005).

[3] Cardot, H., Ferraty, F. and Sarda, P.: Functional linear model. Statistics and Probability Letters. **45**, 11–22 (1999).

[4] Cox, D.D.: Multivariate smoothing splines functions. SIAM Journal on Numerical Analysis. **21**, 789–813 (1984).

[5] Dejean, S. Martin, P.G.P., Baccini, A. and Besse, P.: Clustering time-series gene expression data using smoothing spline derivatives. EURASIP Journal on Bioinformatics and Systems Biology. 2007:Article ID70561 (2007).

[6] Deville, J.C.: Méthodes statistiques et numériques de l'analyse harmonique. Annales de l'INSEE. **15**(Janvier–Avril), 3–97 (1974).

[7] Ferraty, F. and Vieu, P.: The functional nonparametric model and application to spectrometric data. Computational Statistics. **17**, 515–561 (2002).

[8] Ferraty, F. and Vieu, P.: NonParametric Functional Data Analysis. Springer (2006).

[9] Ferré, L. and Villa, N.: Multi-layer perceptron with functional inputs: an inverse regression approach. Scandinavian Journal of Statistics. **33**(4), 807–823 (2006).

[10] Kimeldorf, G. and Wahba, G.: Some results on Tchebycheffian spline functions. Journal of Mathematical Analysis and Applications. **33**(1), 82–95 (1971).

[11] Lee, H.J.: Functional data analysis: classification and regression. PhD thesis, Department of Statistics, Texas, A&M University (2004).

[12] Park, C., Koo, J.Y., Kim, S., Sohn, I. and Lee, J.W.: Classification of gene functions using support vector machine for time-course gene expression data. Computational Statistics and Data Analysis (2007). Article in Press. doi:10.1016/j.csda.2007.09.002.

[13] Ragozin, D.L.: Error bounds for derivative estimation based on spline smoothing of exact or noisy data. Journal of Approximation Theory. **37**, 335–355 (1983).

[14] Ramsay, J.O. and Silverman, B.W.: Functional Data Analysis. Springer Verlag, New York. (1997).

[15] Rossi, F. and Conan-Guez, B.: Functional multi-layer perceptron: a nonlinear tool for functional data anlysis. Neural Networks. **18**(1), 45–60 (2005).

[16] Rossi, F and Villa, N.: Classification in Hilbert spaces with support vector machines. In ASMDA 2005 proceedings. pages 635–642, Brest, France. (2005).

[17] Rossi, F and Villa, N.: Support vector machine for functional data classification. Neurocomputing. **69**(7-9), 730–742 (2006b).

[18] Rossi, F. and Villa, N.: Consistency of derivative based functional classifiers on sampled data. (2007). Submitted.

[19] Steinwart, I.: On the influence of the kernel on the consistency of support vector machines. Journal of Machine Learning Research. **2**, 67–93 (2001).

[20] Steinwart, I.: Support vector machines are universally consistent. Journal of Complexity. **18**, 768–791 (2002).

[21] Utreras, F.I.: Boundary effects on convergence rates for tikhonov regularization. Journal of Approximation Theory. **54**, 235–249 (1988).

[22] Vapnik, V.: Statistical Learning Theory. Wiley, New York. (1998).

[23] Villa, N. and Rossi, F.: Un résultat de consistance pour des SVM fonctionnels par interpolation spline. Comptes Rendus Mathématique. Académie des Sciences. Paris. **343**(8), 555–560 (2006a).

Chapter 42
Wavelet Thresholding Methods Applied to Testing Significance Differences Between Autoregressive Hilbertian Processes

María Ruiz-Medina

Abstract The philosophy of Fan (1996) and Fan and Lin (1998) is adopted in the formulation of significance tests for comparing autoregressive Hilbertian processes. The discrete wavelet domain is considered to derive the test statistic based on thresholding rules. The results derived are applied to the statistical analysis of spatial functional data (SFD) sequences.

42.1 Introduction

Different testing procedures have been developed in the context of functional data analysis by several authors. For example, in time series theory, test for curves were initially considered in Shumway (1988); Brillinger (1973, 1980), and references therein. In the nonparametric setting, we can mention the paper by Hall and Hart (1990), where a bootstrap test for detecting differences between two mean functions is proposed. In Eubank and Hart (1992) and Eubank and LaRiccia (1992), a goodness-of-fit test is considered based on cross-validation. Data-driven methods for smoothed test have been studied by Inglot and Ledwina (1996), Ledwina (1994), and others. Fan (1996) and Fan and Lin (1998) proposed a methodology based on adaptive Neyman test, and wavelet thresholding techniques for testing significance differences between curves. The wavelet domain is also used in the testing procedures proposed, for functional analysis of variance, by Abramovich and Angelini (2006); Abramovich, Antoniadis, Sapatinas and Vidakovic (2004); among others.

One of the main drawbacks in the formulation of testing methods in the functional context is the dimensionality. Optimal signal compression methods are then needed to address this problem. Usually, orthogonal transforms like

María Ruiz-Medina
Department of Statistics and Operation Researchs, Campus Fuente Nueva s/n, E-18071 Granada, Spain, e-mail: mruiz@ugr.es

the ones involved in Fourier, Principal Component and Wavelet analysis have been incorporated in the development of estimation methods for functional data (see, for example, Ferraty and Vieu, 2006; Ramsay and Silverman, 2005; Vidakovic, 1999). In practice, the application of such orthogonal transformations requires the pre-implementation of binning, interpolation or smoothing methods. In the case of discrete transformations, adaptive procedures are formulated for selection of the most significative coefficients (see, for example, Fan, 1996; Fan and Lin, 1998). However, in the case of continuous transforms, like the continuous wavelet transform, such selection procedures must be formulated in terms of useful windows, patch or areas selected under some optimality criterion (see, for instance, Maraun and Kurths, 2004; Maraun, Kurths and Holschneider, 2007).

In this paper, a sequential testing procedure is formulated for detecting significance differences between two SFD sequences, in the context of Gaussian autoregressive Hilbertian processes. The data are transformed into the discrete wavelet domain in terms of a basis of orthonormal compactly supported spatial wavelets. We consider hard thresholding rules (in terms of the universal threshold) for selection of the most significant wavelet coefficients. The test statistic is then defined, considering a functional formulation of the thresholding test statistic of Fan and Lin (1998), in terms of suitable factorizations of the spatial covariance operators involved, as well as the associated thresholded scalograms.

In Section 2, we describe the preliminary elements involved in the testing procedure proposed. In Section 3, the main results are summarized. References are listed at the end of the paper.

42.2 Preliminaries

In the following development, we will assume that the sequence $\{Y_t(\cdot) : t \geq 0\}$ of SFD to be analyzed obeys the following autoregressive Hilbertian model of order one (ARH(1)) (see Bosq, 2000):

$$Y_t(\mathbf{x}) = \mathcal{A}[Y_{t-1}](\mathbf{x}) + \nu_t(\mathbf{x}), \quad \mathbf{x} \in D,\ t \in \mathbb{N}, \tag{42.1}$$

where ν is a Gaussian strong Hilbertian white noise, that is, a sequence of independent and identically Gaussian distributed Hilbert-valued random variables in H with

$$E[\|\nu_t\|_H^2] = \sigma_\nu^2 < \infty, \tag{42.2}$$

uncorrelated with the random initial condition Y_0. The autocorrelation operator $\mathcal{A}$ is a bounded operator defined on a dense domain in H. Here, H is defined as a Hilbert space of spatial functions on a compact domain $D \subset \mathbb{R}^n$.

For each $t \geq 0$, the spatial wavelet transform of Y_t is defined as

$$\{Y_{\mathbf{k}}(t):\ \mathbf{k}\in\Gamma_0\}\cup\left\{Y_{j,\mathbf{k}}(t):\ \mathbf{k}\in\widetilde{\Gamma}_j,\ j\in\mathbb{N}\right\},$$

where

$$\begin{aligned} Y_{\mathbf{k}}(t) &= \int_D Y_t(\mathbf{z})\phi_{\mathbf{k}}(\mathbf{z})d\mathbf{z}, \quad \mathbf{k}\in\Gamma_0,\\ Y_{j,\mathbf{k}}(t) &= \int_D Y_t(\mathbf{z})\psi_{j,\mathbf{k}}(\mathbf{z})d\mathbf{z}, \quad \mathbf{k}\in\widetilde{\Gamma}_j,\ j\in\mathbb{N}, \end{aligned} \tag{42.3}$$

with $\{\phi_{\mathbf{k}},\ \mathbf{k}\in\Gamma_0\}$ representing an orthogonal system of compactly supported scaling functions generating the space V_0, that is, the space providing, by projection, the random draft of Y_t, and $\left\{\psi_{j,\mathbf{k}},\ \mathbf{k}\in\widetilde{\Gamma}_j\right\}, j\in\mathbb{N}$, representing orthogonal compactly supported wavelet bases generating the spaces W_j, $j\in\mathbb{N}$, that is, the spaces providing, by projection, the local variability properties of Y_t. The Hilbert-valued process Y is assumed to satisfy the necessary regularity and moment conditions to ensure that the integrals in equation (42.3) are well-defined.

For each $t\geq 0$, the wavelet periodogram or scalogram (also referred as wavelet sample spectrum, specially when smoothing is performed in the scale or spatial direction) of Y_t is usually defined as

$$\{|Y_{\mathbf{k}}(t)|^2:\ \mathbf{k}\in\Gamma_0\}\cup\left\{|Y_{j,\mathbf{k}}(t)|^2:\ \mathbf{k}\in\widetilde{\Gamma}_j,\ j\in\mathbb{N}\right\}.$$

We consider hard thresholding procedures, we then have the thresholded scalogram

$$\begin{aligned} \mathcal{S}_{Th}(Y_t) &= \{|Y_{\mathbf{k}}(t)|^2 I(|Y_{\mathbf{k}}(t)|>\delta):\ \mathbf{k}\in\Gamma_0\}\cup\\ &\quad \left\{|Y_{j,\mathbf{k}}(t)|^2 I(|Y_{j,\mathbf{k}}(t)|>\delta):\ \mathbf{k}\in\widetilde{\Gamma}_j,\ j\in\mathbb{N}\right\}, \end{aligned}$$

where I denotes the indicator function. In the results described in the next section, we consider functional adaptations of the hard thresholding parameter values proposed by Fan (1996), and Fan and Lin (1998), related to the universal threshold (see, for example, Vidakovic, 1999).

42.3 Main results

Let $\{Y_t(\cdot): t\geq 0\}$ be defined as in equation (42.1).

(C) Assume that the covariance operator of the Hilbert-valued process Y

$$\mathcal{R}_Y = E\left[Y_t\otimes Y_t\right] = E\left[Y_0\otimes Y_0\right],\quad \forall t\geq 0,$$

given by

$$R_Y(\phi) = E\left[Y_0 \langle Y_0, \phi\rangle\right], \quad \forall \phi \in H,$$

defines an isomorphism from $\widetilde{H}^*$ onto $\widetilde{H}$, with $\widetilde{H}$ a dense subspace of H.

Then, for each $t \geq 0$, Y_t admits an orthogonal decomposition in terms of dual Riesz bases, providing orthonormal bases of the associated RKHS and its dual space, given by dual linear transformations of an orthonormal basis of H. Moreover, R_Y admits the factorization:

$$R_Y = \mathcal{T}_Y \mathcal{T}_Y^*,$$

in terms of an isomorphism $\mathcal{T}_Y$ from H onto $\widetilde{H}$ (see, for example, Ruiz-Medina, Angulo and Anh, 2003). We consider the case where $H = L^2(D)$. Then, $\widetilde{H}$ and $\widetilde{H}^*$ can belong to the scale of fractional Besov spaces (in particular, the scale of fractional Sobolev spaces can be considered). In this case, we can define the dual Riesz bases involved from an orthonormal basis of wavelets of $L^2(D)$ (see Angulo and Ruiz-Medina, 1999; Ruiz-Medina and Angulo, 2002; Ruiz-Medina, Angulo and Fernández-Pascual, 2007). For such a class of Hilbert-valued process the following representation in the wavelet domain can be considered:

$$\mathbf{TWD}(Y_t) = \mathbf{TW2D}(\mathcal{T}_Y)\mathbf{TWD}(\varepsilon_t^Y),$$

where $\mathbf{TWD}(Y_t)$ denotes the spatial wavelet transform

$$\{Y_{\mathbf{k}}(t) : \ \mathbf{k} \in \Gamma_0\} \cup \left\{Y_{j,\mathbf{k}}(t) : \ \mathbf{k} \in \widetilde{\Gamma}_j, \ j \in \mathbb{N}\right\},$$

of Y_t. The same notation is used for $\mathbf{TWD}(\varepsilon_t^Y)$, with ε^Y representing Gaussian $H-$white noise. That is, a Gaussian Hilbert-valued process with

$$R_{\varepsilon^Y}(\phi)(\psi) = \langle \phi, \psi\rangle_H \,.$$

Here, $\mathbf{TW2D}(\mathcal{T}_Y)$ denotes the 2D-wavelet transform of operator $\mathcal{T}_Y$.

Since we have considered an orthonormal wavelet basis of $L^2(D)$, and ε^Y is Gaussian $L^2(D)-$white noise, the spatial wavelet transform

$$\mathbf{TWD}(\varepsilon_t^Y) = \left\{\varepsilon_{\mathbf{k}}^Y(t) : \ \mathbf{k} \in \Gamma_0\right\} \cup \left\{\varepsilon_{j,\mathbf{k}}^Y(t) : \ \mathbf{k} \in \widetilde{\Gamma}_j, \ j \in \mathbb{N}\right\}$$

defines a Gaussian white noise process, that is, for each $t \geq 0$, the random components, defining the wavelet coefficients of ε_t^Y, are independent.

Condition $\mathbf{C}$ is also assumed to be satisfied by the Hilbert-valued innovation process ν. Therefore, for each $t \geq 0$,

$$\mathbf{TWD}(\nu_t) = \mathbf{TW2D}(\mathcal{T}_\nu)\mathbf{TWD}(\varepsilon_t^\nu),$$

with $\mathbf{TWD}(\varepsilon_t^\nu)$ being a white-noise process independent of the white-noise process $\mathbf{TWD}(\varepsilon_t^Y)$. Hence,

$$\begin{aligned}\mathbf{TWD}(Y_t) &= \mathbf{TW2D}(\mathcal{A})\mathbf{TWD}(Y_{t-1}) + \mathbf{TWD}(\nu_t)\\ &= \mathbf{TW2D}(\mathcal{A}\mathcal{T}_Y)\mathbf{TWD}(\varepsilon_{t-1}^Y) + \mathbf{TW2D}(\mathcal{T}_\nu)\mathbf{TWD}(\varepsilon_t^\nu)\end{aligned} \quad (42.4)$$

In the functional formulation of the thresholding test statistic of Fan and Lin (1998), in the next section, we consider the thresholded version of equation (42.4), denoting by $\widetilde{\mathbf{TWD}}(\cdot)$ and $\widetilde{\mathbf{TW2D}}(\cdot)$ the thresholded one- and two-dimensional wavelet transforms.

42.3.1 Comparing two sequences of SFD in the ARH context

Let us consider two independent sequences of SFD $\{Y_t(\cdot) : t = 1, \ldots, T\}$ and $\{Z_t(\cdot) : t = 1, \ldots, T\}$ satisfying condition **(C)** as before. The following hypotheses are then tested

$$\begin{aligned}&(1)\ H_0 : \nu^Y \underset{d}{=} \nu^Z \quad \text{versus} \quad H_1 : \nu^Y \underset{d}{\neq} \nu^Z\\ &(2)\ H_0 : \mathcal{A}_Y = \mathcal{A}_Z \quad \text{versus} \quad H_1 : \mathcal{A}_Y \neq \mathcal{A}_Z.\end{aligned} \quad (42.5)$$

Test (2) is applied if the null hypothesis in Test (1) is accepted. For applying Test (1), consistent estimators of the autocorrelation operators $\mathcal{A}_Y$ and $\mathcal{A}_Z$ must be considered (see, for instance, Guillas, 2001).

The test statistic in (1) is defined in terms of

$$T_1 = \sum_{t=1}^{T} \|\mathbf{D}(t)\|^2,$$

where

$$\mathbf{D}(t) = [\widetilde{\mathbf{TW2D}}(\mathcal{T}_{\nu^Y})]^{-1}\widetilde{\mathbf{TWD}}(\nu_t^Y) - [\widetilde{\mathbf{TW2D}}(\mathcal{T}_{\nu^Z})]^{-1}\widetilde{\mathbf{TWD}}(\nu_t^Z),$$

with $\widetilde{\mathbf{TW2D}}(\mathcal{T}_{\nu^Y})$ and $\widetilde{\mathbf{TW2D}}(\mathcal{T}_{\nu^Z})$ being estimated from the factorization of the thresholded scalograms of ν_t^Y and ν_t^Z. In practice, for each $t = 1, \ldots, T$, the vectors $\widetilde{\mathbf{TWD}}(\nu_t^Y)$ and $\widetilde{\mathbf{TWD}}(\nu_t^Z)$, and the thresholded scalograms of ν_t^Y and ν_t^Z are computed, for each sequence of SFD, in terms of the difference between the thresholded wavelet transform of each SFD and the thresholded wavelet transform of the associated (estimated) autocorrelation operator applied to a previous element of the SFD sequence, and the thresholded scalograms associated with such differences.

In the Test (2), under H_0 ($\mathcal{A}_Y = \mathcal{A}_Z = \mathcal{A}$), after accepting H_0 of Test (1), we have that the differences $\widetilde{\mathbf{TWD}}(Y_t) - \widetilde{\mathbf{TWD}}(Z_t)$, $t = 1, \ldots, T$, obey a multivariate normal distribution with covariance matrix $\widetilde{\mathbf{TW2D}}(\mathcal{A}(R_Y + R_Z)\mathcal{A}^* + 2R_\nu)$. As before, the Fan and Lin (1998) philosophy can then be applied to formulate the test statistic in a functional framework. The $\widetilde{\mathbf{TW2D}}(\mathcal{A}(R_Y+R_Z)\mathcal{A}^*+2R_\nu)$ is estimated from the thresholded scalograms of the differences between the data.

References

[1] Abramovich, F. and Angelini, C.: Testing in mixed effects FANOVA models. Journal of Statistical Planning and Inference. **136**, 4326-4348 (2006).

[2] Abramovich, F., Antoniadis, A., Sapatinas, T. and Vidakovic, B.: Optimal testing in functional analysis of variance models. Int. J. Wavelets Multiresolution Inform. Process. **2**, 323-349.

[2] Angulo, J.M. and Ruiz-Medina, M.D. (1999). Multiresolution approximation to the stochastic inverse problem. Adv. App. Prob. **31**, 1039-1057 (2004).

[3] Bosq, D.: Linear processes in function spaces. Springer-Verlag. (2000).

[4] Brillinger, D. R.: The analysis of time series collected in an experiment design. In Multivariate analysis, III. Krishnaiah, P.R. (ed.), Academic Press, 241-256 (1973).

[5] Brillinger, D. R.: Some aspect of the analysis of evoked response experiments. In Statistics and related topics. Csörgö, M., Dawson, D.A. Rao, J.N.K. and Saleh, A.K. (eds.), North-Holland, 15-168 (1980).

[6] Fan, J.: Test of significance based on wavelet thresholding and neyman 's truncation. Journal of American Statistical Association. **91**, 674-688 (1996).

[7] Fan, J. and Lin, S.J.: Test of significance when data are curves. Journal of American Statistical Association. **93**, 1007-1021 (1998).

[8] Ferraty, F. and Vieu, P.: Nonparameric functional data analysis. Springer. (2006).

[9] Eubank, R.L. and Hart, J.D.: Testing goodness-of-fit in regression via order selection criteria. The Annals of Statistics. **20**, 1412-1425 (1992).

[10] Eubank, R.L. and LaRiccia, V.N.: Asymptotic comparison of Cramér-von Mises and non-parametric function estimation techniques for testing goodness-of-fit. The Annals of Statistics. **20**, 2071-2086 (1992).

[11] Guillas, S.: Rates of convergence of autocorrelation estimates for autoregressive Hilbertian processes. Stat. Prob. Lett. **55**, 281-291 (2001).

[12] Hall, P. and Hart, J.D.: Bootstrap test for difference between means in nonparametric regression. Journal of American Statistical Association. **85**, 1039-1049 (1990).

[13] Inglot, T. and Ledwina, T.: Asymptotic optimality of data-driven Neyman's tests for uniformity. The Annals of Statistics. **24**, 1982-2019 (1996).

[14] Ledwina, T.: Data-driven version of Neyman's smooth test of fit. Journal of American Statistical Association. **89**, 1000-1005 (1994).

[15] Maraun, D. and Kurths, J.: Cross-wavelet analysis: significance testing and pitfalls. Nonlinear Processes in Geophysics. **11**, 505-514 (2004).

[16] Maraun, D., Kurths, J. and Holschneider, M.: Nonstationary Gaussian processes in wavelet domain: Synthesis, estimation, and significance testing. Physical Review E. **5**, 016707-1–016707-14 (2007).

[17] Ruiz-Medina, M.D. and Angulo, J.M.: Spatio-temporal filtering using wavelets. Stoch. Environm. Res. Risk Assess. **16**, 241-266 (2002).

[18] Ruiz-Medina, M.D., Angulo, J.M. and Anh, V.V.: Fractional generalized random fields on bounded domains. Stoch. Anal. Appl. **21**, 465-492 (2003).

[19] Ruiz-Medina, , M.D., Angulo, J.M. and Fernández-Pascual, R.: Wavelet-vaguelette decomposition of spatiotemporal random fields. Stoch. Environm. Res. Risk Assess. **21**, 273-281 (2007).

[20] Ramsay, J.O. and Silverman, B.W.: Functional data analysis. Springer. (2005).

[21] Shumway, R.H.: Applied statistical time series analysis. Prentice-Hall. (1988).
[22] Vidakovic, B.: Statistical modeling by wavelets. John Wiley & Sons. (1999).

Chapter 43
Explorative Functional Data Analysis for 3D-geometries of the Inner Carotid Artery

Laura Maria Sangalli, Piercesare Secchi and Simone Vantini

Abstract We analyze reconstructions of inner carotid arteries, obtained from 3D angiographic images, and we investigate the role of vessel geometry on the pathogenesis of cerebral aneurysms.

43.1 Introduction

Cerebral aneurysms are lesions of cerebral vessels characterized by a bulge of the vessel wall. Many authors believe that the onset, development and possibly rupture of an aneurysm are conditioned by the geometry of the vessel through its effect on blood fluid-dynamics (see, for instance, Hassan *et al.* (2005)). We thus aim at studying possible relations between vessel geometries and this pathology. In particular, we focus here on the analysis of vessel radius and curvature profiles. Indeed, these two geometric features, together with blood density, viscosity and velocity, determine the local hemodynamics.

Our study is part of AneuRisk Project, a joint research program involving MOX Laboratory for Modeling and Scientific Computing (Dip. di Matematica, Politecnico di Milano), Laboratory of Biological Structures (Dip. di In-

Laura Maria Sangalli
MOX Laboratory for Modeling and Scientific Computing, Dipartimento di Matematica Politecnico di Milano, P.zza Leonardo da Vinci, 32, 20133 Milano, Italy, e-mail: laura.sangalli@polimi.it

Piercesare Secchi
MOX Laboratory for Modeling and Scientific Computing, Dipartimento di Matematica Politecnico di Milano, P.zza Leonardo da Vinci, 32, 20133 Milano, Italy, e-mail: piercesare.secchi@polimi.it

Simone Vantini
MOX Laboratory for Modeling and Scientific Computing, Dipartimento di Matematica Politecnico di Milano, P.zza Leonardo da Vinci, 32, 20133 Milano, Italy, e-mail: simone.vantini@polimi.it

gegneria Strutturale, Politecnico di Milano), Istituto Mario Negri (Ranica), Ospedale Niguarda Ca' Granda (Milano), and Ospedale Maggiore Policlinico (Milano). The Project is supported by Fondazione Politecnico di Milano and Siemens-Medical Solutions Italia. The AneuRisk dataset is the largest collection of 3D cerebral angiographies available for the study of the aneurysmal pathology; it includes the spatial coordinates of vessel centerlines and vessel radius profiles for the internal carotid arteries (ICA) of 65 patients with and without cerebral aneurysms. Details about the elicitation of these data from 3D-angiographies are in Piccinelli *et al.* (2007). Figure 43.1 shows the draw of the reconstruction of an ICA with aneurysm.

We perform explorative analyses of this dataset, which support the existence of a strong relationship between vessel geometry and aneurysm location. We first fit raw data and estimate vessel centerlines and their curvature functions, by means of 3D free knot regression splines. Centerlines estimates then undergo a process of registration, that separates their amplitude variability from their phase variability, enabling meaningful comparisons across patients. The main uncorrelated modes of variability, of registered radius and curvature profiles, are thus found by functional principal component analysis. Finally, a quadratic discriminant analysis of principal components scores identifies the optimal number of principal components that discriminate at best patients with aneurysms located in different vascular districts: patients having an aneurysm at or after the terminal bifurcation of the ICA (Upper group), and patients having an aneurysm along the ICA or healthy (Lower group). The quadratic discriminant analysis also allows to select special cases for numerical simulations.

43.2 Efficient estimation of 3D vessel centerlines and their curvature functions by free knot regression splines

For every patient i in our dataset, we know, for each point s_{ij} on a fine grid along a curvilinear abscissa (that goes from the terminal bifurcation of the ICA, towards the heart), the three spatial coordinates x_{ij}, y_{ij} and z_{ij} of vessel centerline, and the vessel radius R_{ij}. Due to measurement and reconstruction errors, reconstructed centerlines may be quite wiggly and thus need to be smoothed, and so do the estimates of their derivatives, in order to obtain sensible estimates of their curvature functions. We do so by means of free knot regression splines, i.e. regression splines where the number and position of knots are not fixed in advance, but chosen in a way to minimize a penalized average squared error criterion. Since our data are 3D, the idea is to fit simultaneously the three spatial coordinates $(x(s), y(s), z(s))$ of the centerline versus the curvilinear abscissa s, looking for the optimal spline knots along the curvilinear abscissa. See Figure 43.3. Optimal knots are searched by an algorithm which is a modification of the algorithm developed by Zhou and Shen (2001) in the 1D case. The first and second derivatives of the fitted centerline are thus used to estimate its curvature. See Figure 43.2.

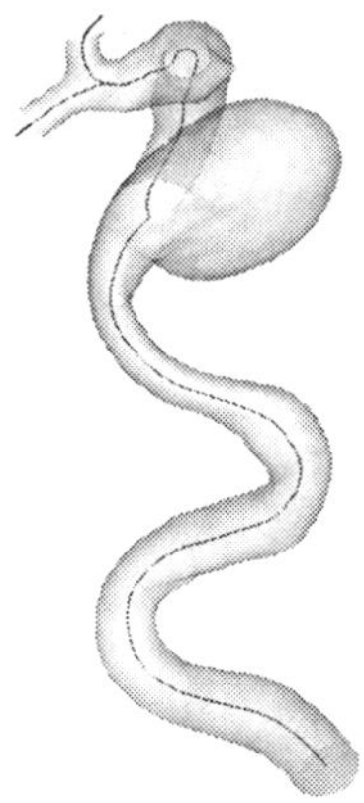

Fig. 43.1 3D image of an internal carotid artery with an aneurysm [patient 1]. The black curve inside the vessel is the centerline.

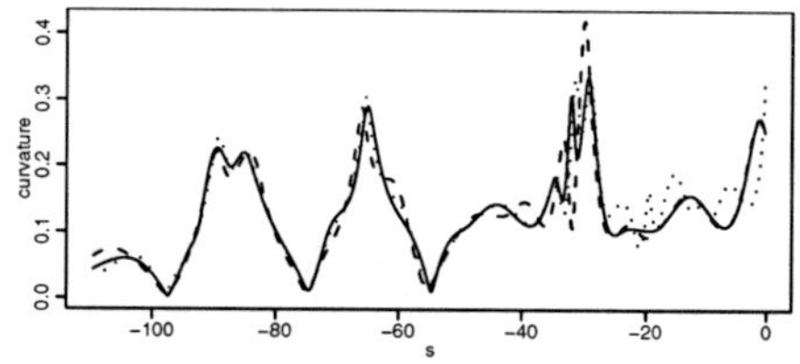

Fig. 43.2 Estimated curvature functions [patient 1], obtained by free knot splines with three different penalizations (FKRS1 dotted, FKRS2 solid, and FKRS3 dashed; see Figure 43.4). The points of approximately zero curvature are the siphon delimiters shown in Figure 43.3.

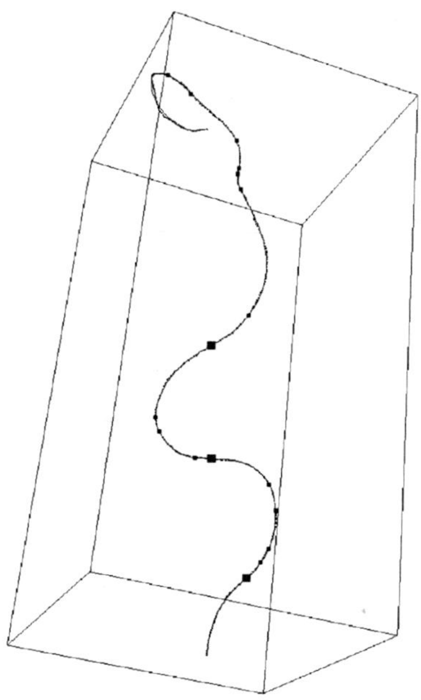

Fig. 43.3 3D image of fitted centerline (the little bullets show the positions of the spline knots), together with rough data [patient 1]. The big squares are the siphon delimiters. See Figure 43.2.

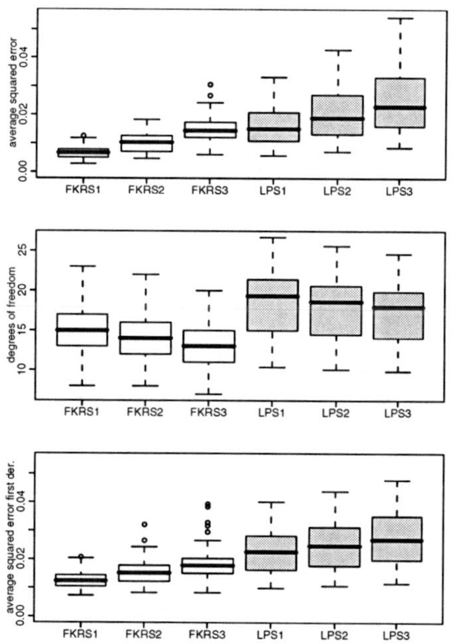

Fig. 43.4 Boxplots of the distribution of average squared error, degrees of freedom, and average squared error on first derivatives, for the fits corresponding to the 65 patients, obtained by free knot regression splines (FKRS1, FKRS2 and FKRS3) and by local polynomial smoothing with different bandwidths (LPS1, LPS2 and LPS3).

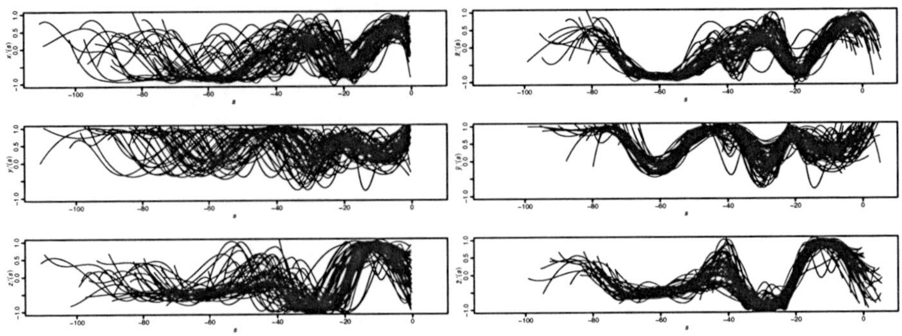

Fig. 43.5 On the left column, from top to bottom, estimated first derivatives $x_i'(s)$, $y_i'(s)$ and $z_i'(s)$ before registration. On the right column, estimated first derivatives $\tilde{x}_i'(s)$, $\tilde{y}_i'(s)$ and $\tilde{z}_i'(s)$ after registration.

The amount of penalization in the penalized average squared error criterion is chosen taking into account the trade off between average squared error and degrees of freedom of the model (computed as the trace of the linear operator that takes from the raw data to the fitted values). Indeed, when the penalization increases, the average squared error increases while the degrees of freedom decrease. See Figure 43.4. Nonetheless, the estimate of the curvature is quite robust with respect to different choices of the penalization. In particular, the points of approximately zero curvature, which are taken as delimiters of different carotid siphons, do not change. See Figures 43.2 and 43.3.

Smoothing by free knot regression splines turns out to be more efficient than classical local polynomial smoothing, in the sense that the former technique attains lower average squared errors, and also lower average squared errors on first derivatives (with respect to first central differences), using less degrees of freedom. See Figure 43.4. The degrees of freedom of local polynomials (i.e. the trace of the corresponding linear operator) are computed according to Zhang (2003) empirical formula. Moreover, by using free knot regression splines we are reducing the dimension of data, a fundamental issue for our highly dimensional dataset. See Sangalli, Secchi, Vantini and Veneziani (2007a) for details.

43.3 Registration

Visual inspection of the first derivatives of estimated centerlines $(x_i'(s), y_i'(s), z_i'(s))$, for the 65 patients, makes evident that data present a phase variability that must be removed to enable meaningful comparisons across patients (Ramsay and Silverman, 2005). See Figure 43.5. This can be achieved by means of a registration procedure that finds the 65 warping functions h_i of the abscissa, that capture this phase variability, leading to the new registered centerlines $(\tilde{x}_i(s), \tilde{y}_i(s), \tilde{z}_i(s))$, where $\tilde{x}_i = x_i \circ h_i^{-1}$, $\tilde{y}_i = y_i \circ h_i^{-1}$, and $\tilde{z}_i = z_i \circ h_i^{-1}$. We look for the optimal warping functions on the space of increasing affine transformations, maximizing the following similarity index between $(\tilde{x}_i(s), \tilde{y}_i(s), \tilde{z}_i(s))$ and a reference centerline $(x_0(s), y_0(s), z_0(s))$:

$$\frac{1}{3}\left[\frac{\int_{S_i} \tilde{x}_i' x_0' ds}{\sqrt{\int_{S_i} \tilde{x}_i'^2 ds}\sqrt{\int_{S_i} x_0'^2 ds}} + \frac{\int_{S_i} \tilde{y}_i' y_0' ds}{\sqrt{\int_{S_i} \tilde{y}_i'^2 ds}\sqrt{\int_{S_i} y_0'^2 ds}} + \frac{\int_{S_i} \tilde{z}_i' z_0' ds}{\sqrt{\int_{S_i} \tilde{z}_i'^2 ds}\sqrt{\int_{S_i} z_0'^2 ds}}\right]$$

where S_i is the support of the i-th centerline. This is analogous to the criterion proposed by Ramsay and Silverman (2005), but suitable for managing curves defined on different supports, as the ones we deal with. A Procrustes fitting criterion is used to estimate both the 65 warping functions and the reference centerline. The iterative procedure converges in few steps. See Sangalli, Secchi, Vantini and Veneziani (2007b) for details.

43.4 Statistical analysis

Many interesting traits emerge from the analysis of registered radius and curvature. Figure 43.6 shows that the mean radius of the vessel gets progressively narrower toward the terminal bifurcation of the ICA (the so-called tapering effect). Moreover, it shows that two peaks of curvature are usually present at about -3.5 and -2.0 cm to the terminal bifurcation. The same figure also displays a density estimate of the location of aneurysms along the ICA. Note that most aneurysms are clustered in two groups, both located in the terminal part of the ICA, where tapering is evident, and one located just after the last peak of curvature. These results provide evidence of a link between morphology and aneurysms onset, induced by hemodynamics. Moreover, some details of the structure of sample autocovariance functions of registered radius and covariance, not shown here, are amenable of an anatomical interpretation. See Sangalli, Secchi, Vantini and Veneziani (2007b) for details.

The main uncorrelated modes of variability, of registered radius and curvature profiles, are found by functional principal component analysis (FPCA). See Ramsay and Silverman (2005). Since the 65 curves are known on different abscissa intervals, these analyses focus on the interval where all curves are available. Figure 43.6 shows the first two eigenfunctions of radius, $\hat{\beta}_{R1}(s)$ and $\hat{\beta}_{R2}(s)$, and curvature, $\hat{\beta}_{C1}(s)$ and $\hat{\beta}_{C2}(s)$. The distributions of FPCA scores for the two groups of patients, the Upper group (composed by patients having an aneurysm at or after the terminal bifurcation of the ICA) and the Lower group (composed by patients having an aneurysm along the ICA or healthy), have significantly different means and/or variances. According to these dif-

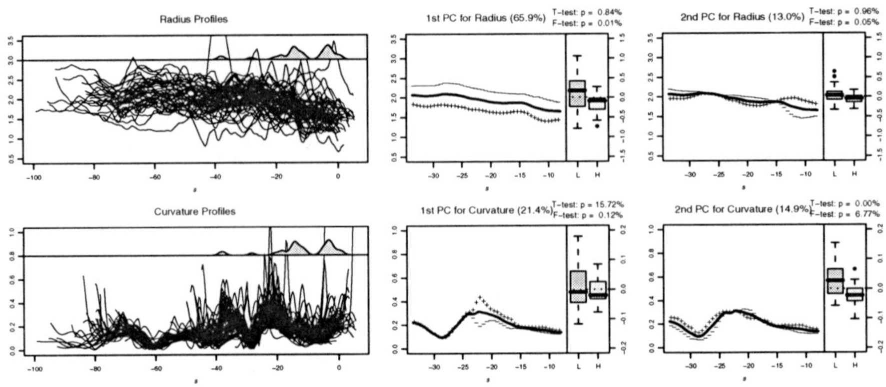

Fig. 43.6 Left: registered radius and curvature profiles (top and bottom respectively), with superimposed density estimate of aneurysms location. Right: first two principal components of registered radius and curvature (top and bottom respectively), with boxplots of corresponding scores: darker boxplots for Lower group patients and lighter boxplots for Upper group patients.

ferences, Upper group patients have on average wider, more tapered and less curved ICA's, with respect to Lower group patients. Moreover Upper group patients display a significantly smaller variance of these geometrical features. The first two eigenfunctions of radius and curvature are in fact the optimal set of eigenfunctions to discriminate the two groups of patients, by means of quadratic discriminant analysis of FPCA scores. Moreover, this analysis allows to select representative geometries for numerical simulations of the hemodynamics in the vascular neighborhood of the aneurysm. More details are in Sangalli, Secchi, Vantini and Veneziani (2007b).

These numerical simulations will generate new functional data relative to the hemodynamics of the ICA, i.e. pressure, velocity and shear stress along the vessel; we hope that, by extending the previous analyses with the inclusion of this new information, we will be able to explore and model the causal relationship between the complex hemodynamics of the ICA and the onset and rupture of cerebral aneurysms.

Acknowledgements Special thanks to Alessandro Veneziani (MOX), leader of the AneuRisk Project, to Edoardo Boccardi (Ospedale Niguarda Ca' Granda), who provided the 3D-angiographies and motivated our reaserch by posing fascinating medical questions, and to Luca Antiga and Marina Piccinelli (Istituto Mario Negri), who performed the image reconstructions.

References

[1] Hand, D. J.: Discrimination and Classification. John Wiley & Sons, London. (1981).

[2] Hassan, T., Timofeev, E. V., Saito, T., Shimizu, H., Ezura, M., Matsumoto, Y., Takayama, K., Tominaga, T., and Takahashi, A.: A proposed parent vessel geometry-based categorization of saccular intracranial aneurysms: computational flow dynamics analysis of the risk factors for lesion rupture, J. Neurosug. **103**, 662–680 (2005).

[3] Piccinelli, M., Bacigaluppi, S., Boccardi, E., Ene-Iordache, B., Remuzzi, E., Veneziani, A., and Antiga, L.: Influence of internal carotid artery geometry on aneurysm location and orientation: a computational geometry study, Available at www.maths.emory.edu. (2007).

[4] Ramsay, J. O. and Silverman, B. W.: Functional Data Analysis. Springer New York NY, 2nd ed. (2005).

[5] Sangalli, L. M., Secchi, P., Vantini, S., and Veneziani, A.: Efficient estimation of 3D centerlines of inner carotid arteries and their curvature profiles by free knot regression splines, Tech. Rep. 23/2007, MOX, Dipartimento di Matematica, Politecnico di Milano. Available at http://mox.polimi.it/it/progetti/pubblicazioni.(2007a).

[6] Sangalli, L. M., Secchi, P., Vantini, S., and Veneziani, A.: A Case Study in Explorative Functional Data Analisys: Geometrical Features of the Internal Carotid Artery, Tech. rep., MOX, Dipartimento di Matematica, Politecnico di Milano. Available at http://mox.polimi.it/it/progetti/pubblicazioni. (2007b).

[7] Zhang, C.: Calibrating the degrees of freedom for automatic data smoothing and effective curve checking, J. Amer. Statist. Assoc. **98**, 609–628 (2003).

[8] Zhou, S. and Shen, X.: Spatially adaptive regression splines and accurate knot selection schemes, J. Amer. Statist. Assoc. **96**, 247–259 (2001).

Chapter 44
Inference on Periodograms of Infinite Dimensional Discrete Time Periodically Correlated Processes

Zohreh Shishebor, Ahmad Reza Soltani and Ahmad Zamani

Abstract In this work we shall consider two classes of periodically correlated processes with values in separable Hilbert spaces: weakly second order and strongly second order. It is proved that the sample Fourier transforms are asymptotically uncorrelated and the periodograms are asymptotically unbiased for corresponding spectral densities.

44.1 Introduction

In a series of papers, following the work of Rozanov (1992), Miamee and Salehi (1971), Miamee (1976) and Salehi and Soltani (1983), basic spectral foundations of infinite dimensional second order stationary processes were established. Such processes are considered as bounded linear transformations on a Hilbert space (or a Banach space) into $L^2(\Omega, \mathcal{F}, \mathcal{P})$, the space of mean zero random variables with finite second moments, possessing certain covariance structure. We refer to such processes as weakly second order (WSO). More description is provided in the next section. Processes with trajectories in metric spaces that the metric of the elements are in $L^2(\Omega, \mathcal{F}, P)$ are also studied intensively by different authors: Gihman and Skorohod (1974), Tala-

Zohreh Shishebor
Projet Department of Statistics, Faculty of Science, Shiraz University, Shiraz 71454, Iran, e-mail: sheshebor@susc.sc.ir

Ahmad Reza Soltani
Department of Statistics and Operations Research, Faculty of Science, Kuwait University, P.O. Box 5969, Safat 13060, Kuwait, e-mail: soltani@kuc01.kuniv.edu.kw

Ahmad Zamani
Projet Department of Statistics, Faculty of Science, Shiraz University, Shiraz 71454, Iran, e-mail: zamani_1127@yahoo.com.

grand (1991), Bosq (2000), among others. We refer to this type of processes with values in Hilbert spaces as strongly second order (SSO) processes.

Recent applications of infinite dimensional process in applied problems has inspired advanced researches in inference of such processes. In this work we consider periodically correlated (PC) processes of type WSO or SSO with values in separable Hilbert spaces. Indeed basic spectral structures of WSO PC processes were established by Soltani and Shishebor (2007). No study on periodogram of infinite dimensional PC processes has yet been carried out. Periodograms are important tools in spectral analysis of time series to highlight hidden frequencies. The work by Soltani and Azimmohseni (2007) and Hurd (1989) are on periodograms of univariate PC processes; also see Pourahmadi and Salehi (1983).
In summery, sample finite Fourier transforms (SFFT) and periodograms for X-valued, X a separable Hilbert space, WSO and SSO processes are introduced, their basic statistical properties are derived, and it is proved that periodograms are asymptotically unbiased for the corresponding spectral densities, and SFFT at distinct frequencies are asymptotically uncorrelated. This article is organized as follows.

44.2 Preliminaries and results

Let X be a Hilbert space, let $L(X)$ stands for the bounded linear operators on X and $L^2(\Omega, \mathcal{F}, \mathcal{P})$ for the linear space of all complex random variables with mean zero and with finite absolute second moments. The inner product on X is denoted by $(\ , \ )_X$.

A random variable $\xi : \Omega \to X$ is said to be second order in the strong sense if $\|\xi(\omega)\|_X \in L^2(\Omega, \mathcal{F}, \mathcal{P})$; and second order in the weak sense (WSO) if $(\xi(\omega), x)_X \in L^2(\Omega, \mathcal{F}, \mathcal{P})$ for every $x \in X$. Since $E| \ \xi(\omega)|^2 \leq E\|\xi(\omega)\|_X^2 \|x\|_X^2$, every SSO random variable is WSO. Also since $\|\xi(\omega)\|_X^2 = \sum_{i=1}^{\infty} |(\xi(\omega), e_i)_X|^2$, where $\{e_i\}$ is an orthonormal basis in X, a WSO random variable is SSO if and only if

$$\sum_{i=1}^{\infty} E|(\xi(\omega), e_i)_X|^2 < \infty.$$

Similarly, WSO and SSO X-valued process $\xi = \{\xi_x^n, \ x \in X, \ n \in \mathbb{Z}\}$ are defined. A WSO as well as a SSO X-valued stochastic process is said to be periodically correlated (PC) if there exists an integer $T > 0$ such that for every $x, y \in X$ and $m, n \in \mathbb{Z}$,

$$E\xi_x^n \overline{\xi_y^m} = E\xi_x^{n+T} \overline{\xi_y^{m+T}}. \tag{2.1}$$

Univariate second order PC processes where introduced and studied by Gladyshev (1961). X-valued WSO PC processes where studied by Soltani and Shishebor (2007). We assume that the spectral density $\frac{d}{ds}\mathcal{F}(ds)$ exists:

$$\mathbf{f}(s) = [f_{p-l}(ds + \frac{2\pi p}{T})]_{l,p=0,\dots,T-1}. \quad s \in [0, \frac{2\pi}{T}). \tag{2.2}$$

Another spectral representation, time dependent, was derived by Soltani and Shishebor (2007), namely $\xi_x^n = \int_0^{2\pi} e^{ins}\Phi(ds)V_n(s)x$, in the sense that

$$E\xi_x^n\overline{\xi_y^m} = \int_0^{2\pi} e^{i(n-m)s}(V_n(s)x, V_m(s)y)_X ds \tag{2.3}$$

in which Φ is an orthogonally scattered random measure, and $V_n(s) = \sum_{k=0}^{T-1} e^{i\frac{2\pi kn}{T}}$ $a_k(s + \frac{2\pi k}{T})$, is the sequence of T-periodic, $L(X)$-valued functions for $s \in [0, 2\pi)$ and $n \in \mathbb{Z}$. Furthermore $\mathbf{f}(s) = \mathbf{A}^*(s)\mathbf{A}(s) \quad s \in [0, \frac{2\pi}{T})$, where $\mathbf{A}(s) = \overline{[a_{j-k}(s + \frac{2\pi j}{T})]_{k\leq j}} \quad k, j = 0, \dots, T-1$. The operator-matrix $\mathbf{A}$ is indeed the Cholesky factor of the spectral density $\mathbf{f}$. It is plain to verify that

$$E(\sum_{i=1}^{\infty} |(\xi^n(\omega), e_i)_X|^2) = \sum_{i=1}^{\infty} \int_0^{2\pi} \|V_n(s)e_i\|_X^2 ds < \infty. \tag{2.4}$$

In summary we conclude that *a matrix-operator* $\mathbf{f}$ *is the spectral density of a SSO PC process if and only if it is a nuclear, i.e. every of its entries is nuclear. Equivalently, the matrix-operator* $\mathbf{A}$ *is Hilbert Schmidt.*

Let $\eta^n = \int_0^{2\pi} e^{ins}\Phi(ds), \quad n \in \mathbb{Z}$,, and let us correspondingly define the following finite Fourier transforms (FFT) and X-valued processes based on a finite segment of the processes ξ and η. Let

$$d_\xi(\lambda) = N^{-1/2} \sum_{t=0}^{N-1} \xi_t e^{it\lambda}, \quad d_\eta(\lambda) = N^{-1/2} \sum_{t=0}^{N-1} \eta_t e^{it\lambda}, \quad \lambda \in [0, 2\pi).$$

Indeed we define the FFT terms to be step functions with jumps at Fourier frequencies $\frac{2\pi k}{N}$, $k = 0, \dots, N-1$. Also let

$$\tilde{\xi}_N^n = N^{-1/2} \sum_{p=0}^{N-1} e^{-it\lambda_p} d_\eta(\lambda_p)V_n(\lambda_p), \quad n \in \mathbb{Z},$$

and let us define the following auxiliary process and its FFT.

$$d_{\tilde{\xi}}^N(\lambda) = N^{-1/2} \sum_{n=0}^{N-1} e^{in\lambda}\tilde{\xi}_N^n, \qquad \lambda \in [0, 2\pi)$$

Clearly If ξ is SSO, then $d_\xi^N(\lambda)$ and $d_{\tilde{\xi}}^N(\lambda)$, $N = 1, 2, \ldots$, $\lambda \in [0, 2\pi)$ are SSO X-valued process indexed by $\{1, 2, \ldots\} \times [0, 2\pi)$.

Our first asymptotic result below exhibits that the auxiliary PC process $\tilde{\xi}$ approximates ξ in mean square. Let for $x \in X$ and $n = 0, \ldots, T-1$, $u_{n,x}(\theta) = \|V_n(\theta)x\|_X^2$, and $v_n(\theta) = \sum_{i=0}^\infty \|V_n(\theta)e_i\|_X^2$, Also $u_{n,x}(\theta, \theta') = (V_n(\theta)x, V_n(\theta'))_X$, and $v_n(\theta, \theta') = \sum_{i=0}^\infty (V_n(\theta)e_i, V_n(\theta')e_i)_X$.

Lemma 44.1. *(i): Let ξ be a WSO PC process for which $u_{n,x}(\theta)$, $n = 0, \ldots, T-1$ are continuous and of bounded variations on $[0, 2\pi)$. Then for each $t \in \mathbb{Z}$,*

$$E|\tilde{\xi}_x^t - \xi_x^t|^2 \to 0, \quad N \to \infty.$$

(ii): Let ξ be a SSO PC process for which $v_n(\theta)$, $n = 0, \ldots, T-1$ are continuous and of bounded variations on $[0, 2\pi)$. Then for each $t \in \mathbb{Z}$,

$$E\|\tilde{\xi}^t - \xi^t\|_X^2 \to 0, \quad N \to \infty.$$

Let us highlight some applications of the Lemma 1. Indeed it can be used to test whether a segment $\xi_1, \ldots, \xi_n$ is a segment of a PC process with given $\{V_0(\cdot), \ldots, V_n(\cdot)\}$. Indeed by replacing $\tilde{\xi}$ by ξ in (2.16), one can solve the resulting linear equations for $\{d_\eta(\lambda_p),\ p = 0, \ldots, N-1\}$. Then test whether the solutions are the FFT of a white noise process. If the process ξ is generated by independent innovations, i.e., the process η has independent values then $d_\eta(\lambda_p)x$ will have normal distribution with mean zero and variance $2\pi\|x\|^2$. Also it is easy to see that that since

$$E|\tilde{\xi}_N^n x|_X^2 = 2\pi \sum_{p=0}^{N-1} u_{n,x}(\lambda_p), \quad n \in \mathbb{Z}, \quad N = 1, 2, \ldots.$$

and

$$E\|\tilde{\xi}_N^n\|_X^2 = 2\pi \sum_{p=0}^{N-1} v_n(\lambda_p) \quad n \in \mathbb{Z}, \quad N = 1, 2, \ldots.$$

Therefore for large N the contribution of every Fourier frequency to the variances of a WSO or SSO PC process can be measured by solving a number of simultaneous equations. We note that the variances are periodic too.
We also define

$$h_{k,x}(\theta) = \|a_k(\theta + \frac{2\pi k}{T})x\|_X^2, \qquad x \in X,$$

$$g_k(\theta) = \sum_{i=0}^\infty \|a_k(\theta + \frac{2\pi k}{T})e_i\|_X^2, \qquad n = 0, \ldots, T-1$$

Also

$$h_{k,k',x}(\theta,\theta') = (a_k(\theta + \frac{2\pi k}{T})x, a_{k'}(\theta' + \frac{2\pi k'}{T})_X, \qquad x \in X,$$

$$v_{k,k'}(\theta,\theta') = \sum_{i=0}^{\infty}(a_k(\theta + \frac{2\pi k}{T})e_i, a_{k'}(\theta' + \frac{2\pi k'}{T})e_i)_X, \qquad n = 0,\dots,T-1$$

In the following lemma we prove that the mean square deviation between $d_{\tilde{\xi}}^N(\lambda)$ and $d_{\xi}^N(\lambda)$ goes to zero as N tends to infinity, under the mild assumption of continuity of the Cholesky factor.

Lemma 44.2. *(i): Let ξ be a WSO PC process for which $h_{n,x}(\theta)$, $n = 0,\dots,T-1$ are continuous on $[0,2\pi)$. Then at every Fourier frequency λ*

$$E|d_{\tilde{\xi}}(\lambda)x - d_{\xi}(\lambda)x|^2 \to 0, \quad N \to \infty.$$

(ii): Let ξ be a SSO PC process for which $g_n(\theta)$, $n = 0,\dots,T-1$ are continuous on $[0,2\pi)$. Then at every Fourier frequency

$$E\|d_{\tilde{\xi}}(\lambda) - d_{\xi}(\lambda)\|^2 \to 0, \quad N \to \infty.$$

Let us introduce periodograms for WSO and SSO PC processes. Let

$$\mathbf{d}_{\xi}^T(\lambda) = (d_{\xi}(\lambda),\dots,d_{\xi}(\lambda + \frac{2\pi(T-1)}{T}))', \qquad \lambda \in [0,\frac{2\pi}{T}).$$

The periodogram for a WSO as well as SSO PC process is defined to be

$$\mathbf{I}_{\xi}^T(\lambda) = [I_{k,\ell}(\lambda)]_{k,\ell=0,\cdots,T-1}, \quad \lambda \in [0,\frac{2\pi}{T}),$$

$$(I_{k,\ell}(\lambda)x, y) = d_{\xi}(\lambda + \frac{2\pi k}{T})x\overline{d_{\xi}(\lambda + \frac{2\pi\ell}{T})y}, \quad x, y \in X.$$

Indeed each $I_{k,\ell}(\lambda)$ is a random $L(X)$ valued function on $[0,\frac{2\pi}{T})$. For WSO or SSO processes, respectively:

$$E|(I_{k,\ell}(\lambda)x,y)| < \infty, \qquad E\|I_{k,\ell}(\lambda)x\|^2 < \infty, \quad k,\ell = 0,\cdots T-1.$$

Corollary 44.1. *Under the assumption of Lemma 3.2, for Fourier frequencies λ, λ',*

$$E|d_{\tilde{\xi}}(\lambda)x\overline{d_{\tilde{\xi}}(\lambda')y} - d_{\xi}(\lambda)x\overline{d_{\xi}(\lambda')y}| \to 0, \quad N \to \infty, \quad x \in X,$$

for WSO PC processes and

$$E|\sum_{i=0}^{\infty}\{d_{\tilde{\xi}}(\lambda)e_i\overline{d_{\tilde{\xi}}(\lambda')x} - d_{\xi}(\lambda)e_i\overline{d_{\xi}(\lambda')x}\}| \to 0, \quad N \to \infty, x \in X,$$

for SSO PC processes.

Let us give the main result of this article. The proof is based on Lemma 2 and the following crucial fact.

$$\mathbf{d}_\xi^T(\lambda) = \mathbf{A}(\lambda)\mathbf{d}_\eta^T(\lambda),$$

Theorem 44.1. *Let ξ be an X-valued PC process with the spectral density $\mathbf{f}(\lambda)$, $\lambda \in [0, 2\pi)$. Let $\mathbf{A}(\lambda)$, $\lambda \in [0, 2\pi)$ be the Cholesky factor of f. Assume for every $x \in X$, $\mathbf{A}(\lambda)x$ is continuous in λ w.r.t. the norm in X. Also $\mathbf{d}_\xi^T(\lambda)$ and $\mathbf{I}_\xi^T(\lambda)$ be the corresponding SFFT and Periodogram.*

(i) If ξ is WSO then for $k, \ell = 0, \cdots, T-1$,

$$E(I_{k,\ell}(\lambda)x, y)_X \longrightarrow (f_{k,\ell}(\lambda)x, y)_X, \quad N \to \infty, \quad x, y \in X.$$

(ii) If ξ is SSO then for $k, \ell = 0, \cdots, T-1$,

$$E\|I_{k,\ell}(\lambda)x - f_{k,\ell}(\lambda)x\|_X^2 \longrightarrow 0, \quad N \to \infty, \quad x \in X.$$

(iii) For arbitrary frequencies $\lambda_1, \cdots, \lambda_J$ in $[0, \frac{2\pi}{T})$, SFFT $\mathbf{d}_\xi^T(\lambda_1), \cdots, \mathbf{d}_\xi^T(\lambda_J)$ are asymptotically uncorralated with mean zero and covariance operators $\mathbf{f}(\lambda_1), \cdots, \mathbf{f}(\lambda_J)$.

References

[1] Bosq, D. Linear Processes in Function Spaces. Theory and Applications. Lecture Notes in Statistics, **149**, Springer, Berlin. (2000).

[2] Gladyshev, E. G.: Periodically correlated random sequences. Soviet Math. Dokl., 2 385-388 (1961).

[3] Gihman,I. I. and Skorohod, A. V. The Theory of Stochastic Processes. Springer-Verlag, Berlin. (1974).

[4] Hurd, H. L.: Representation of strongly harmonizable periodically corrolated processes and their covariance. J. Mult. Anal **29**, 53-67 (1989).

[5] Ledoux, M and Talagrand, M.: Probability in Banach spaces. Springer-Verlag, Berlin. (1991).

[6] Makagon, A., Miamee, A. G. Salehi, H. and A.R. Soltani (2007). On spectral domain of periodically correlated processes. Theor Prob. Their Appl. **52** (2), 1-12, 2007.

[7] Mandrekar, V., and Salehi, H.: On Singularity and Lebesgue type decomposition for Operator-Valued measures. J. Multivariate Anal. **2**, 167-185. (1971).

[8] Miamee, A.G.: On B(X, K) - Valued Stationary Stochastic Processes. Indiana University Mathematics Journal, **25** (10), 921-932 (1976).

[9] Pourahmadi, M. and Salehi, H.: On subordination and linear transformation of harmonizable and periodically correlated processes, in probability Theory on Vector Spaces, III (Lublin), Berline-New York, Springer, 195-213 (1983).

[10] Rozanov, Yu. A.: Some approximation problems in the theory of stationary processes. J. Multivariate Anal. **2**, 135-144 (1972).

[11] Rudin, W.: Functional analysis. New York, McGraw Hill. (1976).

[12] Soltani, A. R. and Azimmohseni, M.: Periodograms Asymptotic Distributions in Periodically orrelated Processes and Multivariate Stationary Processes: An Alternative Approach. J. Statistical Planning and Inference, **137** (4), 1236-1242 (2007).
[13] Soltani, A. R. and Shishebor, Z.: A spectral representation for weakly periodic sequences of bounded linear transformations. Acta Math. Hungary. **80**, 265-270 (1998).
[14] Soltani, A. R. and Shishebor, Z.: On infinite dimentional discrete time dependent periodically correlated processes. Rocky Mountain Math J, **37** (3), 1043-1058 (2007).

Printed in the United States
116191LV00001B/71-74/P

9 783790 820614